Classics in Mathematics

Richard Courant • Fritz John    Introduction to Calculus and Analysis
Volume II/2

Springer
Berlin
Heidelberg
New York
Barcelona
Hong Kong
London
Milan
Paris
Singapore
Tokyo

Richard Courant · Fritz John

# Introduction
## to Calculus and Analysis

Volume II/2
Chapters 5 - 8
Reprint of the 1989 Edition

Springer

Originally published in 1974 by Interscience Publishers, a division
of John Wiley and Sons, Inc.
Reprinted in 1989 by Springer-Verlag New York, Inc.

Mathematics Subject Classification (1991): 26xx, 26-01

Cataloging-in-Publication Data applied for

Die Deutsche Bibliothek - CIP-Einheitsaufnahme
Courant, Richard:
Introduction to calculus and analysis / Richard Courant; Fritz John.- Reprint.- Berlin; Heidelberg;
New York; Barcelona; Hong Kong; London; Milan; Paris; Singapore; Tokyo: Springer
(Classics in mathematics)
Vol.2. / With the assistance of Albert A. Blank and Alan Solomon 2. Chapter 5-8.- Reprint of
the 1989 ed.- 2000
ISBN 3-540-66570-6

Photograph of Richard Courant from: C. Reid, *Courant in Göttingen
and New York. The Story of an Improbable Mathematician,*
Springer New York, 1976

Photograph of Fritz John by kind permission of The Courant Institute
of Mathematical Sciences, New York

ISSN 1431-0821
ISBN 3-540-66570-6 Springer-Verlag Berlin Heidelberg New York

© Springer-Verlag Berlin Heidelberg 2000
Printed in Germany

The use of general descriptive names, registered names, trademarks etc. in this publication does not imply,
even in the absence of a specific statement, that such names are exempt from the relevant protective laws and
regulations and therefore free for general use.

SPIN 11562054      41/3111-5 4 3 2 – Printed on acid-free paper

# Richard Courant   Fritz John

# Introduction to Calculus and Analysis

## Volume II

With the assistance of
Albert A. Blank and Alan Solomon

With 120 Illustrations

Springer

Richard Courant (1888 - 1972)                    Fritz John
Courant Institute of Mathematical Sciences
New York University
New York, NY 10012

Originally published in 1974 by Interscience Publishers, a division of John Wiley and Sons, Inc.

---

Mathematical Subject Classification: 26xx, 26-01

---

Printed on acid-free paper.

Printed and bound by Edwards Brothers, Inc., Ann Arbor, MI.
Printed in the United States of America.

9  8  7  6  5  4  3

ISBN 0-387-97152-1 Springer-Verlag New York Berlin Heidelberg
ISBN 0-540-97152-1 Springer-Verlag Berlin Heidelberg New York   SPIN 10691322

# *Preface*

Richard Courant's Differential and Integral Calculus, Vols. I and II, has been tremendously successful in introducing several generations of mathematicians to higher mathematics. Throughout, those volumes presented the important lesson that meaningful mathematics is created from a union of intuitive imagination and deductive reasoning. In preparing this revision the authors have endeavored to maintain the healthy balance between these two modes of thinking which characterized the original work. Although Richard Courant did not live to see the publication of this revision of Volume II, all major changes had been agreed upon and drafted by the authors before Dr. Courant's death in January 1972.

From the outset, the authors realized that Volume II, which deals with functions of several variables, would have to be revised more drastically than Volume I. In particular, it seemed desirable to treat the fundamental theorems on integration in higher dimensions with the same degree of rigor and generality applied to integration in one dimension. In addition, there were a number of new concepts and topics of basic importance, which, in the opinion of the authors, belong to an introduction to analysis.

Only minor changes were made in the short chapters (6, 7, and 8) dealing, respectively, with Differential Equations, Calculus of Variations, and Functions of a Complex Variable. In the core of the book, Chapters 1–5, we retained as much as possible the original scheme of two roughly parallel developments of each subject at different levels: an informal introduction based on more intuitive arguments together with a discussion of applications laying the groundwork for the subsequent rigorous proofs.

The material from linear algebra contained in the original Chapter 1 seemed inadequate as a foundation for the expanded calculus structure. Thus, this chapter (now Chapter 2) was completely rewritten and now presents all the required properties of $n$th order determinants and matrices, multilinear forms, Gram determinants, and linear manifolds.

The new Chapter 1 contains all the fundamental properties of linear differential forms and their integrals. These prepare the reader for the introduction to higher-order exterior differential forms added to Chapter 3. Also found now in Chapter 3 are a new proof of the implicit function theorem by successive approximations and a discussion of numbers of critical points and of indices of vector fields in two dimensions.

Extensive additions were made to the fundamental properties of multiple integrals in Chapters 4 and 5. Here one is faced with a familiar difficulty: integrals over a manifold $M$, defined easily enough by subdividing $M$ into convenient pieces, must be shown to be independent of the particular subdivision. This is resolved by the systematic use of the family of Jordan measurable sets with its finite intersection property and of partitions of unity. In order to minimize topological complications, only manifolds imbedded smoothly into Euclidean space are considered. The notion of "orientation" of a manifold is studied in the detail needed for the discussion of integrals of exterior differential forms and of their additivity properties. On this basis, proofs are given for the divergence theorem and for Stokes's theorem in $n$ dimensions. To the section on Fourier integrals in Chapter 4 there has been added a discussion of Parseval's identity and of multiple Fourier integrals.

Invaluable in the preparation of this book was the continued generous help extended by two friends of the authors, Professors Albert A. Blank of Carnegie-Mellon University, and Alan Solomon of the University of the Negev. Almost every page bears the imprint of their criticisms, corrections, and suggestions. In addition, they prepared the problems and exercises for this volume.[1]

Thanks are due also to our colleagues, Professors K. O. Friedrichs and Donald Ludwig for constructive and valuable suggestions, and to John Wiley and Sons and their editorial staff for their continuing encouragement and assistance.

FRITZ JOHN
NewYork
September 1973

[1]In contrast to Volume I, these have been incorporated completely into the text; their solutions can be found at the end of the volume.

# Contents

ix

# Chapter 2 Vectors, Matrices, Linear Transformations

*Chapter 3 Developments and Applications of the Differential Calculus*

**APPENDIX**

# Chapter 4 Multiple Integrals

## APPENDIX: DETAILED ANALYSIS OF THE PROCESS OF INTEGRATION

# Chapter 5 *Relations Between Surface and Volume Integrals*

## *Chapter 6 Differential Equations*

# Introduction to Calculus and Analysis
## Volume II

# CHAPTER
# 5

# *Relations Between Surface and Volume Integrals*

The multiple integrals discussed in the previous chapter are not the only possible extension of the concept of integral to more than one independent variable. Other generalizations arise from the fact that regions of several dimensions may contain manifolds of fewer dimensions and that we can consider integrals over such manifolds. Thus, for two independent variables, we considered not only the integrals over two-dimensional regions but also integrals along curves, which are one-dimensional manifolds. With three independent variables, besides integrals over three-dimensional regions and integrals along curves, we encounter integrals over curved surfaces. In the present chapter we shall introduce surface integrals and discuss the mutual relations between integrals over manifolds of varying dimensions.[1]

## 5.1 Connection Between Line Integrals and Double Integrals in the Plane (The Integral Theorems of Gauss, Stokes, and Green)

For functions of a single independent variable the fundamental

---

[1]We use the term *manifold* without precise definition as a generic name for sets of an unspecified number of dimensions. In this book we deal exclusively with manifolds that are subsets of some euclidean space, such as the curves, two-dimensional surfaces, hypersurfaces, and four-dimensional regions in four-dimensional euclidean space. More generally, manifolds can be defined without reference to a surrounding euclidean space. Such manifolds locally resemble deformed portions of euclidean space, while their over-all structure can be much more complicated than that of euclidean space.

formula stating the relation between differentiation and integration (cf. Volume I, p. 190) is

(1) $$\int_{x_0}^{x_1} f'(x)\, dx = f(x_1) - f(x_0).$$

An analogous formula—*Gauss's theorem,* also called the *divergence theorem*—holds in two dimensions. Here again, the integral of a derivative of functions

$$\iint_R f_x(x,y)\, dx\, dy \qquad \text{or} \qquad \iint_R g_y(x,y)\, dx\, dy$$

is transformed into an expression that depends on the values of the functions themselves on the boundary. We regard here the boundary $C$ of the set $R$ as an *oriented* curve $+C$, choosing as positive sense on $C$ the one for which the region $R$ remains on the "left" side as we describe the boundary curve $C$.[1] Gauss's theorem then states that

(2) $$\iint_R [f_x(x,y) + g_y(x,y)]\, dx\, dy = \int_{+C} [f(x,y)\, dy - g(x,y)\, dx].$$

This theorem contains as a special case our previous formula expressing the area $A$ of the set $R$ as a line integral over the boundary $C$ of $R$. We put $f(x, y) = x$, $g(x, y) = 0$ and at once obtain

$$A = \iint_R dx\, dy = \int_{+C} x\, dy.$$

In exactly the same way, for $f(x, y) = 0$ and $g(x, y) = y$, we obtain

$$A = \iint_R dx\, dy = -\int_{+C} y\, dx$$

in agreement with Volume I (p. 367).

The divergence theorem becomes particularly suggestive in the notation of the calculus of differential forms, as explained on pp. 307–324. In (2), the line integral has the integrand

$$L = f(x,y)\, dy - g(x,y)\, dx,$$

a first-order differential form. Indeed, $L$ can be identified with the most general first-order form $a(x, y)dx + b(x, y)dy$ if we take $f = b$, $g = -a$. By the definition on p. 313 the derivative of this form is

---

[1] Assuming that the $x$, $y$-coordinate system is right-handed.

$$dL = df\, dy - dg\, dx = (f_x\, dx + f_y\, dy)\, dy - (g_x\, dx + g_y\, dy)\, dx$$

$$= f_x\, dx\, dy - g_y\, dy\, dx = (f_x + g_y)\, dx\, dy,$$

which is just the integrand of the double integral in (2). Hence, formula (2) takes the form[1]

(2a) $$\iint_R dL = \int_{+C} L.$$

In the proof we restrict ourselves to the case in which $R$ is an open set whose boundary $C$ is a simple closed curve consisting of a finite number of smooth arcs; moreover, we assume that every parallel to one of the coordinate axes intersects $C$ in at most two points.[1] We require $f$ and $g$ to be continuous and to have continuous first derivatives in the closure of $R$ (consisting of $R$ and of its boundary $C$).

We first assume that the function $g$ vanishes identically. Then the double integral of $f_x$ over $R$ exists and can be written as a repeated integral[2]

(3) $$\iint_R f_x(x,y)\, dx\, dy = \int dy \int f_x(x,y)\, dx.$$

On each parallel to the $x$-axis, the variable $y$ is constant. The parallels to the $x$-axis intersecting $R$ correspond to $y$-values forming an open interval $\eta_0 < y < \eta_1$, the projection of $R$ onto the $y$-axis.[3] For

---

[1]The process of forming the boundary of a set $R$ presents formal analogies with differentiation. For that reason one frequently uses the symbol $\partial R$ for the boundary $+C$ of $R$, writing (2a) as

(2b) $$\iint_R dL = \int_{\partial R} L.$$

This formula actually applies much more generally to differential forms integrated over manifolds in $n$-dimensional space (see p. 624).

[1]In the Appendix the theorem (and its generalizations in higher dimensions) is proved under the assumption that $R$ is the closure of an open set bounded by a simple curve that is smooth everywhere.

[2]The set $R$ is bounded by the union of a finite number of smooth arcs and, hence, (see p. 521) is Jordan-measurable. The integral of the continuous function $f_x$ over $R$ exists then and is defined as the integral of $\phi_R f_x$ over the whole plane, where $\phi_R$ is the characteristic function of the set $R$ (that is, $\phi_R$ is 1 in the points of $R$ but is 0 in all other points). The reduction of the double integral to a repeated integral is permitted (see p. 531) since the function $\phi_R f_x$ can be integrated over each parallel to the $x$-axis; indeed, each parallel to the $x$-axis meets $R$ in either an open interval or nowhere, so that the integral of $\phi_R f_x$ over a parallel to the $x$-axis is either the integral of the continuous function $f_x$ over an open interval or zero.

[3]The projection of $R$ is an open interval because $R$ is open and its boundary is a simple closed curve and, hence, connected.

each $y$ in that interval the corresponding parallel to the $x$-axis cuts out of $R$ an interval $x_0(y) < x < x_1(y)$ whose end points are the abscissas of the two points of intersection of the parallel with $C$ (see Fig. 5.1). Formula (3) asserts more precisely that

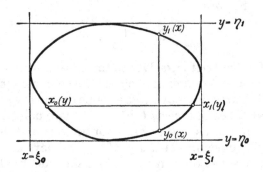

**Figure 5.1**

$$\iint_R f_x \, dx \, dy = \int_{\eta_0}^{\eta_1} h(y) \, dy,$$

where

$$h(y) = \int_{x_0(y)}^{x_1(y)} f_x(x, y) \, dx = f(x_1(y), y) - f(x_0(y), y).$$

Hence,

$$(4) \qquad \iint_R f_x \, dx \, dy = \int_{\eta_0}^{\eta_1} f(x_1(y), y) \, dy - \int_{\eta_0}^{\eta_1} f(x_0(y), y) \, dy.$$

We introduce the two simple oriented arcs $+C_1$, $+C_0$ given parametrically, respectively, by

$$+C_1 \colon x = x_1 \, t \,, \, y = t, \qquad \text{for} \qquad \eta_0 \leq t \leq \eta_1$$

$$+C_0 \colon x = x_0 \, t \,, \, y = t, \qquad \text{for} \qquad \eta_0 \leq t \leq \eta_1,$$

where in each case the sense of increasing $t$ corresponds to the orientation of the arc. Formula (4) can then be written as

$$\iint_R f_x \, dx \, dy = \int_{+C_1} f \, dy - \int_{+C_0} f \, dy.$$

Now $C_1$ and $C_0$ form respectively the right and left portions of $C$, where, however, $+C_1$ has the same orientation as $C$ and $+C_0$ the opposite one. Denoting by $-C_0$ the arc obtained by reversing the orientation of $C_0$, we obtain (see p. 94)

$$\iint_R f \, dx \, dy = \int_{+C_1} f \, dy + \int_{-C_0} f \, dy = \int_{+C} f \, dy.$$

We can similarly decompose $+C$ into an "upper" arc

$$+\Gamma_1: x = t, \qquad y = y_1(t), \qquad \text{for} \qquad \xi_0 \leq t \leq \xi_1$$

and "lower" arc

$$+\Gamma_0: x = t, \qquad y = y_0(t), \qquad \text{for} \qquad \xi_0 \leq t \leq \xi_1,$$

oriented according to the sense of increasing $t$. Here the interval $\xi_0 < x < \xi_1$ represents the projection of $R$ onto the $x$-axis. Then,

$$\iint_R g_y \, dx \, dy = \int_{\xi_0}^{\xi_1} dx \int_{y_0(x)}^{y_1(x)} g_y \, dy$$

$$= \int_{\xi_0}^{\xi_1} g(x, y_1(x)) \, dx - \int_{\xi_0}^{\xi_1} g(x, y_0(x)) \, dx$$

$$= \int_{+\Gamma_1} g \, dx - \int_{+\Gamma_0} g \, dx$$

$$= -\int_{-\Gamma_1} g \, dx - \int_{+\Gamma_0} g \, dx$$

$$= -\int_{+C} g \, dx$$

since here $\Gamma_0$ has the same orientation as $C$ and $\Gamma_1$ the opposite one. Adding the two identities obtained, we arrive at the general formula (2).

We can now extend our formula to more general open sets $R$ bounded by a simple closed curve $C$, provided $C$ can be decomposed into a finite number of simple arcs $C_1, \ldots, C_n$ each of which is intersected in at most one point by any parallel to one of the coordinate axes.[1] In order to prove that here also

(5) $$\iint_R f_x \, dx \, dy = \int_{+C} f \, dy,$$

we draw parallels to the $y$-axis through all of the end points of the simple arcs $C_i$ (see Fig. 5.2). In this way $R$ is decomposed into a finite

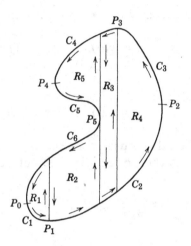

**Figure 5.2**

number of sets $R_1, \ldots, R_N$ each of which is bounded laterally by straight segments parallel to the $y$-axis and above and below by simple subarcs of two of the arcs $C_i$. We can apply the formula

$$\iint_{R_i} f_x \, dx \, dy = \int_{+\Gamma_i} f \, dy$$

to each of the sets $R_i$ with boundary $\Gamma_i$, since $\Gamma_i$ is intersected by each parallel to the $x$-axis in at most two points. Here the orientation of the boundary curve $+\Gamma_i$ agrees with that of $+C$ in the nonvertical portions and is that of increasing $y$ on the right-hand boundary and of decreasing $y$ on the left-hand one. Adding up the formulae

---

[1]This assumption is not always satisfied. The boundary curve $C$ may, for example, consist in part of the curve $y = x^2 \sin(1/x)$, which is cut by the $x$-axis in an infinite number of points and can not be decomposed into a finite number of arcs cut in only one point.

for $i = 1, \ldots, N$ the double integrals over the $R_i$ yield the double integral over $R$. In the line integrals over the $+\Gamma_i$ the contributions over the vertical auxiliary segments cancel out, since each segment is traversed twice, once upward, once downward. Hence, the line integrals over the curves $+\Gamma_i$ add up to that over the whole curve $+C$, and one obtains formula (5). In the same way one proves that

$$\iint_R g_y \, dx \, dy = - \int_{+C} g \, dx$$

by dividing $R$ by parallels to the $x$-axis through all of the end points of the arcs $C_i$.

The same arguments also show that we can dispense with the assumption that the boundary $C$ of $R$ consists of a *single* closed curve $C$. The divergence theorem (2) applies just as well when $C$ consists of several closed curves, as long as $C$ can be decomposed into a finite number of simple arcs each intersected in at most one point by parallels to the axes. In taking the integral over $+C$ we have to give each of the closed components of $C$ the orientation corresponding to leaving $R$ on the left-hand side. Decomposition by parallels to the $y$-axis still results then in regions whose boundary is intersected in at most two points by any parallel to the $x$-axis (see Fig. 5.3).

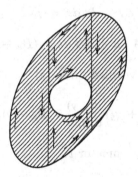

**Figure 5.3**

In this manner we prove the divergence theorem for more general regions $R$ by *decomposing* $R$ into regions for which the theorem has already been proved. Often, we can instead *transform* $R$ into a region to which the theorem is known to apply. Writing the divergence theorem as

$$\iint_R dL = \int_{+C} L,$$

we notice that the differential forms $dL$ and $L$ are defined independently of coordinates, as explained in Section 3.6d, p. 322. Let

$$x = x(u, v), \quad y = y(u, v)$$

be a continuously differentiable 1–1 transformation, with positive Jacobian, that takes $R$ into a set $R^*$ with boundary $C^*$ in the $u, v$-plane. Then,

$$L = f \, dy - g \, dx = f(y_u \, du + y_v \, dv) - g(x_u \, du + x_v \, dv)$$

$$= (fy_u - gx_u) \, du + (fy_v - gx_v) \, dv$$

$$= A \, du + B \, dv,$$

where

$$A = fy_u - gx_u, \qquad B = fy_v - gx_v.$$

The derivative of $L$ computed in either $x, y$ or $u, v$ variables is given by

$$dL = df \, dy - dg \, dx = (f_x + g_y) \, dx \, dy$$

$$= dA \, du + dB \, dv = (B_u - A_v) \, du \, dv,$$

so that (as can also be verified directly)

$$(f_x + g_y) \frac{d(x,y)}{d(u,v)} = B_u - A_v.$$

Let $C$ be referred to a parameter $t$:

$$x = x(t), \qquad y = y(t) \qquad\qquad a \le t \le b,$$

where the orientation of $+C$ corresponds to increasing $t$. Using for the corresponding points of $+C^*$ the same parameter value $t$, we have for the line integrals of $L$ over $C$ and $C^*$ the common value

$$\int L = \int \frac{L}{dt} \, dt = \int_{+C} \left( f \frac{dy}{dt} - g \frac{dx}{dt} \right) dt = \int_{+C^*} \left( A \frac{du}{dt} + B \frac{dv}{dt} \right) dt.$$

Similarly, we have the same value for the area integrals in the two planes:

$$\iint dL = \iint_{R} (f_x + g_y) \, dx \, dy$$

$$= \iint_{R^*} (f_x + g_y) \frac{d(x,y)}{d(u,v)} \, du \, dv$$

$$= \iint_{R^*} (B_u - A_v) \, du \, dv.$$

Hence, the divergence theorem for $R$

$$\iint_{R} (f_x + g_y) \, dx \, dy = \int_{C} (f \, dy - g \, dx)$$

will follow from the corresponding formula for $R^*$,

$$\iint_{R^*} (B_u - A_v) \, du \, dv = \int_{+C^*} (A \, du + B \, dv).$$

For the validity of the theorem for a region $R$, it is sufficient that $R$ can be transformed into a region whose boundary consists of simple arcs intersected by parallels to the axes in, at most, one point. If, for example, the boundary $C$ or $R$ is a polygon, we can always rotate the figure in such a way that none of the sides of the polygon is parallel to one axis, and the divergence theorem will apply.

## 5.2 Vector Form of the Divergence Theorem. Stokes's Theorem

Gauss's theorem can be stated in a particularly simple way if we make use of the notations of vector analysis. For this purpose we consider the two functions $f(x, y)$ and $g(x, y)$ as the components of a plane vector field **A**. The integrand of the double integral in formula (2) is denoted by div **A**,

$$\text{div } \mathbf{A} = f_x(x, y) + g_y(x, y)$$

and is called the divergence of the vector **A** (cf. p. 208). In order to obtain a vector expression for the line integral on the right side in the divergence theorem, we introduce the length of arc $s$ of the oriented boundary curve $+C$ (cf. Volume I, p. 352). Here, the sense of increasing $s$ is taken to correspond to the orientation[1] of the curve $+C$. The right side of identity (2) then becomes

$$\int_C [f(x, y)\dot{y} - g(x, y)\dot{x}]\, ds,$$

where we put $dx/ds = \dot{x}$ and $dy/ds = \dot{y}$.

We now recall that the plane vector $\mathbf{t}$ with components $\dot{x}$ and $\dot{y}$ has unit length and has the direction of the tangent in the sense of increasing $s$ and, hence, in the direction given by the orientation of $C$. The vector $\mathbf{n}$ with components $\xi = \dot{y}$ and $\eta = -\dot{x}$ has length 1, is perpendicular to the tangent, and, moreover, has the same position relative to the vector $\mathbf{t}$ as the positive $x$-axis has relative to the positive $y$-axis.[2] If, as usual, a 90° clockwise rotation takes the positive $y$-axis into the positive $x$-axis, the vector $\mathbf{n}$ is obtained by a 90° clockwise rotation from the tangent vector $\mathbf{t}$. Thus, $\mathbf{n}$ is the normal pointing to the "right" side of the oriented curve $C$ (cf. Volume I, p. 346). Since in our case $+C$ is oriented in such a way that the region $R$ lies on the left side of $+C$, it follows that $\mathbf{n}$ is the unit vector in the direction of the outward-drawn normal (see Fig. 5.4). The components $\xi$, $\eta$ of the unit vector $\mathbf{n}$ are the direction cosines of the outward normal:

$$\xi = \cos\theta, \qquad \eta = \sin\theta$$

---

[1]In effect, this convention on $s$ makes the value of a line integral of the form

$$I = \int_C h\, ds$$

independent of the orientation of $C$ as long as the integrand $h$ does not depend on the orientation. If $C$ is represented parametrically in the form $x = x(t)$, $y = y(t)$ for $a \leq t \leq b$ where the sense of increasing $t$ corresponds to a particular orientation of $C$, then

$$I = \int_C h\, ds = \int_a^b h\, \frac{ds}{dt}\, dt,$$

where $ds/dt > 0$. In particular, $I > 0$ whenever the integrand $h$ is positive along the curve.

[2]We see this from considerations of continuity; we may suppose that the tangent to the curve is made to coincide with the $y$-axis in such a way that $\mathbf{t}$ points in the direction of increasing $y$. Then $x = 0$, $y = 1$, so that the vector $\mathbf{n}$ with components $\xi = 1$ and $\eta = 0$ has the direction of the positive $x$-axis.

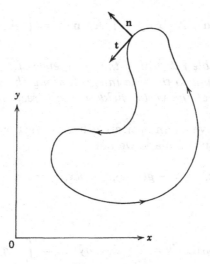

**Figure 5.4**

if $\mathbf{n}$ forms the angle $\theta$ with the positive $x$-axis. It is useful to notice that the components of $\mathbf{n}$ can also be written as directional derivatives of $x$ and $y$ in the direction of $\mathbf{n}$:

$$\xi = \dot{y} = \frac{dx}{dn}, \quad \eta = -\dot{x} = \frac{dy}{dn},$$

since for any scalar $h(x, y)$ the derivative of $h$ in the direction of $\mathbf{n}$ is given by

$$\frac{dh}{dn} = h_x \cos \theta + h_y \sin \theta = \xi h_x + \eta h_y$$

(see p. 44)

Gauss's theorem therefore can be written in the form

$$(6) \qquad \iint_R \operatorname{div} \mathbf{A} \, dx \, dy = \int_C \left( f \frac{dx}{dn} + g \frac{dy}{dn} \right) ds.$$

Here the integrand on the right is the scalar product $\mathbf{A} \cdot \mathbf{n}$ of the vector $\mathbf{A}$ with components $f$, $g$ and the vector $\mathbf{n}$ with components $dx/dn$, $dy/dn$. Since the vector $\mathbf{n}$ has length 1 the scalar product $\mathbf{A} \cdot \mathbf{n}$ represents the component $A_n$ of the vector $\mathbf{A}$ in the direction of $\mathbf{n}$. Consequently, the divergence theorem takes the form

(7) $$\iint_R \operatorname{div} \mathbf{A} \, dx \, dy = \int_C \mathbf{A} \cdot \mathbf{n} \, ds = \int_C A_n \, ds.$$

In words, *the double integral of the divergence of a plane vector field over a set R is equal to the line integral, along the boundary C of R, of the component of the vector field in the direction of the outward-drawn normal.*

In order to arrive at an entirely different vector interpretation of Gauss's theorem in the plane we put

$$a(x, y) = - g(x, y), \qquad b(x, y) = f(x, y).$$

Then, by (2),

(8) $$\iint_R (b_x - a_y) \, dx \, dy = \int_C (a\dot{x} + b\dot{y}) \, ds = \int_{+C} a \, dx + b \, dy.$$

If the two functions $a$ and $b$ are again taken as components of a vector field $\mathbf{B}$ (where at each point $\mathbf{B}$ is obtained from the vector $\mathbf{A}$ by a 90° rotation in the counterclockwise sense), we see that $a\dot{x} + b\dot{y}$ is the scalar product of $\mathbf{B}$ with the tangential unit vector $\mathbf{t}$:

$$a\dot{x} + b\dot{y} = \mathbf{B} \cdot \mathbf{t} = B_t,$$

where $B_t$ is the tangential component of the vector $\mathbf{B}$. The integrand of the double integral in (8) appeared on p. 209 as a component of the curl of a vector in space. In order to apply the concept of curl here we imagine the plane vector field $\mathbf{B}$ continued somehow into $x$, $y$, $z$-space in such a way that in the $x$, $y$-plane the $x$- and $y$-components of $\mathbf{B}$ coincide with $a(x, y)$ and $b(x, y)$, respectively. Then $b_x - a_y$ represents the $z$-component $(\operatorname{curl} \mathbf{B})_z$ of the curl $\mathbf{B}$. The divergence theorem now takes the form

(9) $$\iint_R (\operatorname{curl} \mathbf{B})_z \, dx \, dy = \int_C B_t \, ds.$$

We can formulate the theorem in words as follows:

*The integral of the z-component of the curl of a vector field in space taken over a set R in the x, y-plane is equal to the integral of the tangential component taken around the boundary of R. This statement is Stokes's theorem in the plane.*

If we make use of the vector character of the curl of a vector field in space we can free the Stokes theorem from the restriction that the plane region $R$ lie in the $x$, $y$-plane. Any plane in space can be taken as $x$, $y$-plane of a suitable coordinate system. We thus arrive at the more general formulation of Stokes's theorem:

$$(10) \qquad \iint_R (\text{curl } \mathbf{B})_n \, ds = \int_C B_t \, ds,$$

where $R$ is any plane region in space bounded by the curve $C$, and $(\text{curl } \mathbf{B})_n$ is the component of the vector curl $\mathbf{B}$ in the direction of the normal $\mathbf{n}$ to the plane containing $R$. Here $C$ has to be oriented in such a way that the tangent vector $\mathbf{t}$ points in the counterclockwise direction as seen from that side of the plane toward which $n$ points.

If the complete boundary $C$ of $R$ consists of several closed curves, these formulas remain valid provided that we extend the line integral over each of those curves, oriented properly so as to leave $R$ on its left side.

Of importance is the special case where the functions $a(x, y)$, $b(x, y)$ satisfy the integrability condition

$$(11) \qquad a_y = b_x,$$

that is, where $a \, dx + b \, dy$ is a "closed" torm. Here the double integral over $R$ vanishes and we find from (8) that

$$\int_C a \, dx + b \, dy = 0$$

whenever $C$ denotes the complete boundary of a region $R$ in which (11) holds. This again implies, as we saw on p. 96, that

$$\int a \, dx + b \, dy$$

extended over a simple arc has the same value for all arcs that have the same end points and that can be deformed into each other without leaving $R$ (see p. 104).

## Exercises 5.2

1. Use the divergence theorem in the plane to evaluate the line integral

$$\int_C A \, du + B \, dv$$

for the following functions and paths taken in the counterclockwise sense about the given region

(a) $A = au + bv$, $\quad B = 0$, $\quad u \geq 0$, $\quad v \geq 0$, $\quad \alpha^2 u + \beta^2 v \leq 1$

(b) $A = u^2 - v^2$, $\quad B = 2uv$, $\quad |u| < 1$, $\quad |v| < 1$

(c) $A = v^n$, $\quad B = u^n$, $\quad u^2 + v^2 \leq r^2$.

2. Derive the formula for the divergence theorem in polar coordinates:

$$\int_{+C^*} f(r, \theta) \, dr + g(r, \theta) \, d\theta = \iint_{R^*} \frac{1}{r} \left\{ \frac{\partial g}{\partial r} - \frac{\partial f}{\partial \theta} \right\} dS.$$

3. Assuming the conditions for the divergence theorem hold, derive the following expressions in polar coordinates for the area of a region $R$ with boundary $C$,

$$\frac{1}{2} \int_{+C^*} r^2 \, d\theta, \qquad - \int_{+C^*} r\theta \, dr,$$

where in the second formula we assume that $R$ does not contain the origin.

4. Apply Stokes's theorem in the $x$, $y$-plane to show that

$$\iint_{R^*} \frac{d(u, v)}{d(x, y)} \, dS = \int_{+C^*} u(\text{grad } v) \cdot \mathbf{t} \, ds,$$

where $\mathbf{t}$ is the positively oriented unit tangent vector for $C$.

## 5.3 Formula for Integration by Parts in Two Dimensions. Green's Theorem

The divergence theorem

(12) $$\iint_R (f_x + g_y) \, dx \, dy = \int_C \left( f \frac{dx}{dn} + g \frac{dy}{dn} \right) ds$$

[see formula (6)] combined with the rule for differentiating a product immediately yields a formula for *integration by parts* that is basic in the theory of partial differential equations. Let $f(x, y) = a(x, y) u(x, y)$ and $g(x, y) = b(x, y) v(x, y)$, where the functions $a$, $u$, $b$, $v$ have continuous first derivatives. Since here

$$f_x + g_y = (au_x + bv_y) + (a_x u + b_y v),$$

we can write formula (12) in the form

(13)
$$\iint_R (au_x + bv_y) \, dx \, dy = \int_C \left( au \frac{dx}{dn} + bv \frac{dy}{dn} \right) ds$$

$$- \iint_R (a_x u + b_y v) \, dx \, dy.$$

To obtain *Green's first theorem* we apply this formula to the case where $v = u$ and where $a$ and $b$ are of the form $a = w_x$ and $b = w_y$. (We assume that $u$ has continuous first derivatives and $w$ continuous second derivatives in the closure of $R$.) We obtain the equation

$$\iint_R (u_x w_x + u_y w_y) \, dx \, dy = \int_C u \left( w_x \frac{dx}{dn} + w_y \frac{dy}{dn} \right) ds$$

$$- \iint_R u(w_{xx} + w_{yy}) \, dx \, dy.$$

Using the symbol $\Delta$ for the *Laplace operator* (p. 211), we write

$$w_{xx} + w_{yy} = \Delta w.$$

Moreover, $dx/dn$ and $dy/dn$ are the direction cosines of the outward normal of the boundary $C$ of $R$ (see p. 552); thus, we have in

$$w_x \frac{dx}{dn} + w_y \frac{dy}{dn} = \frac{dw}{dn}$$

the directional derivative of $w$ taken in the direction of the outward normal to $C$.[1] In this notation *Green's first theorem* becomes

(14)
$$\iint_R (u_x w_x + u_y w_y) \, dx \, dy = \int_C u \frac{dw}{dn} \, ds - \iint_R u \Delta w \, dx \, dy$$

If in addition $u$ has continuous second derivatives, we obtain from (14) by interchanging the roles of $u$ and $v$ the formula

$$\iint_R (w_x u_x + w_y u_y) \, dx \, dy = \int_C w \frac{du}{dn} \, ds - \iint_R w \Delta u \, dx \, dy$$

Subtracting the two relations yields an equation symmetric in $u$ and $w$ and known as *Green's second theorem:*

---

[1] Usually $dw/dn$ is called, for short, *the normal derivative of w.*

$$(15) \qquad \iint_R (u\Delta w - w\Delta u)\, dx\, dy = \int_C \left(u\frac{dw}{dn} - w\frac{du}{dn}\right)\, ds.$$

The two theorems of Green are basic in the study of the solutions of the partial differential equation $u_{xx} + u_{yy} = 0$ (Laplace equation).[1]

## 5.4  The Divergence Theorem Applied to the Transformation of Double Integrals

### a.  *The Case of 1-1 Mappings*

The divergence theorem yields a new proof for the fundamental rule for transformation of double integrals to new independent variables (see p. 403). The divergence theorem for a region $R$ with boundary $C$ can be stated in the form

$$(16) \qquad \int_R dL = \int_{+C} L$$

[see formula (2a), p. 545].[2] Here, putting $f = b$, $g = -a$,

$$(17a) \qquad L = a(x, y)\, dx + b(x, y)\, dy$$

$$(17b) \qquad dL = (b_x - a_y)\, dx\, dy.$$

If the curve $C$ has a parametric representation

$$x = x\,(t), \qquad y = y\,(t), \qquad \alpha \leqq t \leqq \beta,$$

where the sense of increasing $t$ corresponds to the orientation of $+C$, we can write the line integral in (16) as the ordinary integral

$$(17c) \qquad \int_{+C} L = \int_{+C} a\, dx + b\, dy = \int_a^\beta \frac{L}{dt}\, dt$$

with the integrand

---

[1]See the section on potential theory (p. 713).
[2]Here and in what follows we always assume tacitly that the assumptions used in the proof of the divergence theorem are satisfied; that is, that $R$ is an open set whose boundary $C$ consists of a finite number of smooth arcs, each of which is intersected in at most one point by parallels to the axes. The coefficients of the linear form $L$ are assumed to have continuous first derivatives in the closure of $R$.

$$\frac{L}{dt} = a\,\frac{dx}{dt} + b\,\frac{dy}{dt}$$

(see p. 307).
We now consider a mapping defined by functions

(18a) $$u = u(x, y), \qquad v = v(x, y).$$

We assume that the mapping is 1–1 in the closure of $R$ and that the Jacobian $d(u, v)/d(x, y)$ is positive throughout. Let $R$ be mapped onto the set $R'$ in the $u$, $v$-plane and $C$ onto the boundary $C'$ of $R'$. Moreover, $C'$ also shall consist of a finite number of smooth arcs, each of which is intersected in, at most, one point by any parallel to a coordinate axis. Since the Jacobian is positive, the orientation is preserved; that is, for increasing $t$ the point $(u, v)$ given by

$$u = u(x(t), y(t)), \qquad v = v(x(t), y(t))$$

describes the curve $C'$ in such a way that we leave the set $R'$ to our left. Referred to the coordinates $u$, $v$ we have

$$L = A\,du + B\,dv = A\,(u_x\,dx + u_y\,dy) + B\,(v_x\,dx + v_y\,dy) = a\,dx + b\,dy,$$

where the coefficients $A$, $B$ in the $u$, $v$-system are connected with the coefficients $a$, $b$ in the $x$, $y$-system by the relations

$$a = Au_x + Bv_x, \qquad b = Au_y + Bv_y.$$

Along $C'$

$$\frac{L}{dt} = a\,\frac{dx}{dt} + b\,\frac{dy}{dt} = A\,\frac{du}{dt} + B\,\frac{dv}{dt},$$

so that by (17c)

(18b) $$\int_{+C} L = \int_a^\beta \frac{L}{dt}\,dt = \int_a^\beta A\,du + B\,dv = \int_{+C'} L.$$

Applying the divergence theorem (16) to the region $R'$ in the $u$, $v$-plane, we find that

(18c) $$\int_{C'} L = \iint_{R'} dL,$$

where, in analogy to (17b),

$$dL = (B_u - A_v) \, du \, dv.$$

One verifies immediately that[1]

$$
\begin{aligned}
b_x - a_y &= (Au_y + Bv_y)_x - (Au_x + Bv_x)_y \\
&= (A_u u_x + A_v v_x)u_y + (B_u u_x + B_v v_x)v_y - (A_u u_y + A_v v_y)u_x \\
&\qquad - (B_u u_y + B_v v_y)v_x \\
&= (B_u - A_v)(u_x v_y - u_y v_x).
\end{aligned}
$$

Thus, we conclude from (18b, c) and (16) that

$$
(19) \quad \iint_{R'} dL = \iint_{R'} (B_u - A_v) \, du \, dv = \iint_{R} dL
$$

$$
= \iint_{R} (b_x - a_y) \, dx \, dy = \iint_{R} (B_u - A_v) \frac{d(u, v)}{d(x, y)} \, dx \, dy.
$$

This formula contains the general *law of transformation*

$$
(20) \quad \iint_{R'} f(u, v) \, du \, dv = \iint_{R} f(u\,(x, y),\, v\,(x, y)) \frac{d(u, v)}{d(x, y)} \, dx \, dy
$$

*for double integrals* [see (16b), p. 403]. We only have to choose the functions $A$, $B$ in (19) in such a way that $A = 0$ and $B_u = f(u, v)$. This means that for fixed $v$ the function $B$ shall be some indefinite integral of $f(u, v)$ as a function of $u$ alone:

$$
B(u, v) = \int_{g(v)}^{u} f(w, v) \, dw + h(v),
$$

where $h(v)$ is arbitrary and $g(v)$ is chosen in such a way that the point $(g(v),\, v)$ lies in $R'$. For the special function $f = 1$, formula (20) yields an expression for the area of the image region as a double integral:

---

[1]This formula follows without any algebraic computations if we use the fact proved on p. 322 that $dL$ can be formed for a form $L$ without reference to any particular coordinate system; hence, by (56c), p. 308,

$$
b_x - a_y = \frac{dL}{dx \, dy} = \frac{dL}{du \, dv} \frac{d(u, v)}{d(x, y)} = (B_u - A_v) \frac{d(u, v)}{d(x, y)}
$$

(20a)
$$\iint_{R'} du\, dv = \iint_{R} \frac{d(u,\, v)}{d(x,\, y)}\, dx\, dy$$

*Essentially formula (20) expresses the fact that the double integral of a second-order differential form* $\omega = f\, du\, dv$ *does not change under changes of the independent variables.* This fact is proved here by expressing $\omega$ as derivative $dL$ of a first-order form $L$, reducing the double integral to a line integral by means of the divergence theorem, and making use of the invariance of a line integral $\int L$.

### b. Transformation of Integrals and Degree of Mapping

It is interesting to observe what happens to the transformation formula (20) when the mapping

$$u = u(x,\, y), \quad v = v(x,\, y)$$

is no longer 1–1 and when its Jacobian is not necessarily positive. First, we look at the case where the mapping of $R$ onto $R'$ is 1–1, but the Jacobian is negative throughout the closure of $R$. The only difference in the argument leading to (20) is that now $+C$ and $+C'$ have opposite orientations: if increasing parameter values $t$ on $C'$ means leaving $R'$ on the left, then increasing $t$ on $C$ means leaving $R$ on the right. In applying the divergence theorem (16) we assume that the boundary of the two-dimensional region is oriented in such a way that the region lies on the positive (left) side of the boundary. The result is that formula (20)[1] has to be replaced by

(20b)
$$\iint_{R'} f\, du\, dv = - \iint_{R} f\frac{d(u,\, v)}{d(x,\, y)}\, dx\, dy.$$

We can combine formulae (20) and (20b) into a single formula valid whenever the mapping from $(x,\, y)$ onto $(u,\, v)$ is 1–1 and the Jacobian is of constant sign:

---

[1]Formula (20) applies unchanged if the two-dimensional regions $R$ and $R'$ themselves are considered as *oriented* manifolds. In that case, the sign of an integral over the manifold changes when the orientation of the manifold is reversed. A negative Jacobian for the mapping implies that $R$ and $R'$ have opposite orientations, so that formula (20) persists if written as

$$\iint_{+R'} f\, du\, dv = \iint_{+R} f\, \frac{d(u,\, v)}{d(x,\, y)}\, dx\, dy.$$

Instead of orienting the regions, we can also replace the Jacobian by its absolute value as in formula (16b) on p. 403.

(21) $$\iint f\varepsilon_R \, du \, dv = \iint_R f \frac{d(u, v)}{d(x, y)} \, dx \, dy.$$

Here the integral on the left side is to be extended over the whole $u, v$-plane, and the function $\varepsilon_R = \varepsilon_R(u, v)$ is defined as

$$\varepsilon_R(u, v) = \begin{cases} 0 & \text{if } (u, v) \text{ is not the image of a point of } R \\ \text{sign } \dfrac{d(u, v)}{d(x, y)} & \text{if } (u, v) \text{ is the image of a point of } R. \end{cases}$$

More generally we consider the case where the mapping of $R$ is not necessarily 1-1. We assume that we can divide $R$ into subsets $R_i$, each of which is mapped 1-1 and in each of which the Jacobian is of constant sign $\varepsilon_{R_i}$. Then

$$\iint_R f \frac{d(u, v)}{d(x, y)} \, dx \, dy = \sum_i \iint_{R_i} f \frac{d(u, v)}{d(x, y)} \, dx \, dy$$

$$= \sum_i \iint f\varepsilon_{R_i} \, du \, dv = \iint f\chi_R \, du \, dv.$$

Here the last integral is extended over the whole $u, v$-plane, and the function $\chi_R$ stands for

$$\chi_R(u, v) = \sum_i \varepsilon_{R_i}(u, v).$$

Each term $\varepsilon_{R_i}(u, v)$, when $(u, v)$ is image of a point of $R_i$, is equal to the sign of the Jacobian at the point. Hence, the function $\chi_R(u, v)$, *the degree of the mapping of $R$* at the point $(u, v)$, is the excess of the number of points of $R$ with image $(u, v)$ for which $d(u, v)/d(x, y)$ is positive over the number of those points for which $d(u, v)/d(x, y) < 0$. With this definition of $\chi_R(u, v)$ the transformation formula for integrals becomes

(22) $$\iint f(u, v) \, \chi_R(u, v) \, du \, dv = \iint_R f(u(x, y), v(x, y)) \frac{d(u, v)}{d(x, y)} \, dx \, dy.$$

Taking the constant 1 for $f$, we obtain the formula

(23) $$\iint_R \frac{d(u, v)}{d(x, y)} \, dx \, dy = \iint \chi_R(u, v) \, du \, dv,$$

which generalizes formula (20a) to mappings with nonvanishing Jacobian that are not necessarily 1-1.

As an example, consider the mapping

(24a) $$u = e^x \cos y, \qquad v = e^x \sin y,$$

for which

$$\frac{d(u, v)}{d(x, y)} = e^{2x} > 0$$

for all $(x, y)$. Using polar coordinates $r$, $\theta$ in the $u$, $v$-plane defined by $u = r \cos \theta$, $v = r \sin \theta$, we see that the image of the point $(x, y)$ is the point with polar coordinates $r = e^x$, $\theta = y$. Now let $R$ be the rectangle

(24b) $$0 < x < \log 2, \qquad -\frac{3}{2}\pi < y < \frac{3}{2}\pi.$$

The image points lie in the annulus $1 < r < 2$ (see Fig. 5.5) The points of the annulus with $u < 0$ are covered twice by the image of $R$ (they can be assigned polar angles between $\pi/2$ and $3\pi/2$ or between $-\pi/2$ and $-3\pi/2$). The other points of the annulus are covered once.

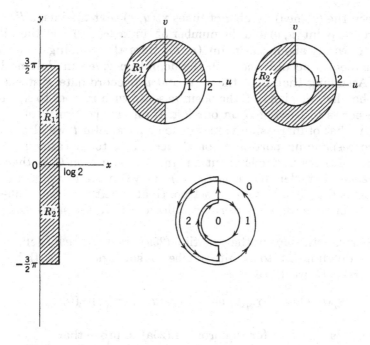

**Figure 5.5** Degree of the mapping $u = e^x \cos y$, $v = e^x \sin y$ applied to the rectangle $0 < x < \log 2$, $|y| < 3/2\,\pi$.

Hence,

$$\chi_R(u, v) = \begin{cases} 0 & \text{for} \quad 0 \leq r \leq 1 \quad \text{or} \quad r \geq 2 \\ 2 & \text{for} \quad 1 < r < 2 \quad \text{and} \quad u < 0 \\ 1 & \text{for} \quad 1 < r < 2 \quad \text{and} \quad u \geq 0. \end{cases}$$

Here, since each half of the annulus $1 < r < 2$ has area $3\pi/2$, we have

$$\iint \chi_R(u, v) \, du \, dv = 2\left(\frac{3}{2}\pi\right) + \frac{3}{2}\pi = \frac{9}{2}\pi.$$

Alternatively, by direct calculation,

$$\iint_R \frac{d(u, v)}{d(x, y)} \, dx \, dy = \int_{-3\pi/2}^{3\pi/2} dy \int_0^{\log 2} e^{2x} \, dx = 3\pi \int_0^{\log 2} e^{2x} \, dx = \frac{9}{2}\pi.$$

We have the remarkable identity

(25a) $$\chi_R(u, v) = \mu_C(u, v)$$

between the (signed) number of times $\chi_R(u, v)$ that the image $R'$ of $R$ covers the point $(u, v)$ and the number of times $\mu_C(u, v)$ that the image $C'$ of $C$ winds about the point $(u, v)$. Here the winding number is determined in accordance with the definition given in Volume I (p. 431). Assuming that both the $x$, $y$- and $u$, $v$-coordinate systems are right-handed, we give to $C$ the positive sense with respect to $R$, which corresponds to leaving $R$ on our left. If on any portion $\gamma$ of $C$ this sense is that of increasing values of some parameter $t$, we also orient the corresponding portion $\gamma'$ of $C'$ according to increasing $t$. The number of times $C'$ winds about a point $(u_0, v_0)$ not on $C'$ is then the difference—here denoted by $\mu_C(u_0, v_0)$—between the number of times $C'$ crosses the ray $u = u_0$, $v > v_0$ from right to left and the number of times $C'$ crosses from left to right, following $C'$ in the sense assigned to it.

Clearly, both sides in the equation (25a) are *additive* by definition; that is, dividing $R$ into a finite number of subregions $R_i$ with boundary curves $C_i$ we have

$$\chi_R(u, v) = \sum_i \chi_{R_i}(u, v), \qquad \mu_C(u, v) = \sum_i \mu_{C_i}(u, v).$$

Hence, it is sufficient for the proof of (25a) to prove that

(25b) $$\chi_{R_i}(u, v) = \mu_{C_i}(u, v)$$

for any portion $R_i$ of $R$ that is mapped 1–1 into the $u$, $v$-plane and in which the Jacobian $d(u, v)/d(x, y)$ has a constant sign $\varepsilon_{R_i}$. Let $R_i$ have the boundary curve $C_i$, and let $R_i'$ be the image of $R_i$, $C_i'$ that of $C_i$. Obviously, for any $(u, v)$ not on $C_i$

$$\chi_{R_i}(u,\, v) = \begin{cases} \varepsilon_{R_i} & \text{for } (u,\, v) \text{ in } R_i \\ 0 & \text{for } (u,\, v) \text{ exterior to } R_i. \end{cases}$$

Moreover, $C_i$ is a simple closed curve whose orientation is counter-clockwise for $\varepsilon_{R_i} > 0$, clockwise for $\varepsilon_{R_i} < 0$ (see Section 3.3e, p. 260). Hence, the number of times $C_i$ winds about a point $(u, v)$ also is $\varepsilon_{R_i}$ for $(u, v)$ inside $C_i$ and is 0 for $(u, v)$ outside $C_i$, which proves (25b).

For the example on p. 563 the identity of $\chi_R(u,\, v)$ and $\mu_C(u,\, v)$ is immediate by inspection (see Fig. 5.5).

### 5.5 Area Differentiation. Transformation of $\Delta u$ to Polar Coordinates

On p. 387 we defined the notion of *space differentiation* of a triple integral. In two dimensions we deal with the corresponding concept of *area differentiation* of a double integral

$$(26) \qquad M(R) = \iint_R \rho(x,\, y)\, dx\, dy.$$

We assume here that $\rho(x,\, y)$ is a continuous function defined in an open set $S$ of the $x$, $y$-plane. With any (Jordan-measurable and closed) subset $R$ of $S$ we can then associate through formula (26) a value $M = M(R)$. We denote by $A(R)$ the area of $R$:

$$A(R) = \iint_R dx\, dy.$$

From the mean value theorem (p. 384) we know that the quotient

$$\frac{M(R)}{A(R)}$$

lies between the supremum and the infimum of $\rho(x,\, y)$ in $R$. It follows that at a point $(x_0,\, y_0)$ of $S$

$$(27) \qquad \rho(x_0,\, y_0) = \lim_{n \to \infty} \frac{M(R_n)}{A(R_n)},$$

where the $R_n$ are any sequence of subsets of $S$ that have an area $A(R_n)$, contain the point $(x_0, y_0)$ and have diameters tending to 0 for $n \to \infty$. The limit is analogous to differentiation in one dimension. We call $\rho$ the *area derivative* of $M$ with respect to $A$.

Physically, we can interpret the differential form $\rho(x, y)\, dx\, dy$ (at least for $\rho > 0$) as the element of mass of a certain mass-distribution in the plane, the integral $M(R)$ representing the *total mass* contained in the set $R$. Equation (27) then shows the $\rho(x, y)$ can be obtained as the limit of the masses of the sets $R_n$ divided by their areas as the $R_n$ shrink into the point $(x, y)$. Calling $M(R_n)/A(R_n)$ the *average density* of mass-distribution in the set $R_n$, we define $\rho(x, y)$ as the *density* at $(x, y)$, or as the *mass per unit area*. In a different physical interpretation not restricted to positive $\rho$, we can think of $\rho\, dx\, dy$ as *element of electric charge*, of $M(R)$ as the *total charge in R*, and of $\rho(x, y)$ as the *charge density* or *charge per unit area*.

In a mapping

$$\bar{x} = \bar{x}(x, y), \quad \bar{y} = \bar{y}(x, y)$$

of points $(x, y)$ of the plane onto points $(\bar{x}, \bar{y})$ the area of the image $\bar{R}$ of a set $R$ is given by

$$A(\bar{R}) = \iint_{\bar{R}} d\bar{x}\, d\bar{y} = \iint_{R} \frac{d(\bar{x}, \bar{y})}{d(x, y)}\, dx\, dy$$

[see formula (20a)]. Here clearly the Jacobian

$$\frac{d(\bar{x}, \bar{y})}{d(x, y)} = \lim_{n \to \infty} \frac{A\,(\bar{R}_n)}{A\,(R_n)}$$

is the *area derivative of the area of the image region with respect to the area of the original region.*

Imagine now that the plane is covered by a deformable elastic material where $(x, y)$ is the position of a particle of the material at a certain time $t$ and that $(\bar{x}, \bar{y})$ is the position of the same particle at a later time $\bar{t}$. Let $\rho(x, y)$ denote the density of the material at the position $(x, y)$ at the time $t$ and $\bar{\rho}(\bar{x}, \bar{y})$ that at the time $\bar{t}$ at $(\bar{x}, \bar{y})$. If we postulate that the total mass of the particles filling the set $R$ at time $t$ is the same as that of the same particles at the time $\bar{t}$ when they fill the set $\bar{R}$, then

$$M(\bar{R}) = \iint_{\bar{R}} \bar{\rho}\, d\bar{x}\, d\bar{y} = M(R) = \iint_{R} \rho\, dx\, dy$$

It follows that

$$\bar{\rho} = \lim_{n \to \infty} \frac{M(\bar{R}_n)}{A(\bar{R}_n)} = \lim_{n \to \infty} \frac{M(\bar{R}_n)}{A(R_n)} \frac{A(R_n)}{A(\bar{R}_n)} = \frac{\rho}{d(\bar{x}, \bar{y})/d(x, y)}$$

Hence, mass-densities in mappings $(\bar{x}, \bar{y}) \to (x, y)$ transform according to the rule

$$(28) \qquad \rho = \bar{\rho} \, \frac{d(\bar{x}, \bar{y})}{d(x, y)}.$$

This equation, written as a relation between differential forms (see p. 308), just states the *law of conservation of elements of mass:*

$$(28a) \qquad \rho \, dx \, dy = \bar{\rho} \, d\bar{x} \, d\bar{y}.$$

Applying the notion of area differentiation enables us to transform the expression $\Delta u = u_{xx} + u_{yy}$ to new coordinates, for example, to polar coordinates $(r, \theta)$. For this purpose we use the formula

$$\iint_R \Delta u \, dx \, dy = \int_C \frac{du}{dn} \, ds,$$

which arises from Green's theorem [see (15), p. 558] if we put $w = 1$. If we carry out area differentiation using a sequence of sets $R_n$ with boundaries $C_n$ shrinking into the point $(x, y)$, we find

$$(29) \qquad \Delta u = \lim_{u \to \infty} \frac{1}{A(R_n)} \int_{C_n} \frac{du}{dn} \, ds$$

In order to transform $\Delta u$ to other coordinates, we therefore have only to apply the corresponding transformation to the simple line integral $\int (du/dn) \, ds$, divide by the area, and perform a passage to the limit. The advantage over the direct calculation is that we need not carry out the somewhat complicated calculation of the *second* derivatives of $u$, since only the first derivatives occur in the line integral.

As an important example, we shall work out the transformation of $\Delta u$ to polar coordinates $(r, \theta)$. For $R_n$ we choose a small mesh of the polar coordinate net,[1] say that between the circles $r$ and $r + h$ and the lines $\theta$ and $\theta + k$, whose area, as we know, has the value

$$A(R_n) = kh \left( r + \frac{1}{2} h \right).$$

---

[1] Here $h$ and $k$ are supposed to tend to 0 as $n \to \infty$.

The first derivatives transform according to the formulae

$$u_r = \frac{\partial}{\partial r} u \, (r \cos \theta, \, r \sin \theta) = \frac{1}{r} \, (xu_x + yu_y)$$

$$u_\theta = \frac{\partial}{\partial \theta} u \, (r \cos \theta, \, r \sin \theta) = - \, yu_x + xu_y.$$

On a circle $r =$ constant the direction cosines of the normal (pointing in the direction of increasing $r$) are $x/r$, $y/r$, and hence, $du/dn = u_r$, while $ds = r \, d\theta$. On a ray $\theta =$ constant the direction cosines of the normal (pointing in the direction of increasing $\theta$) are $-y/r$, $x/r$, and hence, $du/dn = u_\theta/r$ while $ds = dr$. Thus, taking the integral of the derivative of $u$ in the direction of the *outward normal* along the boundary $C_n$ of $R_n$, we find

$$\int_{C_n} \frac{du}{dn} \, ds = \int_\theta^{\theta+k} [(r + h)u_r \, (r + h, \, \theta) - ru_r \, (r, \, \theta)] \, d\theta$$

$$+ \int_r^{r+h} \frac{1}{r} \, [u_\theta(r, \, \theta + k) - u_\theta(r, \, \theta)] \, dr$$

$$= \int_\theta^{\theta+k} d\theta \int_r^{r+h} [ru_r(r, \, \theta)]_r \, dr$$

$$+ \int_r^{r+h} dr \int_\theta^{\theta+k} \left[ \frac{1}{r} \, u_\theta(r, \, \theta) \right]_\theta d\theta$$

$$= \iint_{R_n} \left[ \frac{1}{r} \, (ru_r)_r + \frac{1}{r} \left( \frac{1}{r} \, u_\theta \right)_\theta \right] r \, dr \, d\theta.$$

Since here by the formula for area in polar coordinates (p. 000)

$$A(R_n) = \iint_{R_n} r \, dr \, d\theta$$

we find from (29) that

$$(30) \quad \Delta u = \frac{1}{r} \, (ru_r)_r + \frac{1}{r} \left( \frac{1}{r} \, u_\theta \right)_\theta = u_{rr} + \frac{1}{r} + u_r + \frac{1}{r^2} \, u_{\theta\theta},$$

which is the required transformation formula.

This formula suggests some important special solutions of the Laplace differential equation $\Delta u = 0$. From (30) solutions of this equation that depend on $r$ alone—that is, that are of the form $u = f(r)$—must satisfy the condition

$$\frac{1}{r}\,[rf'(r)]_r = 0$$

which leads to $rf'(r) = \text{constant} = a$ or to

(31a) $\qquad u = f(r) = a \log r + b = a \log \sqrt{x^2 + y^2} + b,$

where $a$ and $b$ are constants. Similarly, we find that the general solution of Laplace's equation that depends on $\theta$ alone has the form

(31b) $\qquad u = c\theta + d = c \arctan \dfrac{y}{x} + d,$

with constants $c$ and $d$.

## 5.6 Interpretation of the Formulae of Gauss and Stokes by Two-Dimensional Flows

Our integral theorems find their most natural interpretation in terms of the motion of a liquid moving in the $x, y$-plane. The motion shall be described at every moment by its velocity field.[1] The particle that occupies the location $(x, y)$ at the time $t$ shall have the velocity vector $\mathbf{v} = (v_1, v_2)$.

If the velocity of the liquid were independent of $x, y, t$, the liquid that crosses a line segment $I$ during the time interval from $t$ to $t + dt$ fills at the time $t + dt$ a parallelogram of area $(\mathbf{v} \cdot \mathbf{n})\, s\, dt$, where $s$ is the length of $I$ and $\mathbf{n}$ is the unit normal vector to $I$ pointing to the side of $I$ to which the liquid crosses (see Fig. 5.6).[2] If instead we arbitrarily choose for $\mathbf{n}$ any one of the two unit normal vectors to $I$, then $(\mathbf{v} \cdot \mathbf{n})s\, dt$ is the area filled by the liquid crossing $I$ in the time interval from $t$ to $t + dt$, counted positive if the liquid crosses toward the side to which $\mathbf{n}$ points, and negative otherwise. If $\rho$ is the density of the

---

[1] The motion in the $x, y$-plane may be thought of as part of a motion in $x, y, z$-space, in which the velocity of any particle is parallel to the $x, y$-plane and is independent of the $z$-coordinate.

[2] The parallelogram is formed by the points $(\bar{x}, \bar{y})$ for which the segment with end points $(\bar{x}, \bar{y})$ and

$$(x, y) = (\bar{x} - v_1\, dt, \bar{y} - v_2\, dt)$$

has points in common with $I$.

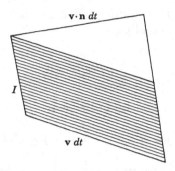

**Figure 5.6** Amount of liquid
crossing segment $I$ in time $dt$
for uniform flow of velocity **v**.

liquid, then $(\mathbf{v} \cdot \mathbf{n})\, \rho\, s\, dt$ is the *mass* of the liquid that crosses $I$ toward
the side to which **n** points.

Let $C$ be a curve in the $x$, $y$-plane. Along $C$ we arbitrarily select
one of the two possible unit normal vectors and denote it by **n**. In
a flow with velocity and density depending on $x$, $y$, $t$ the integral

$$(32a) \qquad \int_C (\mathbf{v} \cdot \mathbf{n})\rho \, ds$$

represents the mass of the liquid crossing $C$ in unit time toward that
side of $C$ pointed to by **n**. This follows immediately by approximating
$C$ by a polygon and the flow by one for which the velocity is constant
across each side of the polygon.

If $C$ is the boundary of a region $R$ and if **n** is the outward drawn
normal the integral represents the mass of the liquid *leaving* $R$ in unit
time.[1] Applying the divergence theorem in the form (7), p. 554, we
can express the flow through $C$ as a double integral:

$$(32b) \qquad \int_C (\mathbf{v} \cdot \mathbf{n}) \, \rho \, ds = \int_C (\rho\mathbf{v}) \cdot \mathbf{n} \, ds = \iint_R \operatorname{div} (\rho\mathbf{v}) \, dx \, dy.$$

We can compare this flow of mass through $C$ out of $R$ with the
change of mass contained in $R$. The total mass of the liquid contained
in the region $R$ at the time $t$ is[2]

---

[1] This will be a negative quantity if the net flow is *into* $R$.
[2] This generally is a function of $t$, since $\rho = \rho(x, y, t)$ is permitted to vary with $t$. The
region $R$ and its boundary $C$ are held fixed in the present consideration.

$$\iint_R \rho \, dx \, dy.$$

Thus, in unit time there is a loss of mass contained in $R$ by the amount

$$-\frac{d}{dt}\iint_R \rho(x, y, t) \, dx \, dy = -\iint \rho_t(x, y, t) \, dx \, dy.$$

If we assume that mass is preserved, then mass can only be lost to $R$ by passing through the boundary $C$. Hence, by (32b), we must have

(32c) $$\iint_R \text{div} \, (\rho\mathbf{v}) \, dx \, dy = -\iint_R \rho_t \, dx \, dy.$$

This identity holds for arbitrary regions $R$. Dividing by the area of $R$ and shrinking $R$ into a point (that is, by area differentiation), we find in the limit that

(33) $$\rho_t + \text{div} \, (\rho\mathbf{v}) = 0$$

(cf. Section 4.6, Exercise 15). This differential equation[1] and the integral relation (32c) express the *law of conservation of mass* in the flow. In terms of the components $v_1$, $v_2$ of the velocity vector we can write (33) as

(33a) $$\frac{\partial\rho}{\partial t} + v_1\frac{\partial\rho}{\partial x} + v_2\frac{\partial\rho}{\partial y} + \rho\left(\frac{\partial v_1}{\partial x} + \frac{\partial v_2}{\partial y}\right) = 0.$$

An important special case of this equation arises when we deal with an *incompressible homogeneous* medium in which $\rho$ has a constant value independent of location and time. In that case equations (33) or (33a) reduce to an equation for the velocity vector alone:

(34) $$\text{div} \, \mathbf{v} = \frac{\partial v_1}{\partial x} + \frac{\partial v_2}{\partial y} = 0.$$

It follows from (32b) that the total amount of an incompressible liquid crossing a closed curve $C$ in unit time is 0:

(35) $$\int_C \mathbf{v} \cdot \mathbf{n} \, ds = 0.$$

---

[1] In mechanics often referred to as the *continuity equation*.

Stokes's theorem (9), p. 554, applied to the vector **v** also has an interpretation in terms of fluid flow. The integral extended over a closed oriented curve $C$

$$\int_C \mathbf{v} \cdot \mathbf{t} \, ds,$$

where **t** is the unit tangent vector corresponding to the orientation of $C$, is called the *circulation* of the fluid around $C$. By Stokes's theorem the circulation is equal to the double integral

$$\iint_R (\text{curl } \mathbf{v})_z \, dx \, dy$$

over the enclosed region $R$. Hence, the quantity

(36) $$(\text{curl } \mathbf{v})_z = \frac{\partial v_2}{\partial x} - \frac{\partial v_1}{\partial y},$$

which is called the *vorticity* of the motion, measures the *density of circulation* at the point $(x, y)$ in the sense that the area integral of the vorticity gives the circulation around the boundary.

A flow is called *irrotational* if the vorticity vanishes everywhere, that is, if

(37) $$\frac{\partial v_2}{\partial x} - \frac{\partial v_1}{\partial y} = 0.$$

By Stokes's theorem the circulation around a closed curve $C$ vanishes if $C$ is the boundary of a region where the motion is irrotational. Since (37) is the condition for $v_1 \, dx + v_2 \, dy$ to be an exact differential (see p. 104), there exists for an irrotational flow in every simply connected region a function $\varphi = \varphi(x, y, t)$ such that

(38) $$v_1 = - \varphi_x, \qquad v_2 = - \varphi_y.$$

The scalar $\varphi$ (which is determined within a constant) is called a *velocity potential*. In vector notation (38) can be replaced by the single equation

(38a) $$\mathbf{v} = - \text{grad } \varphi.$$

The *irrotational motion of an incompressible homogeneous liquid* satisfies both equations (37) and (34). Substituting for $v_1$ and $v_2$ in (34) their expressions from (38), we find that the *velocity potential is a solution of Laplace's equation:*

$$\Delta\varphi = \varphi_{xx} + \varphi_{yy} = 0.$$

As an example, we consider the flow that corresponds to the solution

$$\varphi = a \, \log r = a \, \log \sqrt{x^2 + y^2}$$

of the Laplace equation [cf. (31a), p. 569]. By (38) the velocity vector **v** has components

$$v_1 = -\frac{ax}{r^2}, \qquad v_2 = -\frac{ay}{r^2}$$

and is singular at the origin (see Fig. 5.7a). All velocity vectors point towards the origin for $a > 0$, away from the origin for $a < 0$. In this example the velocity of the liquid at a given location does not change with time, although we have different velocities at different points; we speak of a *steady flow*. The circulation around any closed curve $C$ not passing through the origin vanishes, since

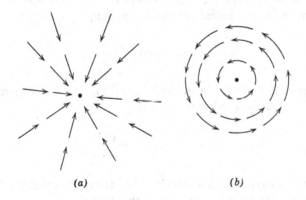

(a)             (b)

**Figure 5.7**    (a) Flow with sink. (b) Flow with vortex.

$$\int_C \mathbf{v} \cdot \mathbf{t} \, ds = \int_C v_1 \, dx + v_2 \, dy = -\int_C d\varphi = 0.$$

On the other hand, the amount of liquid passing outward through the closed curve $C$ in unit time is

$$\rho \int_C \mathbf{v} \cdot \mathbf{n} \, ds = \rho \int_C \left( v_1 \frac{dx}{dn} + v_2 \frac{dy}{dn} \right) ds = \rho \int_C v_1 \, dy - v_2 \, dx$$

$$= - a\rho \int_C \frac{x \, dy - y \, dx}{x^2 + y^2} = - a\rho \int_C d\theta,$$

where $\theta$ is the polar angle from the origin. Since (see p. 354)

$$\frac{1}{2\pi} \int_C d\theta$$

is an integer that measures the number of times $C$ winds around the origin, we see that if the closed curve $C$ is simple, does not pass through the origin, and is oriented counterclockwise,

$$\rho \int_C \mathbf{v} \cdot \mathbf{n} \, ds = \begin{cases} 0 \text{ if } C \text{ does not enclose the origin} \\ -2\pi a\rho \text{ if } C \text{ encloses the origin.} \end{cases}$$

Thus, the same amount of mass flows in unit time through every simple closed curve $C$ enclosing the origin. For $a > 0$ the origin is a *sink,* where mass disappears at the rate of $2\pi a\rho$ units in unit time. For $a < 0$ we have a *source* of mass at the origin.

The opposite behavior is encountered if we consider the steady flow with velocity potential [see (31b), p. 569]

$$\varphi = c\theta = c \arctan \frac{y}{x}.$$

While $\varphi$ itself is a multiple valued function, the corresponding velocity field has univalued components

$$v_1 = \frac{cy}{r^2}, \quad v_2 = - \frac{cx}{r^2}.$$

The vector $\mathbf{v}$ is perpendicular to the radii from the origin. (Fig. 5.7b). Again the velocity field is singular at the origin.

The circulation around a closed curve $C$ has the value

$$\int_C v_1 \, dx + v_2 \, dy = - \int_C d\varphi = - c \int_C d\theta.$$

Hence, the circulation is zero for a simple closed curve not enclosing the origin. For a simple closed curve running around the origin in the counterclockwise sense we find the value $-2\pi c$ for the circulation. This corresponds to a *vortex of strength* $-2\pi c$ concentrated at the origin. On the other hand, the flow of mass in unit time through any closed curve $C$ not passing through the origin is 0, since here

$$\rho \int_C \mathbf{v} \cdot \mathbf{n} \, ds = \rho \int_C v_1 \, dy - v_2 \, dx$$

$$= c\rho \int_C \frac{x \, dx + y \, dy}{x^2 + y^2}$$

$$= c\rho \int_C \frac{dr}{r} = 0.$$

Thus, the origin is not a source or sink of mass.

## 5.7   Orientation of Surfaces

The theory of integration for three independent variables includes not only triple integrals and line integrals, which we have discussed previously, but also the concept of *surface integral*. In order to explain the latter, we begin with considerations of a general nature, which at the same time will serve to refine our previous ideas relating to double integrals. In treating integrals of a differential over a curve $C$ in the plane or in space (p. 89), we found it necessary not just to consider $C$ as a set of points in space but to assign to it a certain *sense,* or *orientation.* The same holds when we consider integrals of differential forms over surfaces in space of three or more dimensions. Similarly, the definition of integrals of third-order differential forms over three-dimensional manifolds requires a definition of orientation for such manifolds. In discussing this topological concept of orientation we shall restrict ourselves to the simplest situations of curves, surfaces, and such lying in a euclidean space of any dimension and possessing smooth parametric representations in a sufficiently small neighborhood of any point.

### a.   Orientation of Two-Dimensional Surfaces in Three Space

In Section 3.4, we described surfaces in three-dimensional space by means of their parametric representations. In what follows we use a somewhat refined notion of a surface, as a set of points in space that exists independently of any particular parametric representation and that for its complete description may even require several systems of parameters. We define a two-dimensional surface $S$ as a set of points

$x$, $y$, $z$-space with *regular local* representations by means of two parameters. That is, in a neighborhood of any point $P_0$ of $S$ the position vectors $\mathbf{X} = \overrightarrow{OP} = (x, y, z)$ of the points $P$ of $S$ are representable in the form

(39a) $$\mathbf{X} = \mathbf{X}(u, v)$$

where the parameters $u$, $v$ range over an open set $\gamma$ in the $u$, $v$-plane and different $(u, v)$ correspond to different points on $S$. We require, moreover, the representation (39a) to be *regular* in the sense that the vector $\mathbf{X}(u, v)$ has derivatives $\mathbf{X}_u = (x_u, y_u, z_u)$ and $\mathbf{X}_v = (x_v, y_v, z_v)$ with respect to $u$, $v$ in $\gamma$ that are continuous and linearly independent.[1] Independence of the vectors $\mathbf{X}_u$, $\mathbf{X}_v$ is expressed algebraically by the condition [see formula (40d) p. 279]

(39b) $$\mathbf{X}_u \times \mathbf{X}_v \neq 0$$

or by

(39c)    $$\Gamma(\mathbf{X}_u, \mathbf{X}_v) = \begin{vmatrix} \mathbf{X}_u \cdot \mathbf{X}_u & \mathbf{X}_u \cdot \mathbf{X}_v \\ \mathbf{X}_v \cdot \mathbf{X}_u & \mathbf{X}_v \cdot \mathbf{X}_v \end{vmatrix} = |\mathbf{X}_u \times \mathbf{X}_v|^2 > 0$$

where $\Gamma$ denotes the Gram determinant of the vectors $\mathbf{X}_u$, $\mathbf{X}_v$ [see p. 191 and formula (45a), p. 284].

The vectors $\mathbf{X}_u(u, v)$ and $\mathbf{X}_v(u, v)$ at a point $P = \mathbf{X}(u, v)$ of $S$ with parameters $u$, $v$ are tangential to $S$ at $P$ and "span" the tangent plane $\pi(P)$ of $S$ at $P$; that is, every point of the tangent plane has a position vector of the form

$$\mathbf{X}(u, v) + \lambda \mathbf{X}_u(u, v) + \mu \mathbf{X}_v(u, v)$$

with suitable constants $\lambda$, $\mu$ (see p. 144). We *orient* the surface $S$ by assigning an orientation to each of the tangent planes of $S$ *in a continuous manner.* We shall give a precise meaning to this statement.

---

[1]Even for as simple a surface as a sphere we cannot hope to find a *single* regular parametric representation for the whole surface. For that reason we only require existence of *local* representations for $S$. Incidentally, we exclude surfaces that have edges and corners, where no *regular* local representation is possible (for example, cubes).

More generally, a (simple) *m*-dimensional surface in *n*-dimensional $x_1, \ldots, x_n$-space is defined as a set of points with local parametric representations of the form
$$\mathbf{X} = \mathbf{X}(u_1, \ldots, u_m),$$
where the first derivatives of the vector $\mathbf{X}$ with respect to the variables $u_k$ are continuous and linearly independent.

An oriented tangent plane $\pi^*(P)$ is obtained from the plane $\pi(P)$ by specifying an *ordered* pair of independent vectors $\xi(P)$ and $\eta(P)$ in $\pi(P)$. The orientation of $\pi^*$ is then that of the ordered pair $\xi$, $\eta$ or, symbolically,[1]

(40a) $$\Omega(\pi^*(P)) = \Omega(\xi(P), \eta(P)).$$

Any other ordered pair of independent tangential vectors $\xi'$, $\eta'$ at $P$ determines the same orientation if

(40b) $$[\xi, \eta; \xi', \eta'] = \begin{vmatrix} \xi \cdot \xi' & \xi \cdot \eta' \\ \eta \cdot \xi' & \eta \cdot \eta' \end{vmatrix} > 0;$$

(see p. 196). More generally,

(40c) $$\Omega(\xi, \eta) = \text{sgn } [\xi, \eta; \xi', \eta'] \, \Omega(\xi', \eta')$$

The orientation $\Omega(\pi^*)$ can be described more easily in terms of the unit vector (see Fig. 5.8)

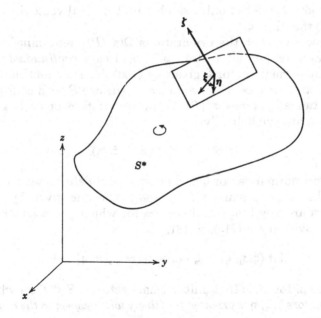

**Figure 5.8**

---

[1]We can picture $\Omega(\pi^*(P))$ as a sense of rotation in the plane $\pi(P)$; namely, as the sense of that rotation by an angle less than 180° that takes the direction of the vector $\xi$ into that of $\eta$.

(40d)
$$\zeta = \frac{\xi \times \eta}{|\xi \times \eta|}.$$

which is normal to $\xi$ and $\eta$ and, hence, to the tangent plane $\pi(P)$. The vector $\zeta$ does not depend on the individual pair of tangential vectors $\xi$, $\eta$ but only on the orientation determined by the vectors. This follows from the general identity for vector products[1]

(40e)
$$(\xi \times \eta) \cdot (\xi' \times \eta') = \begin{vmatrix} \xi \cdot \xi' & \xi \cdot \eta' \\ \eta \cdot \xi' & \eta \cdot \eta' \end{vmatrix} = [\xi, \eta; \xi', \eta'].$$

If here the ordered pairs of tangential vectors $\xi$, $\eta$ and $\xi'$, $\eta'$ give the same orientation to $\pi$, then by (40b) the corresponding unit normals $\zeta$ and $\zeta'$ satisfy

(40f)
$$\zeta \cdot \zeta' = \frac{[\xi, \eta; \xi', \eta']}{|\xi \times \eta| \, |\xi' \times \eta'|} > 0.$$

Since $\zeta$ and $-\zeta$ are the only possible unit normal vectors, it follows from (40f) that $\zeta' = \zeta$.

We now say that the orientations $\Omega(\pi^*(P))$ determined by (40a) from pairs of tangential vectors $\xi(P)$, $\eta(P)$ *vary continuously* with P if the unit normal vector $\zeta$ given by (40d) depends continuously on P. An *oriented surface* $S^*$ is defined as a surface $S$ with continuously oriented tangent planes $\pi^*(P)$. If the orientation of $\pi^*$ is given by (40a), we write symbolically

(40g)
$$\Omega(S^*) = \Omega(\pi^*) = \Omega(\xi, \eta).$$

Any unit normal vector $\zeta$ at a point $P$ of $S$ determines an orientation of the tangent plane $\pi(P)$, namely, the one given by $\Omega(\xi, \eta)$, where $\xi$, $\eta$ are any tangential vectors for which $\xi \times \eta$ has the direction of $\zeta$. By formula (71c), p. 181,

(40h)
$$\det(\xi, \eta, \zeta) = \zeta \cdot (\xi \times \eta) = |\xi \times \eta| > 0.$$

Hence (see p. 186), $\zeta$ is that unit normal vector of $S$ at $P$ for which the *triple of vectors $\zeta$, $\xi$, $\eta$ is oriented positively with respect to the coordinate axes;* that is,

---

[1]The identity can be verified directly by writing it in terms of the components of the vectors involved; see also Exercise 9b, Section 2.4, p. 203. Formula (39c) is the special case $\xi = \xi'$, $\eta = \eta' = X_v$.

$(40i)^1$ $$\Omega(\zeta, \xi, \eta) = \Omega(x, y, z).$$

An orientation of $S$ consists then in choosing in a continuous fashion a unit normal vector $\zeta$ at all points of $S$. Here $\zeta$ is given by (40d) whenever $\Omega\,(S^*) = \Omega(\xi, \eta)$ for the oriented surface $S^*$. We say that $\zeta$ is the unit normal vector *pointing to the positive side* of the oriented surface $S^*$ or is the *positive unit normal* of $S^*$.[2]

Let $S$ be a *connected* surface, that is, one with the property that any two points of $S$ can be joined by a curve lying on $S$. It is then easy to see that either $S$ cannot be oriented at all or that there are exactly two different ways of orienting $S$.[3] For two orientations of $S$ correspond to two choices $\zeta\,(P)$ and $\zeta'(P)$ of unit normal vectors on $S$. Here, necessarily, $\zeta' = \varepsilon\zeta$, where $\varepsilon = \varepsilon(P)$ has one of the values $+1$ or $-1$. Since, by assumption, the vectors $\zeta$ and $\zeta'$ vary continuously with $P$, the same holds for the scalar $\varepsilon(P) = \zeta \cdot \zeta'$. Thus, $\varepsilon$ is a continuous function on $S$ assuming only the values $+1$ or $-1$. If $\varepsilon(P) \neq \varepsilon(Q)$ for any two points $P$, $Q$ on $S$, it would follow from the intermediate value theorem that $\varepsilon = 0$ somewhere along a curve on $S$ joining $P$ and $Q$, contrary to the definition of $\varepsilon$. Consequently, $\varepsilon$ has the same value at all points of $S$. Thus, any orientation of $S$ is either the one described by the normal $\zeta(P)$ or the one described by $-\zeta\,(P)$. If $S^*$ is the oriented surface with positive normal $\zeta$, we write $-S^*$ for the one with the other orientation of $S$, so that

$(40j)$ $$\Omega(-S^*) = -\Omega(S^*).$$

Obviously, the orientation of the positive normal $\zeta$ to a connected surface $S$ at a single point $P$ uniquely determines the positive normal at any other point $Q$ and, hence, determines the orientation of $S$. We

_____

[1]Formula (40i) shows that the sense of rotation of the plane $\pi$ associated with $\Omega(\xi, \eta)$ appears counterclockwise when viewed from that side of $\pi$ to which $\zeta$ points, provided the $x$, $y$, $z$-coordinate system is right-handed. Notice that the connection between $\Omega(\xi, \eta)$ and the direction of $\zeta$ depends on the orientation of the coordinate system used, since the vector product $\xi \times \eta$ depends on that orientation.

[2]More generally, any nontangential vector $\zeta$ with initial point $P$ is said to *point to the positive side* of $S^*$ if (40i) holds. For a "material" oriented surface, say a thin metal sheet, the two sides of the surface can be painted in distinctive colors. The pigment layer on the positive side would then only occupy points that can be reached by starting at a point $P$ of the surface and moving a short distance in the direction of the positive normal to the surface.

[3]The assumption that $S$ is *connected* is essential. For a surface consisting of several disjoint connected components, the individual components might be oriented independently of each other. That there exist surfaces that cannot be oriented at all will be shown on p. 583.

only need to connect $Q$ to $P$ by a curve $C$ on $S$ and define a unit normal to $S$ along $C$ that coincides with $\zeta$ at $P$ and varies continuously along $C$; the normal then also coincides at $Q$ with the positive normal.

It is particularly simple to orient a surface $S$ that forms the boundary of a three-dimensional region $R$ of space (here $S$ need not be connected, as in the case of a spherical shell $R$). At each point $P$ of $S$ we can distinguish an *interior normal* pointing into $R$ and an *exterior normal* pointing away from $R$, both varying continuously with $P$. Taking the exterior normal as positive normal defines an orientation for $S$. We call the corresponding oriented surface $S^*$ *oriented positively with respect* to $R$.[1]

If, for example, $R$ is the spherical shell

$$\text{(40k)} \qquad a \leqq |\mathbf{X}| \leqq b,$$

the positive oriented boundary $S^*$ of $R$ has the positive unit normal

$$\text{(40l)} \quad \zeta = -\mathbf{X}/a \quad \text{for} \quad |\mathbf{X}| = a \quad \text{and} \quad \zeta = \mathbf{X}/b \quad \text{for} \quad |\mathbf{X}| = b.$$

Let a portion of the oriented surface $S^*$ have a regular parametric representation $\mathbf{X} = \mathbf{X}(u, v)$ for $(u, v)$ varying over an open set $\gamma$ of the $u, v$-plane. Then,

$$\text{(40m)} \qquad \mathbf{Z} = \frac{\mathbf{X}_u \times \mathbf{X}_v}{|\mathbf{X}_u \times \mathbf{X}_v|}$$

defines a unit normal vector for $(u, v)$ in $\gamma$. If $\zeta$ is the positive unit normal of $S^*$, we have

$$\text{(40n)} \qquad \zeta = \varepsilon \mathbf{Z}$$

---

[1]As defined here, the positive orientation of the boundary $S$ of a region $R$ depends on the orientation of the $x, y, z$-coordinate system or on the orientation of three-space determined by that system. It is often more convenient to think of $R$ also as oriented and to define unambiguously the oriented boundary $S^*$ of the oriented connected region $R^*$ in three-space. Here the "orientation" of $R^*$ consists of a particular choice of $x, y, z$-coordinate system, which then is "oriented positively with respect to $R$" by definition:

$$\Omega(R^*) = \Omega(x, y, z).$$

The positively oriented boundary surface $S^*$ of $R^*$ (usually denoted by $\partial R^*$) is defined such that

$$\Omega(\zeta, \xi, \eta) = \Omega(R^*)$$

whenever $\xi, \eta$ are tangential vectors at a point $P$ of $S$ with $\Omega(S^*) = \Omega(\xi, \eta)$, and $\zeta$ is the exterior normal unit vector at $P$.

with $\varepsilon = \varepsilon(u, v) = \pm 1$. Since both $\zeta$ and $\mathbf{Z}$ are continuous, it follows that $\varepsilon$ is continuous and, hence, constant in any connected part of $\gamma$. For $\varepsilon = 1$, that is, for

(40o) $$\Omega(S^*) = \Omega(\mathbf{X}_u, \mathbf{X}_v),$$

we say that $S^*$ is *oriented positively with respect to the parameters* $u, v$ and write

(40p) $$\Omega(S^*) = \Omega(u, v).$$

If the same portion of $S^*$ has a second regular parametric representation in terms of parameters $u'$, $v'$ varying over a region $\gamma'$, we have by formula (42), p. 283,

(40q) $$\mathbf{X}_u \times \mathbf{X}_v = \left( \frac{d(y, z)}{d(u, v)}, \frac{d(z, x)}{d(u, v)}, \frac{d(x, y)}{d(u, v)} \right)$$

$$= \frac{d(u', v')}{d(u, v)} (\mathbf{X}_{u'} \times \mathbf{X}_{v'}).$$

Hence, the unit normals $\mathbf{Z}$ and $\mathbf{Z}'$ corresponding to the two parametric representations are related by

(40r) $$\mathbf{Z} = \operatorname{sgn} \frac{d(u', v')}{d(u, v)} \mathbf{Z}'.$$

Thus, if $S^*$ is oriented positively with respect to the parameters $u, v$, then it is also positively oriented with respect to the parameters $u'$, $v'$, provided

(40s) $$\frac{d(u', v')}{d(u, v)} > 0.$$

In illustration, we consider the unit sphere $S^*$ with center at the origin, oriented positively with respect to its interior. Using $u = x$, $v = y$ as parameters for $z \neq 0$, we have

(40t) $$\mathbf{X} = (u, v, \varepsilon \sqrt{1 - u^2 - v^2}), \quad \text{where } \varepsilon = \operatorname{sgn} z.$$

The corresponding normal vector $\mathbf{Z}$ defined by (40m) becomes here

$$\mathbf{Z} = (\varepsilon x, \varepsilon y, \varepsilon z) = \varepsilon \zeta,$$

where $\zeta$ is the exterior unit normal. Hence, $S^*$ is oriented positively with respect to the parameters $x$, $y$ for $z > 0$ and negatively for $z < 0$ (see Fig. 5.9).

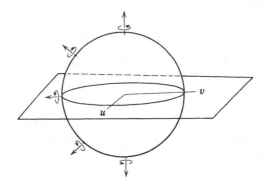

**Figure 5.9**

A surface in three-space for which no distinction between the sides can be made or along which we cannot select a continuously varying unit normal cannot be orientable. The simplest example of a "one-sided" surface of this type, shown in Fig. 5.10(a) is called a *Möbius*

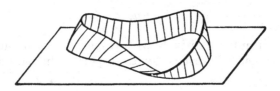

**Figure 5.10(a)** Möbius band.

*band* after its discoverer. We can easily make such a surface out of a rectangular strip of paper by fastening the ends of the strip together after rotating one end through an angle of 180°. If we start out with the rectangle $0 < u < 2\pi$, $-a < v < a$ (where $0 < a < 1$) in the $u$, $v$-plane, we arrive at a Möbius band if we move each segment $u =$ constant rigidly in such a way that its center moves to the point $(\cos u, \sin u, 0)$ of the unit circle in the $x$, $y$-plane and such that it becomes perpendicular to that circle and makes the angle $u/2$ with the positive $z$-axis (the assumption $a < 1$ keeps the surface from intersecting itself). The resulting band $S$ has the parametric representation

(40u)    $$\mathbf{X} = \left( \left(1 + v \sin \frac{u}{2}\right) \cos u, \left(1 + v \sin \frac{u}{2}\right) \sin u, v \cos \frac{u}{2} \right)$$

with $v$ restricted to the interval $-a < v < a$. The points $(u, v)$, $(u + 4\pi, v)$, $(u + 2\pi, -v)$ in the $u$, $v$-plane correspond to the same point on the surface. If for an arbitrary point $P_0$ of $S$ we make one possible choice $u_0$, $v_0$ of parameters, formula (40u) yields a regular local parametric representation of $S$ for $u$, $v$ restricted to the rectangle $\gamma$ given by

$$u_0 - \pi < u < u_0 + \pi, \qquad -a < v < a.$$

Along the center line $v = 0$ of the surface, equation (40m) defines a unit normal vector

$$\mathbf{Z} = \left( \cos u \cos \frac{u}{2}, \sin u \cos \frac{u}{2}, -\sin \frac{u}{2} \right)$$

that varies continuously with $u$. Starting out with the unit normal $\mathbf{Z} = (1, 0, 0)$ at the point $(1, 0, 0)$ of $S$ corresponding to $u = 0$ and letting $u$ increase from 0 to $2\pi$, we describe a complete circuit along the center line of the surface returning to the same point but with the opposite unit normal $\mathbf{Z} = (-1, 0, 0)$. We would find similarly that carrying during our motion a small oriented tangential curve we return to the same point with the orientation reversed. Thus, it is not possible to choose a continuously varying unit normal, or a side of $S$, or to choose a sense of rotation on $S$ in a consistent way. The one-sidedness of the Möbius band is strikingly illustrated by the insects crawling along the band in the drawing by M.C. Escher, reproduced in Fig. 5.10(b). We see that a surface does not automatically enjoy the property of *orientability*.

We oriented a surface by orienting its tangent planes in a continuous manner. The orientation of the tangent planes $\pi^*(P)$ was described by a suitable pair of independent tangential vectors $\xi(P)$, $\eta(P)$. When it came to defining "continuity" of $\Omega(\pi^*) = \Omega(\xi, \eta)$, we made use of the normal vector $\zeta$ formed according to (40d) and required $\zeta$ to be continuous. It is desirable to define continuity of the orientations $\Omega(\xi(P), \eta(P))$ without recourse to normal vectors or cross products. This is of particular importance when it comes to defining orientation for manifolds in higher-dimensional spaces, say, for a two-dimensional surface $S$ in four-dimensional euclidean space. Here again, orientation of each tangent plane can be described by an ordered pair of independent tangential vectors $\xi$, $\eta$. But there is no

**Figure 5.10(b)** *Band Van Möbius II*, by M. C. Escher (Escher Foundation, Haags Gemeentemuseum, The Hague, Netherlands).

unique unit normal vector or "side" of $S$ we can associate with $S$. We also cannot require the tangential vectors $\xi(P)$, $\eta(P)$ describing

$\Omega$ ($\pi^*$) to be defined and continuous for all $P$ on $S$.[1] We discuss shortly two definitions of orientation of surfaces in three-space equivalent
to the one given before, but not involving normals and, hence, capable
of generalization to higher dimensions.

Any regular parametric representation $\mathbf{X} = \mathbf{X}(u, v)$ of a portion
of a surface of $S$ in three-space determines a continuously varying
unit normal $\mathbf{Z}$ on that portion by means of formula (40m). Let there
be given a number of regular parametric representations for different
portions of $S$. They will then define a continuously varying unit
normal on all of $S$ and, hence, an orientation of $S$, provided at least
one of the representations is valid near any point $P$ of $S$ and provided
any two representations valid at $P$ lead to the same unit normal vector
$\mathbf{Z}$. By (40r) the latter condition simply requires that

(41a) $$\frac{d(u', v')}{d(u, v)} > 0$$

wherever two of the representations with parameters $u$, $v$ and $u'$, $v'$
hold. The surface is then oriented positively with respect to each of
the given parametric representations.

For instance, various portions of the unit sphere $S$ have the regular
parametric representations

(41b) $$\mathbf{X} = (\sin u \cos v, \sin u \sin v, \cos u)$$

$$\text{for} \quad 0 < u < \pi, \quad v_0 - \pi < v < v_0 + \pi$$

(41c) $$\mathbf{X} = (u', v', \sqrt{1 - u'^2 - v'^2}) \quad \text{for} \quad u'^2 + v'^2 < 1$$

(41d) $$\mathbf{X} = (v'', u'', -\sqrt{1 - u''^2 - v''^2}) \quad \text{for} \quad u''^2 + v''^2 < 1.$$

It is easily seen that all of these representations define an orientation
of $S$. For example, both (41b) and (41d) apply on the hemisphere $z < 0$,
and there

$$\frac{d(u'', v'')}{d(u, v)} = \frac{d(\sin u \sin v, \sin u \cos v)}{d(u, v)} = -\sin u \cos u > 0.$$

The unit normal $\mathbf{Z}$ obtained from all these parametric representations
is the exterior normal, and the orientation of $S$ is the one that is
positive with respect to the interior.

---

[1]Even for as simple a surface as a sphere in three-space no nonvanishing tangential
vectors $\xi(P)$ can be found that are continuous at all points of the surface. We can,
however, always choose the vectors $\xi(P)$, $\eta(P)$ in such a way that they vary continuously *in a neighborhood of a given point*.

The second method to be mentioned expresses the condition of continuity of $\Omega(\xi(P), \eta(P))$ directly in terms of the vectors $\xi$, $\eta$. Let $\zeta(P)$ be the unit normal vector associated with $\xi$, $\eta$ by (40d). In a neighborhood of a given point $P_0$ of $S$, a regular parametric representation $\mathbf{X} = \mathbf{X}(u, v)$ holds, defining a continuously varying normal vector $\mathbf{Z}$ by (40m). Then $\zeta(P) = \varepsilon(P)\,\mathbf{Z}(P)$ with a certain $\varepsilon(P) = \pm 1$. Continuity of the vector $\zeta(P)$ at $P_0$ obviously is equivalent to the condition $\varepsilon(P) = $ constant near $P_0$ or to the condition

$$\zeta(P) \cdot \zeta(P_0) = \varepsilon(P)\,\varepsilon(P_0)\,\mathbf{Z}(P) \cdot \mathbf{Z}(P_0) > 0$$

for all $P$ sufficiently close to $P_0$. Now, using the identity (40e), we find that

$$\zeta(P) \cdot \zeta(P_0) = \frac{[\xi(P), \eta(P); \xi(P_0), \eta(P_0)]}{|\xi(P) \times \eta(P)|\,|\xi(P_0) \times \eta(P_0)|}.$$

Consequently, the orientations $\Omega(\xi, \eta)$ vary continuously and define an orientation of the surface $S$ if for every $P_0$ on $S$

(41e)[1]                $[\xi(P), \eta(P); \xi(P_0), \eta(P_0)] > 0$

for all points $P$ on $S$ sufficiently close to $P_0$.

For example, let $S$ be the unit sphere $x^2 + y^2 + z^2 = 1$. For any point $(x, y, z)$ on $S$ that is not one of the poles $(0, 0, \pm 1)$, the vectors

$$\xi = (xz, yz, z^2 - 1), \quad \eta = (-y, x, 0)$$

are independent and tangential, since they are perpendicular to the position vector $\mathbf{X} = (x, y, z)$. With the additional choice of

$$\xi = (1, 0, 0), \quad \eta = (0, \varepsilon, 0)$$

at the pole $(0, 0, \varepsilon)$, where $\varepsilon = \pm 1$, the orientations $\Omega(\xi, \eta)$ are continuous at every point $P_0$ of $S$. This is clear when $P_0$ is not one of the poles, since then $\xi$ and $\eta$ themselves are continuous and not zero. Thus, one only has to verify condition (41e) when $P_0$ is a pole. For example, for the "north pole" $P_0 = (0, 0, 1)$ and for any point $P = (x, y, z)$ in the "northern hemisphere"

---

[1]One can deduce directly from formula (85c), p. 199, that (41e) is a relation between $\Omega(\pi^*(P))$ and $\Omega(\pi^*(P_0))$ alone and does not depend on the particular vectors $\xi(P)$, $\eta(P)$, $\xi(P_0)$, $\eta(P_0)$ used to represent the orientations of those tangent planes.

$$[\xi(P), \eta(P); \xi(P_0), \eta(P_0)] = \begin{vmatrix} \xi(P) \cdot \xi(P_0) & \xi(P) \cdot \eta(P_0) \\ \eta(P) \cdot \xi(P_0) & \eta(P) \cdot \eta(P_0) \end{vmatrix}$$

$$= \begin{vmatrix} xz & yz \\ -y & x \end{vmatrix} = (x^2 + y^2) z > 0$$

except for $P = P_0$. But, of course, also

$$[\xi(P_0), \eta(P_0); \xi(P_0), \eta(P_0)] = \begin{vmatrix} 1 & 0 \\ 0 & 1 \end{vmatrix} = 1 > 0.$$

### b. *Orientation of Curves on Oriented Surfaces*

We saw that it is possible to distinguish a positive and negative side of an oriented surface $S^*$ lying in a space with a certain orientation of the coordinate system. In the same way, we can define the positive and negative sides of an oriented curve $C^*$ lying on an oriented surface $S^*$. Let $\xi$ be a vector tangential to the curve at a point $P$ and pointing in the direction determined by the orientation of $C^*$:[1]

(41f) $$\Omega(\xi) = \Omega(C^*).$$

Let $\eta$ be a vector tangential to the surface at $P$ and linearly independent of $\xi$. We say that $\eta$ points to the positive side of $C^*$ if

(41g) $$\Omega(\eta, \xi) = \Omega(S^*).$$

Conversely, we can orient a curve $C$ lying on an oriented surface $S^*$ by requiring that a given vector $\eta$ not tangential to $C$ point to the positive side of $C$.[2]

There is a natural way to orient a curve $C$ when $C$ forms part of the boundary of a region $\sigma$ lying on an oriented surface $S^*$ if we require $\sigma$ to lie on the negative side of the oriented curve $C^*$. More precisely,

---

[1] If $\mathbf{X} = \mathbf{X}(t)$ is a parametric representation of $C^*$ and $\Omega(C^*)$ corresponds to increasing $t$, the vector $\xi$ is to have the same orientation as $d\mathbf{X}/dt$.

[2] In order to achieve greater consistency for higher dimensions the notation for *positive* and *negative* sides of a curve has been changed from the one used in Volume I (p. 342). Consider the special case, where $S^*$ is the plane with the usual counterclockwise orientation when viewed from a certain side. If $C^*$ is an oriented arc with the tangent vector $\xi$ pointing in the direction given by the orientation of $C^*$, then by (41g) a vector $\eta$ points to the *positive* side of $C^*$ if a counterclockwise rotation by an angle less than 180° takes $\eta$ into $\xi$; that is, $\eta$ points to the *right* side of $C^*$ if we look in the direction of $\xi$.

we call $C^*$ oriented positively with respect to $\sigma$ if a vector $\eta$ tangential to $S^*$ at a point $P$ of $C^*$ and pointing away from $\sigma$ points to the positive side of $C^*$. Conversely, we can indicate the orientation of a surface $S^*$ graphically by taking a region $\sigma$ on $S^*$ and marking the positive orientation of its boundary curve (see Fig. 5.11).[1]

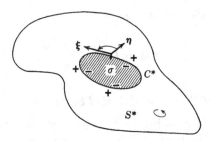

**Figure 5.11**   Oriented curve $C^*$
on oriented surface $S^*$.

If an oriented surface $S^*$ is divided into portions $S_1, S_2, \ldots, S_n$, then any arc $C$ that separates a portion $S_i$ from a portion $S_k$ receives opposite orientations when oriented positively with respect to those portions. This follows immediately from the fact that any vector $\eta$ tangential to $S$ at a point $P$ of $C$ and pointing into $S_i$ points away from $S_k$ (see Fig. 5.12).

**Figure 5.12**

## Exercises 5.7

1. Let $S$ be the two-dimensional surface ("product of two circles") in four-space given by

---

[1]In this manner of indicating orientation of a surface $S^*$ by that of a curve $C^*$ on it, we have to specify clearly the set $\sigma$ with respect to which the curve $C^*$ is to have positive orientation. Ordinarily, $C^*$ is a "small" simple closed curve dividing $S$ into two portions, exactly one of which is also small and which is then taken for $\sigma$.

$$\mathbf{X} = (\cos u, \sin u, \cos v, \sin v).$$

Prove that the vectors

$$\xi = (-x_2, x_1, -x_4, x_3), \quad \eta = (-x_2, x_1, x_4, -x_3)$$

determine an orientation on $S$.

2. Let $S^*$ be the torus with the parametric representation given in Chapter 3 (p. 286) and oriented positively with respect to the parameters $\theta$, $\phi$. Prove that $S^*$ is oriented positively with respect to its interior.

3. Let $S$ be the Möbius band represented parametrically as in (40u).
   (a) Show that the line $v = a/2$ divides $S$ into an orientable and a nonorientable set.
   (b) Show that the line $v = 0$ does not divide $S$, that is, that the set $S_1$ of points obtained by removing from $S$ all points with $v = 0$ is still connected.
   (c) Show that $S_1$ is orientable.

4. Let $\xi$, $\eta$, be independent vectors in the plane $\pi$. Put $a = |\xi|^2$, $b = \xi \cdot \eta$, $c = |\eta|^2$ and form for any $t$ the vector

$$\mathbf{R}(t) = \left(\cos t - \frac{b}{\sqrt{ac - b^2}} \sin t\right) \xi + \frac{a \sin t}{\sqrt{ac - b^2}} \eta.$$

Prove that $\mathbf{R}(t)$ is obtained by rotating the vector $\xi$ in the plane $\pi$ by an angle $t$ in the sense given by the orientation $\Omega(\xi, \eta)$.

## 5.8  Integrals of Differential Forms and of Scalars over Surfaces

### a.  *Double Integrals over Oriented Plane Regions*

In the original definitions of single and multiple integrals, say as limits of Riemann sums, *orientation* plays no role. The integral of a function $f$ is based on the use of length, areas, volumes, and so on, of elementary figures that, naturally enough, are given positive values. The use of signed quantities, amounting to the introduction of orientations, however, imposes itself right away if we want to have simple rules of operating with integrals.[1] Thus, the definite integral

$$\int_a^b f(x)\, dx$$

---

[1]Generally, mathematics would become intolerably clumsy if we restricted ourselves to using only positive quantities, for example, to *positive* distances instead of *signed* distances as coordinates. This would necessitate inumerably many distinctions between different cases in the proof and statement of simple theorems. Positivity is an essential element in the formulation of *inequalities* between mathematical objects but complicates the formulation of most *identities,* which are based usually on unrestricted algebraic manipulation of quantities.

is defined as limit of Riemann sums for $a < b$. If we want the additivity rule

$$\int_a^b f(x) \; dx + \int_b^c f(x) \; dx = \int_a^c f(x) \, dx$$

to hold without restricting the relative positions of $a$, $b$, $c$, we have to define

$$\int_a^b f \, dx$$

as well for $a \geqq b$ by the formula

(42a) $$\int_a^b f(x) \; dx = - \int_b^a f(x) \, dx$$

(see Volume I, p. 136). Geometrically, the ordered pair of numbers $a$, $b$ determines an oriented interval $I^*$ on the $x$-axis with "initial" point $a$ and "final" point $b$. Here the value of

(42b) $$\int_a^b f \, dx = \int_{I^*} f \, dx$$

is the one given by the limit of Riemann sums (which is positive for positive $f$) when the orientation of $I^*$ corresponds to the sense of increasing $x$, that is, for $a < b$. It is the negative of that limit for $a > b$. Interchanging the end points of $I^*$ converts $I^*$ into the interval $-I^*$, with the opposite orientation, so that formula (42a) can also be written as

(42c) $$\int_{-I^*} f \, dx = - \int_{I^*} f \, dx,$$

A similar situation holds for the integral over an oriented (Jordan-measurable) set $R^*$ in the $x\,y$,-plane.[1] When $R^*$ is oriented positively with respect to $x, y$-coordinates, $\Omega\,(R^*) = \Omega\,(x, y)$, the double integral

---

[1]Orientation of $R^*$ is defined here in accordance with the general definition of orientation of surfaces. It is determined by associating with each point of $R^*$ an orientation (described, for example, by a pair of vectors), the orientations varying continuously from point to point. For a connected set only two distinct orientations are possible.

$$\iint_{R^*} f(x, y)\, dx\, dy$$

is to be understood in the sense defined in Chapter 4. That is, the integral is the limit of sums obtained from subdivisions of the plane into squares of area $2^{-2n}$. The integral will have a nonnegative value for nonnegative $f$. In case $\Omega(R^*) = -\Omega(x, y) = \Omega(y, x)$, we define the integral of $f$ over $R^*$ by

$$\iint_{R^*} f\, dx\, dy = - \iint_{R^*} f\, dy\, dx,$$

where now

$$\int_{R^*} f\, dy\, dx$$

has the ordinary meaning as the limit of sums. As a consequence, we have the rule that

(43) $$\iint_{-R^*} f\, dx\, dy = - \iint_{R^*} f\, dx\, dy,$$

where $-R^*$ is obtained by changing the orientation of $R^*$. With this convention the substitution rule [see (16b), p. 403], in the form

(43a) $$\iint_{R^*} f(x, y)\, dx\, dy = \iint_{T^*} f(\phi(u, v),\, \psi(u, v)) \frac{d(x, y)}{d(u, v)}\, du\, dv,$$

holds for smooth 1–1 mappings

$$x = \phi(u, v),\, y = \psi(u, v)$$

of $T^*$ onto $R^*$ as long as the Jacobian $d(x, y)/d(u, v)$ is either positive throughout $T^*$ or negative throughout $T^*$. Here the orientation of $T^*$ has to be the one corresponding to that of $R^*$ under the mapping.[1] If, for example, $\Omega(R^*) = -\Omega(x, y)$ and if $d(x, y)/d(u, v) < 0$,

---

[1] In order to find that orientation, we form, in accordance with (40 o, p), the vectors
$$\mathbf{X}_u = (x_u, y_u),\, \mathbf{X}_v = (x_v, y_v)$$

and put

$$\Omega(R^*) = \varepsilon\, \Omega(\mathbf{X}_u, \mathbf{X}_v) = \varepsilon \left( \operatorname{sgn} \begin{vmatrix} x_u & x_v \\ y_u & y_v \end{vmatrix} \right) \Omega(x, y).$$

where $\varepsilon = \pm 1$ has the value determined by
$$\Omega(R^*) = \Omega(T^*) = \varepsilon\Omega(u, v).$$

then $\Omega(T^*) = \Omega(u, v)$. We might say that the orientation of $R^*$ attributes a certain sign to the differential form $dx\, dy$: the positive sign if the $x$, $y$-coordinate system has the orientation of $R^*$, the negative one otherwise. The sign attributed by the orientation of $T^*$ to the form $du\, dv$ is then the one that agrees with the relationship

$$dx\, dy = \frac{d(x, y)}{d(u, v)}\, du\, dv.$$

In the same way we can define triple integrals

$$\iiint_{R^*} f(x, y, z)\, dx\, dy\, dz$$

over oriented sets in $x$, $y$, $z$-space and similarly in higher dimensions.

### b. Surface Integrals of Second-Order Differential Forms

We can now give a general definition for the integral of any second-order differential form $\omega$ over an oriented surface $S^*$ in space. Let $\omega$ be given by the expression

(44)     $\omega = a(x, y, z)\, dy\, dz + b(x, y, z)\, dz\, dx + c(x, y, z)\, dx\, dy.$

Assume first that the whole surface $S^*$ under consideration can be represented parametrically in the form

(45)     $x = x(u, v), \quad y = y(u, v), \quad z = z(u, v),$

with $(u, v)$ varying over a set $R^*$ in the $u$, $v$-plane. Here $R^*$ has a certain orientation determined by that of $S^*$ (see p. 581).[1]

We can write $\omega$ in the form

$$\omega = K\, du\, dv,$$

where

(46)     $K = \dfrac{\omega}{du\, dv} = a\, \dfrac{d(y, z)}{d(u, v)} + b\, \dfrac{d(z, x)}{d(u, v)} + c\, \dfrac{d(x, y)}{d(u, v)}$

and define

---

[1]The rule for orienting $R^*$ is as follows: $\Omega(R^*) = \varepsilon\Omega(u, v)$ with $\varepsilon = \pm 1$ if $\Omega(S^*) = \varepsilon\Omega(\mathbf{X}_u, \mathbf{X}_v)$, where $\mathbf{X} = (x, y, z)$ is the position vector.

(46a)
$$\iint_{S^*} \omega = \iint_{R^*} K \, du \, dv$$

$$= \iint_{R^*} \left( a \, \frac{d(y, z)}{d(u, v)} + b \, \frac{d(z, x)}{d(u, v)} + c \, \frac{d(x, y)}{d(u, v)} \right) du \, dv.$$

The value obtained in this way for the integral of $\omega$ over the oriented surface $S^*$ is independent of the particular parametric representation for $S^*$. If the surface can also be referred to parameters $u'$, $v'$, we have (see p. 308)

$$\omega = K' \, du' \, dv'$$

where

$$K' = K \, \frac{d(u, v)}{d(u', v')}.$$

The orientation of the region of integration $R'^*$ in the $u'$, $v'$-plane is then such that the substitution rule (43a) applies and

$$\iint_{R^*} K \, du \, dv = \iint_{R'^*} K \, \frac{d(u, v)}{d(u', v')} \, du' \, dv' = \iint_{R'^*} K' \, du' \, dv'.$$

Let, for example, $S^*$ be representable nonparametrically in the form $z = f(x, y)$ with $(x, y)$ varying over the vertical projection $R^*$ of $S^*$ onto the $x$, $y$-plane. The orientation of $S^*$ determines an orientation for $R^*$. The orientation of $S^*$ can be described by specifying the normal of $S^*$ that points to the positive side of $S^*$, when the orientation of space is that of the $x$, $y$, $z$-coordinate system. When that normal forms an acute angle with the positive $z$-axis, the orientation of $R^*$ is that of the $x$, $y$-system, otherwise that of the $y$, $x$-system.[1] In either case we have

$$\iint_{S^*} \omega = \iint_{S^*} (a \, dy \, dz + b \, dz \, dx + c \, dx \, dy)$$

$$= \iint_{R^*} (c - af_x - bf_y) \, dx \, dy.$$

It is now easy to get rid of the special assumption that the whole surface $S^*$ can be represented by means of a single parametric repre-

---

[1] See p. 578. In the first case with $S^*$ referred to the parameters $x$, $y$ the positive normal $\zeta$ has the direction of the vector $(-f_x, -f_y, 1)$, and thus, det $(\zeta, \mathbf{X}_u, \mathbf{X}_v) > 0$.

sentation. We assume that the oriented surface $S^*$ can be divided into a finite number of oriented portions $S_1^*, S_2^*, \ldots, S_N^*$, in such a way that each portion has a parametric representation of the kind discussed. We form the surface integral of the form $\omega$ for each of the portions according to the definition above, and define the integral of $\omega$ over $S^*$ as the sum of the integrals over the $S_i^*$. One has to show, of course, that the integral over $S^*$ defined in this way does not depend on the particular subdivision of $S^*$ into portions $S_i^*$. For the exact assumptions needed for this to be true and the proof, see the Appendix to this chapter.

### c. *Relation Between Integrals of Differential Forms over Oriented Surfaces to Integrals of Scalars over Unoriented Surfaces*

In Chapter 4 (p. 424) we introduced the area $A$ of a surface $S$ in space without any reference to its orientation. If $S$ has the parametric representation

$$x = x(u, v), \ y = y(u, v), \ z = z(u, v)$$

and if $\xi$, $\eta$, $\zeta$ denote the components of the normal vector

(46b) $$\xi = \frac{d(y, z)}{d(u, v)}, \quad \eta = \frac{d(z, x)}{d(u, v)}, \quad \zeta = \frac{d(x, y)}{d(u, v)}$$

[see (30a) p. 428], the area of $S$ is given by

$$A = \iint_R \sqrt{\xi^2 + \eta^2 + \zeta^2} \, du \, dv.$$

Here the integral is extended over the set $R$ in the $u$, $v$-plane corresponding to $S$. The integral is understood in the original sense of a double integral in which the surface element

$$dS = \sqrt{\xi^2 + \eta^2 + \zeta^2} \, du \, dv$$

is treated as a positive quantity or, equivalently, in which $R$ is given the positive orientation with respect to the $u$, $v$-system.[1] Orientability

---

[1] If we introduce the position vector $\mathbf{X} = (x, y, z)$, the quantity $\sqrt{\xi^2 + \eta^2 + \zeta^2}$ represents the length of the vector product of the vectors $\mathbf{X}_u$ and $\mathbf{X}_v$. By (30b), p. 428, it can also be written as

$$\sqrt{EG - F^2} = \sqrt{(\mathbf{X}_u \cdot \mathbf{X}_u)(\mathbf{X}_v \cdot \mathbf{X}_v) - (\mathbf{X}_u \cdot \mathbf{X}_v)^2} = \sqrt{[\mathbf{X}_u, \mathbf{X}_v; \mathbf{X}_u, \mathbf{X}_v]}.$$

The differential $dS$ has the same invariance properties as a second order alternating differential form under parametric substitutions with *positive* Jacobian but changes sign under substitutions with negative Jacobian.

of $S$ is not essential for the definition of $A$. The reader can, for example, easily express as an integral the total area of the unorientable Möbius band with the parametric representation given on p. 583.

More generally, for a function $f(x, y, z)$ defined on the surface $S$, we can form the integral of $f$ over the surface:

$$(47a) \qquad \iint_S f \, dS = \iint_R f \, \sqrt{\xi^2 + \eta^2 + \zeta^2} \, du \, dv.$$

The value of the integral is independent of the particular parameter representation used for $S$ and does not involve any orientation of $S$. It is positive for positive $f$.

In order to relate the integral of a second-order differential form

$$\omega = a(x, y, z) \, dy \, dz + b(x, y, z) \, dz \, dx + c(x, y, z) \, dx \, dy$$

over an oriented surface $S^*$ to the surface integrals of functions over the unoriented surface $S$ as defined just now, we introduce the direction cosines of the positive normal of $S^*$

$$\cos \alpha = \frac{\varepsilon \xi}{\sqrt{\xi^2 + \eta^2 + \zeta^2}}, \; \cos \beta = \frac{\varepsilon \eta}{\sqrt{\xi^2 + \eta^2 + \zeta^2}}, \; \cos \gamma = \frac{\varepsilon \zeta}{\sqrt{\xi^2 + \eta^2 + \zeta^2}}$$

where $\xi$, $\eta$, $\zeta$ are given by (46b), and $\varepsilon = \pm 1$, $\Omega(S^*) = \varepsilon \Omega(\mathbf{X}_u, \mathbf{X}_v)$. Then, by (46),

$$K = \frac{\omega}{du \, dv} = \varepsilon \, (a \cos \alpha + b \cos \beta + c \cos \gamma) \sqrt{\xi^2 + \eta^2 + \zeta^2}.$$

Now, by (46a),

$$\iint_{S^*} \omega = \iint_{R^*} K \, du \, dv = \varepsilon \iint_R K \, du \, dv.$$

Consequently, (47a) yields the identity

$$(47b) \qquad \iint_{S^*} \omega = \iint_{S^*} a \, dy \, dz + b \, dz \, dx + c \, dx \, dy$$

$$= \iint_S (a \cos \alpha + b \cos \beta + c \cos \gamma) \, dS$$

$$= \iint_R (a \cos \alpha + b \cos \beta + c \cos \gamma) \sqrt{\xi^2 + \eta^2 + \zeta^2} \, du \, dv,$$

which expresses the integral of the differential form $\omega$ over the oriented surface $S^*$ as an integral over the unoriented surface $S$ or over the unoriented region $R$ in the parameter plane. Here, however, the *integrand* depends on the orientation of $S^*$, since cos $\alpha$, cos $\beta$, cos $\gamma$ are the direction cosines of that normal $\mathbf{n}$ of $S^*$ that points to the positive side of $S^*$ (using a positive space orientation with respect to $x$, $y$, $z$-coordinates).

If the oriented surface $S^*$ consists of several portions $S_k^*$ each of which permits a parametric representation of the form (45), we apply identity (47b) to each portion and, by addition over the different portions, obtain the same identity for the integral of $\omega$ over the whole surface $S^*$.

The direction cosines of the normal $\mathbf{n}$ pointing to the positive side of $S^*$ can be identified with the derivatives of $x$, $y$, $z$ in the direction of $\mathbf{n}$:

$$\cos \alpha = \frac{dx}{dn}, \qquad \cos \beta = \frac{dy}{dn}, \qquad \cos \gamma = \frac{dz}{dn}.$$

Thus,

$$(47c) \qquad \iint_{S^*} \omega = \iint_S \left( a\frac{dx}{dn} + b\frac{dy}{dn} + c\frac{dz}{dn} \right) dS.$$

In vector notation the formula reduces to

$$(47d) \qquad \iint_{S^*} \omega = \iint_S \mathbf{V} \cdot \mathbf{n}\, dS,$$

where $\mathbf{n} = (\cos \alpha, \cos \beta, \cos \gamma)$ is the unit normal vector on the positive side of $S^*$, and $\mathbf{V}$ the vector with components $a$, $b$, $c$.

The concept of surface integral can be interpreted intuitively in terms of the flow of an incompressible fluid (this time in three dimensions) whose density we take as unity. Let the vector $\mathbf{V} = (a, b, c)$ be the velocity vector of this flow. Then at each point of the surface $S^*$ the product $\mathbf{V} \cdot \mathbf{n}$ gives the component of the velocity of flow in the direction of the normal $\mathbf{n}$ to the surface. The expression

$$\mathbf{V} \cdot \mathbf{n}\, dS = (a \cos \alpha + b \cos \beta + c \cos \gamma)\, dS$$

can therefore be identified with the amount of fluid that flows in unit time across the element of surface $dS$ from the negative side of $S^*$

to the positive side (this quantity may, of course, be negative).[1] The surface integral

(48) $$\iint_{S^*} (a \, dy \, dz + b \, dz \, dx + c \, dx \, dy) = \iint_S \mathbf{V} \cdot \mathbf{n} \, dS$$

therefore represents the total amount of fluid flowing across the surface $S^*$ from the negative to the positive side in unit time. We notice here that an important part is played in the mathematical description of the motion of fluid by the distinction between the positive and negative sides of a surface, that is, by the introduction of orientation.

In other physical applications the vector $\mathbf{V}$ denotes the force due to a field acting at a point $(x, y, z)$. The direction of the vector $\mathbf{V}$ then gives the direction of the *lines of force* and its magnitude gives the *magnitude* of the force. In this interpretation the integral

$$\iint_{S^*} (a \, dy \, dz + b \, dz \, dx + c \, dx \, dy)$$

is called the total *flux of force* across the surface from the negative to the positive side.

## 5.9 Gauss's and Green's Theorems in Space

### a. Gauss's Theorem

The concept of surface integral leads to an extension to three dimensions of Gauss's theorem, which we proved on p. 545 for two dimensions. The essential point in the statement of the theorem in two dimensions is that an integral over a plane region is reduced to a line integral taken around the boundary of the region. We now consider a closed bounded three-dimensional region $R$ in $x$, $y$, $z$-space bounded by a surface $S$ that is intersected by every parallel to one of the coordinate axes in, at most, two points. This last assumption will be removed later.

Let the three functions $a(x, y, z)$, $b(x, y, z)$, $c(x, y, z)$ and their first partial derivatives be continuous in $R$. We consider the integral

---

[1]See the analogous two-dimensional interpretation on. p 570. We think here of the surface in the neighborhood of a point as approximated by a plane piece of area $\Delta S$ and of the velocity vector $\mathbf{V}$ as replaced by a constant vector. A suitable passage to the limit furnishes the integral representation for the amount of liquid crossing $S^*$.

$$\iiint_R \frac{\partial c(x, y, z)}{\partial z} \, dx \, dy \, dz$$

taken over the region $R$, oriented positively with respect to $x$, $y$, $z$-coordinates. The region $R$ can be described by inequalities

$$z_0(x, y) \leqq z \leqq z_1(x, y),$$

where $(x, y)$ varies over the projection $B$ of $R$ onto the $x$, $y$-plane. We assume that $B$ has an area and that the functions $z_0(x, y)$ and $z_1(x, y)$ are continuous and have continuous first derivatives in $B$. We can transform the volume integral over $R$ by means of the formula (see p. 531)

$$\iiint_R f \, dx \, dy \, dz = \iint_B dx \, dy \int_{z_0}^{z_1} f \, dz.$$

Since here $f = \partial c/\partial z$ the integration with respect to $z$ can be carried out, yielding

$$\int_{z_0}^{z_1} \frac{\partial c}{\partial z} \, dz = c(x, y, z_1) - c(x, y, z_0) = c_1 - c_0,$$

so that

$$\iiint_R \frac{\partial c(x, y, z)}{\partial z} \, dx \, dy \, dz = \iint_B c_1 \, dx \, dy - \iint_B c_0 \, dx \, dy.$$

If we assume that the boundary $S$ is positively oriented with respect to the region $R$, then the portion of the oriented boundary surface $S^*$ consisting of the points of entry $z = z_0(x, y)$ has a negative orientation with respect to $x$, $y$-coordinates when projected on the $x$, $y$-plane,[1] while the portion $z = z_1(x, y)$ consisting of the points of exit has a positive orientation. Hence, the last two integrals combine to form the integral

$$\iint_{S^*} c(x, y, z) \, dx \, dy$$

taken over the whole surface $S^*$. We thus obtain the formula

$$\iiint_R \frac{\partial c(x, y, z)}{\partial z} \, dx \, dy \, dz = \iint_{S^*} c(x, y, z) \, dx \, dy.$$

---

[1]See p. 593. On $z = z_0(x, y)$ the positive normal (the one exterior to $R$) points downward.

The formula remains valid if $S^*$ contains cylindrical portions perpendicular to the $x$, $y$-plane, for these contribute nothing to the integral. If, for example, such a portion $S'^*$ of $S^*$ has the representation $y = \phi(x)$, we have for $S'^*$ the parameter representation

$$x = u, \qquad y = \phi(u), \qquad z = v$$

and, thus, indeed

$$\iint_{S^*} c \; dx \; dy = \iint c \, \frac{d(x, y)}{d(u. \, v)} \, du \; dv = \iint c \begin{vmatrix} 1 & 0 \\ \phi' & 0 \end{vmatrix} du \; dv = 0.$$

If we derive the corresponding formulae for the components $a$ and $b$ and add the three formulae, we obtain the general formula

$$(49) \qquad \iiint_R \left[ \frac{\partial a(x, \, y, \, z)}{\partial x} + \frac{\partial b(x, \, y, \, z)}{\partial y} + \frac{\partial c(x, \, y, \, z)}{\partial z} \right] dx \; dy \; dz$$

$$= \iint_{S^*} [a(x, y, z) \, dy \; dz + b(x, y, z) \, dz \; dx + c(x, y, z) \, dx \; dy],$$

which is known as *Gauss's theorem*. Using formula (47b) of p. 595, we can also write this in the form

$$(50) \qquad \iiint_R (a_x + b_y + c_z) \, dx \; dy \; dz$$

$$= \iint_S (a \cos \alpha + b \cos \beta + c \cos \gamma) \, dS$$

$$= \iint_S \left( a \frac{dx}{dn} + b \frac{dy}{dn} + c \frac{dz}{dn} \right) dS.$$

Here, corresponding to the positive orientation of $S^*$ with respect to $R$, we have in $\alpha$, $\beta$, $\gamma$ the angles the *outward-drawn normal* **n** makes with the positive coordinate axes.

This formula can easily be extended to more general regions. We have only to require that the region $R$ be capable of being subdivided by a finite number of portions of surfaces with continuously turning tangent planes, into subregions $R_i$ each of which has the properties assumed above (in particular, that each $R_i$ has a boundary consisting of surfaces that are either intersected by every parallel to a coordinate axis in, at most, two points or are portions of cylinders with generators parallel to one of the coordinate axes). Gauss's theorem holds for each region $R_i$. On adding, we obtain on the left a triple integral over the whole region $R$; on the right, some of the surface integrals combine to form the integral over the oriented surface $S$, while the others (namely, those taken over the surfaces by which $R$ is subdivided) cancel one another, as we have already seen in the case of the plane (p. 549).[1]

As a special case of Gauss's theorem, we obtain the formula for the volume of a region $R$ bounded by a surface $S^*$ oriented positively with respect to $R$. If, for example, we put in (49) $a = 0$, $b = 0$, $c = z$, we immediately obtain the expression

$$V = \iiint_R dx\,dy\,dz = \iint_{S^*} z\,dx\,dy$$

for the volume. In the same way, we find[2] that

$$V = \iint_{S^*} x\,dy\,dz = \iint_{S^*} y\,dz\,dx.$$

If $\mathbf{A}$ is the vector with components $a$, $b$, $c$, we have in $a_x + b_y + c_z$ the divergence of $\mathbf{A}$, and in

---

[1]The proof for general $R$ that we have given here makes use of a definition of integral over a closed surface $S$ that has actually not been shown to be independent of the particular way in which $S$ is divided into portions with simple parameter representations. The proof that for *smooth* $S$ the integral over $S$ is independent of the subdivision will be given in the Appendix, p. 635. In the extension of Gauss's theorem to more general regions $R$ given above, however, we necessarily make use of subregions $R_i$ bounded by surfaces $S_i$ that have *edges* and are not perfectly smooth. For that reason, it is more convenient to use a quite different technique of proof that does not involve decomposition of $R$ into *disjoint* subsets $R_i$, which cannot possibly have smooth boundaries. This is achieved by the method of *partition of unity,* in which, effectively, $R$ is represented as union of *overlapping* regions $R_i$ with smooth boundaries, to each of which the theorem applies directly. See the Appendix to this chapter, pp. 639–642.

[2]It is noteworthy that *cyclic* interchange of $x$, $y$, $z$ in these expressions for $V$ brings about no change in sign, in contrast to the corresponding formulae for the area of a two-dimensional region bounded by an oriented curve $C^*$:

$$A = \int_{C^*} x\,dy = -\int_{C^*} y\,dx$$

This is so because in two dimensions an interchange of the positive $x$-direction with the positive $y$-direction reverses the orientation of the plane: $\Omega(x, y) = -\Omega(y, x)$, while a cyclic interchange of coordinates in three-space preserves the orientation of space:

$$\Omega(x, y, z) = \Omega(y, z, x) = \Omega(z, x, y).$$

$$a\,\frac{dx}{dn} + b\,\frac{dy}{dn} + c\,\frac{dz}{dn}$$

the scalar product of the vectors $\mathbf{A}$ and $\mathbf{n}$, that is, the normal component $A_n$ of the vector $\mathbf{A}$. Hence, in vector notation Gauss's theorem becomes[1]

$$(52) \qquad \iiint_R \operatorname{div} \mathbf{A}\, dx\, dy\, dz = \iint_S \mathbf{A} \cdot \mathbf{n}\, dS = \iint_S A_n\, dS.$$

More striking is the formulation of the Gauss's theorem (49) in terms of exterior differential forms. The second-order differential form

$$\omega = a(x, y, z)\, dy\, dz + b(x, y, z)\, dz\, dx + c(x, y, z)\, dx\, dy$$

just has as its derivative [see (58c), p. 313] the third-order form

$$d\omega = (a_x + b_y + c_z)\, dx\, dy\, dz.$$

*Denoting by $S^*$ the boundary of $R$ oriented positively with respect to $R$, we have simply*

$$(53) \qquad\qquad \iiint_R d\omega = \iint_{S^*} \omega.$$

Heretofore we have made the assumption that the three-dimensional region $R$ is oriented positively with respect to $x$, $y$, $z$-coordinates. We can free ourselves from this assumption by observing that $\omega$ in (53) stands for an arbitrary second-order differential form and that the relation between $\omega$ and $d\omega$ is independent of coordinates used. Denote by $R^*$ an oriented region in space and by $\partial R^*$ its boundary oriented positively with respect to $R^*$. We can always choose an $x$, $y$, $z$-system with respect to which $R^*$ is oriented positively, so that (53) holds with $S^* = \partial R^*$ (see p. 591). With these conventions we have for any orientation of $R^*$

$$(53a) \qquad\qquad \iiint_{R^*} d\omega = \iint_{\partial R^*} \omega.$$

---

[1]Notice that in the surface integrals the orientation given to $S$ only affects the integrand.

Precisely analogous formulae hold more generally for sets of any number of dimensions, as we shall see.[1]

### Exercises 5.9a

1. Evaluate the surface integral

$$\iint \frac{z}{p} \, dS$$

taken over the half of the ellipsoid $x^2/a^2 + y^2/b^2 + z^2/c^2 = 1$, for which $z$ is positive where $1/p = lx/a^2 + my/b^2 + nz/c^2$, $l$, m, n being the direction cosines of the outward-drawn normal.

2. Evaluate the surface integral

$$\iint H \, dS$$

taken over the sphere of radius unity with center at the origin, where

$$H = a_1x^4 + a_2y^4 + a_3z^4 + 3a_4x^2y^2 + 3a_5y^2z^2 + 3a_6x^2z^2.$$

#### b. Application of Gauss's Theorem to Fluid Flow

As in the case of the plane, we can obtain a physical interpretation fo Gauss's theorem in space by taking the vector $\mathbf{A} = (a, b, c)$ as the *momentum vector* in the flow of a fluid of density $\rho$ whose velocity is given by the vector $\mathbf{V} = (u, v, w)$. Here $\rho$ and the velocity components $u, v, w$ depend on the $(x, y, z)$ and the time $t$ considered. The momentum vector (per unit volume) is defined by $\mathbf{A} = \rho\mathbf{V}$. If $R$ is a fixed region in space bounded by the surface $S$, then the total mass of fluid that in unit time flows across a small portion of $S$ of area $\Delta S$ from the interior to the exterior of $R$ is given approximately by the expression $\rho V_n \, \Delta S$, where $V_n$ is the component of the velocity vector $\mathbf{V}$ in the direction of the outward normal $n$ at a point of the surface element. Accordingly, the total amount of fluid that flows across the boundary $S$ of $R$ from the inside to the outside in unit time is given by the integral

---

[1]Generally, for an $n$-dimensional oriented set $R^*$ in euclidean space of $n$ or more dimensions the symbol $\partial R^*$ denotes the boundary of $R^*$ oriented positively with respect to $R^*$; that is, $\partial R^*$ is oriented in such a way that

$$\Omega(R^*) = \Omega(\mathbf{B}, \mathbf{A}^1, \cdots, \mathbf{A}^{n-1})$$

where $\mathbf{A}^1, \ldots, \mathbf{A}^{n-1}$ are vectors tangential at some point to the boundary of $\partial R^*$, with

$$\Omega(\partial R) = \Omega(\mathbf{A}^1, \mathbf{A}^2, \cdots, \mathbf{A}^{n-1}),$$

and where $\mathbf{B}$ is a vector tangential to and pointing away from $R^*$.

$$\iint_S \rho V_n \, dS = \iint_S A_n \, dS$$

taken over the whole boundary $S$. By Gauss's identity (52) the amount of fluid leaving $R$ in unit time through its boundary is thus:

$$\iiint_R \text{div } \mathbf{A} \, dx \, dy \, dz = \iiint_R \text{div } (\rho \mathbf{V}) \, dx \, dy \, dz.$$

On the other hand, the total mass of fluid contained in $R$ at any one time is given by the triple integral

$$\iiint_R \rho(x, y, z, t) \, dx \, dy \, dz$$

and the decrease in unit time of the mass of fluid contained in $R$ by

$$-\frac{d}{dt} \iiint_R \rho(x, y, z, t) \, dx \, dy \, dz = - \iiint_R \rho_t(x, y, z, t) \, dx \, dy \, dz.$$

If the law of conservation of mass is to hold and if there are no sources or sinks of mass in $R$, then the total amount of mass of fluid leaving $R$ through the surface $S$ must be exactly equal to the loss of mass of fluid contained in $R$. We must then have

$$\iiint_R \text{div } (\rho \mathbf{V}) \, dx \, dy \, dz = - \iiint_R \rho_t \, dx \, dy \, dz$$

at any time $t$ for any region $R$. Dividing both sides of this identity by the volume of $R$ and shrinking $R$ into a point (that is, applying space differentiation), we obtain the three dimensional *continuity equation*

$$\text{div } (\rho \mathbf{V}) = -\rho_t$$

or

(55)
$$\frac{\partial \rho}{\partial t} + \frac{\partial (\rho u)}{\partial x} + \frac{\partial (\rho v)}{\partial y} + \frac{\partial (\rho w)}{\partial z} = 0,$$

which expresses the *law of conservation of mass* for motion of fluids in the form of a differential equation

If the law of conservation of mass is not invoked, the expression

$$\rho_t + \text{div } (\rho \mathbf{V})$$

measures the amount of mass created (or annihilated, when negative) in unit time per unit volume.

Particular interest attaches to the case of a homogeneous and incompressible fluid, for which the density $\rho$ has the same value in all places and is unchanging with time. Since $\rho$ is then constant, we deduce from (55) that

$$(56) \qquad \operatorname{div} \mathbf{V} = \frac{\partial u}{\partial x} + \frac{\partial v}{\partial y} + \frac{\partial w}{\partial z} = 0$$

if mass is to be preserved. It then follows from (52) that

$$(57) \qquad \iint_S \mathbf{V} \cdot \mathbf{n} \, dS = 0$$

whenever the surface $S$ bounds a region $R$. Consider, in particular, two surfaces $S_1$ and $S_2$ bounded by the same oriented curve $C^*$ in space, and together forming the boundary $S$ of a three-dimensional region $R$. We find from (57) that

$$(58) \qquad 0 = \iint_S \mathbf{V} \cdot \mathbf{n} \, dS = \iint_{S_1} \mathbf{V} \cdot \mathbf{n} \, dS + \iint_{S_2} \mathbf{V} \cdot \mathbf{n} \, dS,$$

where, on both $S_1$ and $S_2$, $\mathbf{n}$ denotes the normal pointing away from $R$. We can make both $S_1$ and $S_2$ into oriented surfaces $S_1^*$ and $S_2^*$ in such a way that the orientation of $C^*$ is positive with respect to both $S_1^*$ and $S_2^*$. On both these surfaces, let $\mathbf{n}^*$ be the unit normal pointing to the positive side. (For a right-handed orientation of space, this means that $\mathbf{n}^*$ points to that side of the surface from which the orientation of $C^*$ appears ounterclockwise.) Then, necessarily, $\mathbf{n}^* = \mathbf{n}$ on one of the surfaces $S_1$, $S_2$ and $\mathbf{n}^* = -\mathbf{n}$ on the other.[1] It follows from (58) that

$$(59) \qquad \iint_{S_1} \mathbf{V} \cdot \mathbf{n}^* \, dS = \iint_{S_2} \mathbf{V} \cdot \mathbf{n}^* \, dS.$$

In words, *if the fluid is incompressible and homogeneous and mass is conserved, then the same amount of fluid flows across any two surfaces*

---

[1]The normal $\mathbf{n}$ determines an orientation on the whole surface $S$ if we require, for example, that $\mathbf{n}$ points to the positive side of $S$. Orienting $S_1$ and $S_2$ relative to $\mathbf{n}$, the curve $C$ receives opposite senses if we require it to be oriented positively with respect to $S_1$ or to $S_2$ (see p. 588). However, since $C^*$ has the positive sense with respect to both $S_1^*$ and $S_2^*$, it follows that the orientations given by $\mathbf{n}^*$ and by $\mathbf{n}$ agree only on one of the surfaces.

*with the same boundary curve C\* that together bound a three-dimensional region in space.* This amount of fluid does not depend on the precise form of the surfaces; it is plausible that it must be determined by the boundary curve $C^*$ alone.[1] We then ask how we can express the amount of fluid in terms of the curve $C^*$ alone. This question is answered in the next section (p. 614) by means of Stokes's theorem.

### c. Gauss's Theorem Applied to Space Forces and Surface Forces

The forces acting in a continuum may be regarded either as space forces (such as gravitational attraction, electrostatic forces) or as surface forces (such as pressures, tractions). The connection between these two points of view is given by Gauss's theorem.

We consider only the special case of the force in a fluid of density $\rho = \rho(x, y, z)$, in which there is a pressure $p(x, y, z)$, which in general depends on the point $(x, y, z)$. This means that the force acting on a portion $R$ of the liquid exerted by the remaining part of the liquid can be considered as a force acting at each point of the surface $S$ of $R$ in the direction of the inward drawn normal and of magnitude $p$ per unit surface area. Denoting by $dx/dn$, $dy/dn$, $dz/dn$ the direction cosines of the *outward-drawn normal* at a point of the surface $S$ of $R$, the components of the force per unit area are given by

$$-p\frac{dx}{dn}, \qquad -p\frac{dy}{dn}, \qquad -p\frac{dz}{dn}.$$

Thus, the resultant of the surface forces acting on $R$ is a force with components

$$X = -\iint_S p\frac{dx}{dn}\,dS, \quad Y = -\iint_S p\frac{dy}{dn}\,dS, \quad Z = -\iint_S p\frac{dz}{dn}\,dS.$$

By Gauss's theorem (50), p. 599, we can write $X$, $Y$, $Z$ as volume integrals

$$X = -\iiint_R p_x\,dx\,dy\,dz, \qquad Y = -\iiint_R p_y\,dx\,dy\,dz,$$

$$Z = -\iiint_R p_z\,dx\,dy\,dz.$$

In vector notation the resultant is a force $\mathbf{F}$ given by

---

[1]The amount of fluid crossing a surface bounded by the closed curve $C$ in unit time is independent of time if we make the further assumption that the flow is *steady*, that is, that the velocity vector $\mathbf{V}$ is independent of time.

(60) 
$$\mathbf{F} = - \iiint_R \operatorname{grad} p \, dx \, dy \, dz.$$

We can express this result as follows. The forces in a fluid due to a pressure $p(x, y, z)$ may, on the one hand, be regarded as surface forces (pressure) that act with density $p(x, y, z)$ perpendicular to each surface element through the point $(x, y, z)$ and, on the other hand, as volume forces, that is, as forces that act on every element of volume with volume density $-\operatorname{grad} p$.

If a fluid is in equilibrium under the forces due to pressure and to gravitational attraction, the vector $\mathbf{F}$ must balance the total attractive force $\mathbf{G}$ acting on the liquid contained in $R$:

$$\mathbf{F} + \mathbf{G} = 0.$$

If the gravitational force acting on a unit mass at the point $(x, y, z)$ is given by the vector $\Gamma(x, y, z)$, we have

$$\mathbf{G} = \iiint_R \Gamma \rho \, dx \, dy \, dz.$$

From the relation $\mathbf{F} + \mathbf{G} = 0$, valid for any portion $R$ of the fluid, we conclude by space differentiation that the corresponding relation holds for the integrands, that is, that at each point of the fluid the equation

(61) 
$$-\operatorname{grad} p + \rho \Gamma = 0$$

holds. Since the gradient of a scalar is perpendicular to the level surfaces for that scalar, we conclude that *for a fluid in equilibrium under pressure and gravitational attraction the attraction at each point of a surface of constant pressure p ("isobaric" surface) is perpendicular to the surface*. If we make the customary assumption that the gravitational force per unit mass near the surface of the earth is given by the vector $\Gamma = (0, 0, -g)$, where $g$ is the gravitational acceleration, we find[1] from (61) that

(62) 
$$p_x = 0, \qquad p_y = 0, \qquad p_z = -g\rho.$$

Consider in particular a homogeneous liquid of constant density $\rho$ bounded by a *free surface* of pressure 0. Along this free surface, we

---

[1]This formula was derived in Volume I (p. 226), in the description of the pressure variations in the atmosphere.

have, by (62),

$$0 = dp = p_x\, dx + p_y\, dy + p_z\, dz = -g\rho\, dz.$$

Hence, $dz = 0$, which means that *the free surface has to be a plane* $z = $ constant $= z_0$. For any point $(x, y, z)$ of the liquid the value of the pressure is then

$$p(x, y, z) = -\int_z^{z_0} p_z(x, y, \zeta)d\zeta = g\rho\, (z_0 - z).$$

*Thus, at the depth $z_0 - z = h$ the pressure has the value $g\rho h$.* For a solid partly or wholly immersed in the liquid, let $R$ denote the portion of the solid lying below the free surface $z = z_0$. We apply formula (60) to the region $R$ in order to determine the total pressure force acting on the solid.[1] We find from (60) and (62) that the resultant of the pressure forces acting on the solid is equal to a force (buoyancy) with components

$$X = 0, \qquad Y = 0, \qquad Z = \iiint_R g\rho\, dx\, dy\, dz;$$

this force is directed vertically upward and its magnitude is equal to the weight of the displaced liquid (Archimedes' principle).

### d.  Integration by Parts and Green's Theorem in Three Dimensions

Just as in the case of two independent variables (p. 556), Gauss's theorem (50), p. 599 applied to products $au$, $bv$, $cw$ leads to a *formula for integration by parts:*

(63)
$$\iiint_R (au_x + bv_y + cw_z)\, dx\, dy\, dz$$
$$= \iint_S \left(au\, \frac{dx}{dn} + bv\, \frac{dy}{dn} + cw\, \frac{dz}{dn}\right) dS$$
$$- \iiint_R (a_x u + b_y v + c_z w)\, dx\, dy\, dz.$$

If here $u = v = w = U$ and if $a$, $b$, $c$ are of the form $a = V_x$, $b = V_y$, $c = V_z$ for some scalar $V$, we obtain *Green's first theorem*

---

[1] Any portions of the boundary of $R$ lying in the plane $z = z_0$ make no contribution since there $p = 0$ by assumption.

(64)
$$\iiint_R (U_x V_x + U_y V_y + U_z V_z)\, dx\, dy\, dz$$

$$= \iint_S U \frac{dV}{dn}\, dS - \iiint_R U\, \Delta V\, dx\, dy\, dz.$$

Here we use the familiar symbol $\Delta$ for the *Laplace operator* defined by

$$\Delta V = V_{xx} + V_{yy} + V_{zz}$$

and denote by $dV/dn$ the derivative of $V$ in the direction of the *outward* normal:

$$\frac{dV}{dn} = V_x \frac{dx}{dn} + V_y \frac{dy}{dn} + V_z \frac{dz}{dn}.$$

Interchanging $U$ and $V$ in formula (64) and subtracting from (64) yields *Green's second theorem*

(65)    $$\iiint_R (U\, \Delta V - V\, \Delta U)\, dx\, dy\, dz = \iint_S \left( U \frac{dV}{dn} - V \frac{dU}{dn} \right) dS.$$

### e.  Application of Green's Theorem to the Transformation of $\Delta U$ to Spherical Coordinates

If we set $V = 1$ in Green's theorem (65), we obtain

(66)    $$\iiint_R \Delta U\, dx\, dy\, dz = \iint_S \frac{dU}{dn}\, dS = \iint_S (\text{grad } U) \cdot \mathbf{n}\, dS.$$

Just as in the plane, we can use this formula to transform $\Delta U$ to other coordinate systems, notably to the spherical coordinates $r$, $\phi$, $\theta$ defined by

$$x = r \cos \phi \sin \theta, \qquad y = r \sin \phi \sin \theta, \qquad z = r \cos \theta.$$

We apply formula (66) to a wedge-shaped region $R$ described by inequalities of the form

(67)    $$r_1 < r < r_2, \quad \phi_1 < \phi < \phi_2, \quad \theta_1 < \theta < \theta_2.$$

The boundary $S$ of $R$ consists of six faces along each of which one of the coordinates $r$, $\phi$, $\theta$ has a constant value. Applying the formula for transformation of triple integrals we write the left side of equation (66) in the form

(68)
$$\iiint_R \Delta U \, dx \, dy \, dz = \iiint \Delta U \frac{d\,(x,\,y,\,z)}{d\,(r,\,\theta,\,\phi)} \, dr \, d\theta \, d\phi$$

$$= \iiint \Delta U \, r^2 \sin\theta \, dr \, d\theta \, d\phi,$$

with the integral in $r$, $\theta$, $\phi$-space extended over the region (67). In order to transform the surface integral in (66) we introduce the position vector

$$\mathbf{X} = (x, y, z) = (r \cos\phi \sin\theta, \, r \sin\phi \sin\theta, \, r \cos\theta)$$

and notice that its first derivatives satisfy the relations

(68a)      $\mathbf{X}_r \cdot \mathbf{X}_\theta = 0, \quad \mathbf{X}_\theta \cdot \mathbf{X}_\phi = 0, \quad \mathbf{X}_\phi \cdot \mathbf{X}_r = 0$

(68b)      $\mathbf{X}_r \cdot \mathbf{X}_r = 1, \quad \mathbf{X}_\theta \cdot \mathbf{X}_\theta = r^2, \quad \mathbf{X}_\phi \mathbf{X}_\phi = r^2 \sin^2\theta.$

It follows from these relations that at each point the vector $X_r$ is normal to the coordinate surface $r =$ constant passing through that point, the vector $\mathbf{X}_\theta$ normal to the surface $\theta =$ constant, and the vector $\mathbf{X}_\phi$ normal to the surface $\phi =$ constant. More precisely, on one of the faces $r =$ constant $= r_i$ (where $i$ has either the value 1 or 2) the outward normal unit vector $\mathbf{n}$ is given by $(-1)^i \mathbf{X}_r$. Hence, on those faces

$$(\text{grad } U) \cdot \mathbf{n} = (-1)^i \,(\text{grad } U) \cdot \mathbf{X}_r = (-1)^i \frac{\partial U}{\partial r}.$$

Using, moreover, $\theta$ and $\phi$ as parameters along a face $r = r_i$, we have for the element of area the expression [see (30e), p. 429]

$$dS = \sqrt{EG - F^2} \, d\theta \, d\phi = \sqrt{(X_\theta \cdot X_\theta)\,(X_\phi \cdot X_\phi) - (X_\theta \cdot X_\phi)^2} \, d\theta \, d\phi$$
$$= r^2 \sin\theta \, d\theta \, d\phi.$$

It follows that the contribution of the two faces $r = r_1$ and $r = r_2$ to the integral of $dU/dn$ over $S$ is represented by the expression

$$\iint_{r=r_2} r^2 \sin\theta \, \frac{\partial U}{\partial r} \, d\theta \, d\phi - \iint_{r=r_1} r^2 \sin\theta \, \frac{\partial U}{\partial r} \, d\theta \, d\phi,$$

where the integrations are taken over the rectangle

$$\theta_1 < \theta < \theta_2, \qquad \phi_1 < \phi < \phi_2.$$

We can write the difference of these integrals as the triple integral

$$\iiint \frac{\partial}{\partial r}\left(r^2 \sin\theta \frac{\partial U}{\partial r}\right) dr\, d\theta\, d\phi$$

extended over the region (67).

Similarly, we find that on a face $\theta = \text{constant} = \theta_i$

$$\mathbf{n} = (-1)^i \frac{1}{r}\mathbf{X}_\theta, \qquad dS = r\sin\theta\, d\phi\, dr, \qquad \frac{dU}{dn} = \frac{(-1)^i}{r}\frac{\partial U}{\partial\theta}$$

and on a face $\phi = \text{constant} = \phi_i$

$$\mathbf{n} = (-1)^i \frac{1}{r\sin\theta}\mathbf{X}_\phi, \qquad dS = r\, dr\, d\theta, \qquad \frac{dU}{dn} = \frac{(-1)^i}{r\sin\theta}\frac{\partial U}{\partial\phi}.$$

Here also, combining the contributions of opposite faces $\theta = \text{constant}$ or $\phi = \text{constant}$, we find for the total surface integral the expression

$$\iint_S \frac{dU}{dn}\, dS = \iiint \left[\frac{\partial}{\partial r}\left(r^2 \sin\theta \frac{\partial U}{\partial r}\right) + \frac{\partial}{\partial\theta}\left(\sin\theta \frac{\partial U}{\partial\theta}\right)\right.$$

$$\left. + \frac{\partial}{\partial\phi}\left(\frac{1}{\sin\theta}\frac{\partial U}{\partial\phi}\right)\right] dr\, d\theta\, d\phi.$$

Comparing with the expression (68), dividing by the volume of the wedge $R$, and shrinking the wedge to a point leads to the desired expression for the Laplace operator in spherical coordinates:

$$(69) \qquad \Delta U = \frac{1}{r^2 \sin\theta}\left\{\frac{\partial}{\partial r}\left(r^2 \sin\theta \frac{\partial U}{\partial r}\right) + \frac{\partial}{\partial\theta}\left(\sin\theta \frac{\partial U}{\partial\theta}\right) + \frac{\partial}{\partial\phi}\left(\frac{1}{\sin\theta}\frac{\partial U}{\partial\phi}\right)\right\}.$$

## Exercises 5.9e

1. Let the equations

$$x_i = x_i\,(p_1, p_2, p_3) \qquad\qquad (i = 1, 2, 3)$$

define an arbitrary orthogonal coordinate system $p_1, p_2, p_3$; that is, if we put $a_{ik} = \dfrac{\partial x_i}{\partial p_k}$, then the equations

$$a_{11}a_{21} + a_{12}a_{22} + a_{13}a_{23} = 0$$

$$a_{11}a_{31} + a_{12}a_{32} + a_{13}a_{33} = 0$$

$$a_{21}a_{31} + a_{22}a_{32} + a_{23}a_{33} = 0$$

are to hold.

(a) Prove that

$$\frac{\partial(x_1, x_2, x_3)}{\partial(p_1, p_2, p_3)} = \sqrt{e_1 e_2 e_3} \; ,$$

where

$$e_i = a_{1i}{}^2 + a_{2i}{}^2 + a_{3i}{}^2.$$

(b) Prove that

$$\frac{\partial p_i}{\partial x_k} = \frac{1}{e_i} \frac{\partial x_k}{\partial p_i} = \frac{1}{e_i} a_{ki}.$$

(c) Express $\Delta u = u_{x_1 x_1} + u_{x_2 x_2} + u_{x_3 x_3}$ in terms of $p_1$, $p_2$, $p_3$, using Gauss's theorem.

(d) Express $\Delta u$ in the focal coordinates $t_1$, $t_2$, $t_3$ defined in Exercises 9, Section 3.3d, p. 256.

## 5.10  Stokes's Theorem in Space

### a.  *Statement and Proof of the Theorem*

We have already seen Stokes's theorem in two dimensions (p. 554). The analogous theorem in three dimensions connects the integral of the normal component of the curl of a vector over a curved surface with the integral of the tangential component of the vector over the boundary curve of the surface. While in two dimensions Gauss's theorem and Green's theorem go over into each other by a change in notation, they are essentially different theorems in three dimensions.

Let $S$ be an orientable surface in three-space bounded by a closed curve $C$. The choice of an orientation for $S$ converts $S$ into the oriented surface $S^*$. Let $C^*$ be the boundary curve of $S^*$ oriented positively with respect to $S^*$. Assuming that space is oriented positively with respect to $x$, $y$, $z$-coordinates, let **n** at each point of $S^*$ denote the unit normal vector[1] pointing to the positive side of $S^*$. Let **t** be the unit tangent vector on $C^*$ pointing in the direction corresponding to the orientation of $C^*$. Let $\mathbf{A} = (a, b, c)$ be a vector defined near $S$. Stokes's theorem asserts[2] that

$$(70) \qquad \iint_S (\text{curl } \mathbf{A}) \cdot \mathbf{n} \; dS = \int_C \mathbf{A} \cdot \mathbf{t} \; ds.$$

---

[1] In effect this means that when we move a point of $S^*$ into the origin in such a way that **n** coincides with the positive $z$-axis, the sense of rotation on $S^*$ will be that of the 90° rotation taking the positive $x$-axis into the positive $y$-axis.

[2] Precise regularity assumptions for $S$, $C$, $\mathbf{A}$ under which the theorem can be proved are given in the Appendix to this chapter, p. 643.

Denoting by $dx/dn$, $dy/dn$, $dz/dn$ the components of the vector $\mathbf{n}$ and by $dx/ds$, $dy/ds$, $dz/ds$ those of $\mathbf{t}$, we write Stokes's theorem in the form[1]

$$(71) \qquad \iint_S \left[ (c_y - b_z) \frac{dx}{dn} + (a_z - c_x) \frac{dy}{dn} + (b_x - a_y) \frac{dz}{dn} \right] dS$$

$$= \int_C \left( a \frac{dx}{ds} + b \frac{dy}{ds} + c \frac{dz}{ds} \right) ds.$$

Using formula (47c), p. 596, we have, equivalently,

$$(72) \qquad \iint_{S^*} (c_y - b_z)\, dy\, dz + (a_z - c_x)\, dz\, dx + (b_x - a_y)\, dx\, dy$$

$$= \int_{C^*} a\, dx + b\, dy + c\, dz.$$

Introducing the first-order differential form

$$(73\text{a}) \qquad\qquad\qquad L = a\, dx + b\, dy + c\, dz$$

and

$$(73\text{b}) \qquad \omega = (c_y - b_z)\, dy\, dz + (a_z - c_x)\, dz\, dx + (b_x - a_y)\, dx\, dy,$$

we notice (see p. 313) that $\omega$ is just the derivative of $L$:

$$(73\text{c}) \qquad\qquad\qquad \omega = dL.$$

If $\partial S^*$ is the positively oriented boundary $C^*$ of $S^*$,[2] Stokes's theorem becomes simply

$$(74) \qquad\qquad\qquad \iint_{S^*} dL = \int_{\partial S^*} L.$$

In this form it is completely analogous to Gauss's theorem as written in formula (53), p. 601.

The truth of Stokes's theorem can immediately be made plausible from the fact that the theorem has already been proved for plane surfaces [see formula (10), p. 555]. Consequently, if $S$ is a polyhedral surface composed of plane polygonal surfaces, so that the boundary

---

[1]See (94c), p. 209 for the definition of the curl of a vector.
[2]This accords with the general definition in footnote 2, p. 587, for the case $n = 2$.

curve $C$ is a polygon, we can apply Stokes's theorem to each of the plane portions and add the corresponding formulae. In this process the line integrals along all the interior edges of the polyhedron cancel, and we at once obtain Stokes's theorem for the polyhedral surface. In order to obtain the general statement of Stokes's theorem, we only pass to the limit, leading from approximating polyhedra to arbitrary surfaces $S$ bounded by arbitrary curves $C$.

The rigorous validation of this passage to the limit, however, would be troublesome; therefore, having made these heuristic remarks, we carry out the proof by *transforming the whole surface S into a plane surface* and by observing that the theorem is preserved under such transformations.

We assume that there exists a parametric representation[1]

$$x = \phi(u, v), \qquad y = \psi(u, v), \qquad z = \chi(u, v)$$

for $S$, where $\phi$, $\psi$, $\chi$ are functions with continuous first derivatives for which the vector with components

(75)
$$\xi = \frac{d(y, z)}{d(u, v)}, \qquad \eta = \frac{d(z, x)}{d(u, v)}, \qquad \zeta = \frac{d(x, y)}{d(u, v)}$$

does not vanish. Assume that there is an oriented set $\Sigma^*$ in the $u$, $v$-plane bounded by an oriented closed curve $\Gamma^*$ such that $\Sigma^*$ is mapped bi-uniquely onto the surface $S^*$ and $\Gamma^*$ onto $C^*$.[2]

Now $L$ determines a differential form in $du$ and $dv$:

$$L = a\,(x_u\,du + x_v\,dv) + b\,(y_u\,du + y_v\,dv) + c\,(z_u\,du + z_v\,dv)$$
$$= (ax_u + by_u + cz_u)\,du + (ax_v + by_v + cz_v)\,dv$$

and

$$\int_{C^*} L = \int_{\Gamma^*} L,$$

where on the right side we take $L$ as expressed in terms of $du$ and $dv$. Similarly, $\omega$ gives rise to a second-order form in $du$ and $dv$,

---

[1] In the Appendix to this chapter the theorem will be proved more generally for surfaces $S$ that can be patched together from portions with a parametric representation of the type mentioned.

[2] If the vector $(\xi, \eta, \zeta)$ has the direction of **n**, we have $\Omega(\Sigma^*) = \Omega(u, v)$; if $(\xi, \eta, \zeta)$ has the direction of **-n**, we have $\Omega(\Sigma^*) = -\Omega(u, v)$. The curve $\Gamma^*$ is oriented positively with respect to $\Sigma^*$ in either case. See p. 587.

$$\omega = \frac{\omega}{du\ dv}\,du\ dv$$

$$= [(c_y - b_z)\xi + (a_z - c_x)\eta + (b_x - a_y)\zeta]\,du\ dv,$$

and again [see (46a), p. 593]

$$\iint_{S^*} \omega = \iint_{\Sigma^*} \omega$$

Moreover, as we proved on p. 322, the relation $\omega = dL$ does not depend on the choice of independent variables $x$, $y$, $z$ or $u$, $v$.[1] Consequently, the proof of identity (74) has been reduced to the case, involving a first-order differential form $L$ in $du$ and $dv$ and a region $\Sigma^*$ with boundary $\Gamma^*$ in the $u$, $v$-plane. Since Stokes's theorem is known to hold in the $u$, $v$-plane, it now follows for the curved surface $S$. .

Stokes's theorem answers the question raised on p. 0000. We have seen that for a given vector field $\mathbf{V}(x, y, z)$ with div $\mathbf{V} = 0$, the integral

$$\iint_S \mathbf{V} \cdot \mathbf{n}\,dS$$

over a surface $S$ with unit normal $\mathbf{n}$ depends only on the boundary curve $C$ of $S$ and not on the particular nature of $S$. On the other hand, we found on p. 315 that a vector field $\mathbf{V}$ with vanishing divergence can be represented as the curl of a vector $\mathbf{A} = (a, b, c)$—at least if we restrict ourselves to vector fields defined in a parallelepiped with edges parallel to the coordinate axes. Stokes's theorem now enables us to express

$$\iint_S \mathbf{V} \cdot \mathbf{n}\,dS = \iint_S (\text{curl } \mathbf{A}) \cdot \mathbf{n}\,dS$$

in the form

$$\int_C \mathbf{A} \cdot \mathbf{t}\,ds,$$

which involves only the boundary curve $C$ of $S$.

---

[1]This can also be verified directly by proving the identity

$(c_y - b_z)\xi + (a_z - c_x)\eta + (b_x - a_y)\zeta = (ax_v + by_v + cz_v)_u - (ax_u + by_u + cz_u)_v,$

where $\xi$, $\eta$, $\zeta$ are defined by (75).

## Exercises 5.10a

1. Let

$$I = \iint_{S^*} z \, dx \, dy - x \, dy \, dz$$

where $S^*$ is the spherical cap $x^2 + y^2 + z^2 = 1$, $x > 1/2$, oriented positively with respect to the normal pointing to infinity.

(a) Calculate $I$ directly using $y$, $z$ as parameters on $S^*$.

(b) Calculate $I$ from Stokes's formula (74), p. 612, observing that

$$z \, dx \, dy - x \, dy \, dz = dL$$

with

$$L = -yz \, dx - xy \, dz.$$

### b. Interpretation of Stokes's Theorem

The physical interpretation of Stokes's theorem in three dimensions is similar to that already given (p. 572) in two dimensions. Once again we interpret the vector field $\mathbf{V} = (v_1, v_2, v_3)$ as the velocity field of the flow of a fluid. We call the integral

$$\int_C \mathbf{V} \cdot \mathbf{t} \, ds = \int_{C^*} v_1 \, dx + v_2 \, dy + v_3 \, dz$$

taken for an oriented closed curve $C^*$ the *circulation* of the flow along this curve. Stokes's theorem states that the circulation along $C^*$ is equal to the integral

$$\iint_S (\text{curl } \mathbf{V}) \cdot \mathbf{n} \, dS,$$

where $S$ is any orientable surface bounded by $C$, and $\mathbf{n}$ is the unit normal on $S$ chosen in such a way that the screw determined by $\mathbf{n}$ and the sense of rotation of $C^*$ has the same sense (right-handed or left-handed) as that of the $x$, $y$, $z$-system. Suppose we divide the circulation around $C$ by the area of the surface $S$ bounded by $C$ and pass to the limit by letting $C$ shrink to a point while remaining on the surface. This process of space differentiation gives for the limit of the double integral of the normal component of curl $V$ divided by the area the value of (curl $V$) . $\mathbf{n}$ at the limit point. We therefore see that the component of curl $\mathbf{V}$ in the direction of the normal $\mathbf{n}$ to the surface can be regarded as the *specific circulation* or *circulation density* of the flow in the surface at the corresponding point.[1]

---

[1] These considerations also show that the curl of a vector has a meaning independent of the coordinate system and therefore is itself a vector as long as the orientation of the coordinate system (and, hence, the vector $\mathbf{n}$) is not changed.

The vector curl **V** is called the *vorticity* of the motion of the fluid. Thus, the circulation around a curve $C$ is equal to the integral of the normal component of the vorticity over a surface bounded by $C$. The motion is called *irrotational* if the vorticity vector is 0 at every point occupied by the fluid, that is, if the velocity vector satisfies the relations

$$\frac{\partial v_3}{\partial y} - \frac{\partial v_2}{\partial z} = 0, \qquad \frac{\partial v_1}{\partial z} - \frac{\partial v_3}{\partial x} = 0, \qquad \frac{\partial v_2}{\partial x} - \frac{\partial v_1}{\partial y} = 0.$$

As a consequence of Stokes's theorem the circulation in an irrotational motion vanishes along any curve $C$ that bounds a surface contained in the region filled by the fluid.

If we interpret the vector **V** as the field of a mechanical or electrical force, the line integral

$$\int_{C^*} \mathbf{V} \cdot \mathbf{t} \, ds$$

represents the *work* done by the field on a particle when it is made to describe the curve $C^*$ in the sense indicated by its orientation. By Stokes's theorem the expression for this work is transformed into an integral over the surface $S$ bounded by $C$, the integrand being the normal component of the curl of the field of force. If here the curl of the force field vanishes, the work done on a particle returning to the same point is zero, and the field is called *conservative*.

From Stokes's theorem we obtain a new proof for the main theorem on line integrals in space (p. 104). The chief problem is to describe the nature of the vector field $\mathbf{A} = (a, b, c)$ if the integral

$$\int \mathbf{A} \cdot \mathbf{t} \, ds = \int a \, dx + b \, dy + c \, dz$$

is to vanish around an arbitrary closed curve $C$. Stokes's theorem yields a new proof of the fact that the vanishing of the line integral is ensured if curl $\mathbf{A} = 0$, provided $C$ forms the boundary of a surface $S$ contained in the region where $\mathbf{A}$ is defined. The vanishing of curl $\mathbf{A}$ —or, as we shall say, the *irrotational* nature of $\mathbf{A}$—is therefore a sufficient condition for the vanishing of the line integral of the tangential component of $\mathbf{A}$ around any closed curve that bounds a surface $S$ in the domain of definition of $A$. That the condition also is necessary we know already from p. 97. If the condition curl $A = 0$ is satisfied, we can represent $\mathbf{A}$ as gradient of a function $f(x, y, z)$:

$$\mathbf{A} = \operatorname{grad} f.$$

If we take $\mathbf{A}$ as the velocity vector $\mathbf{V}$ of a fluid flow, irrotationality of the flow, that is, the equation curl $\mathbf{V} = 0$, in a simply connected region implies that there exists a *velocity potential* $f(x, y, z)$ such that

$$\mathbf{V} = \operatorname{grad} f.$$

If, in addition, the fluid is homogeneous and incompressible, we have (see p. 604) the relation

$$\operatorname{div} \mathbf{V} = 0.$$

It follows in this case that the velocity potential $f$ satisfies the equation

$$0 = \operatorname{div} \operatorname{grad} f = \Delta f = f_{xx} + f_{yy} + f_{zz},$$

which is *Laplace's equation,* already met before.

### Exercises 5.10b

1. Let $\varphi$, $a$, and $b$ be continuously differentiable functions of a parameter $t$, for $0 \leqq t \leqq 2\pi$, with $a(2\pi) = a(0)$, $b(2\pi) = b(0)$, $\varphi(2\pi) = \varphi(0) + 2n\pi$ ($n$ a rational integer), and let $x$, $y$ be constants. Interpreting the equations

$$\xi = x \cos \varphi - y \sin \varphi + a, \qquad \eta = x \sin \varphi + y \cos \varphi + b$$

   as the parametric equations (with parameter $t$) of a closed plane curve $\Gamma$, prove that

$$\frac{1}{2} \int_\Gamma (\xi \, d\eta - \eta \, d\xi) = A\,(x^2 + y^2) + Bx + Cy + D$$

   where

$$A = \frac{1}{2} \int d\varphi, \qquad B = \int_\Gamma (a \cos \varphi + b \sin \varphi)\, d\varphi,$$

$$C = \int_\Gamma (-a \sin \varphi + b \cos \varphi) d\varphi, \qquad D = \frac{1}{2} \int_\Gamma (a \, db - b \, da).$$

2. Let a rigid plane $P$ describe a closed motion with respect to a fixed plane $\Pi$ with which it coincides. Every point $M$ of $P$ will describe a closed curve of $\Pi$ bounding an area of algebraic value $S(M)$. Denote by $2n\pi$ ($n$ a rational integer) the total rotation of $P$ with respect to $\Pi$. Prove the following results:
   (a) If $n \neq 0$, there is in $P$ a point $C$ such that for any other point $M$ of $P$ we have

$$S(M) = \pi n \overline{CM^2} + S(C);$$

   (b) If $n = 0$, then two cases may arise: first there is in $P$ an oriented line $\Delta$ such that for every point $M$ of $P$

$$S(M) = \lambda\, d(M),$$

where $d(M)$ is the distance of $M$ from $\Delta$ and $\lambda$ is a constant positive factor; or, second, $S(M)$ has the same value for all the points $M$ of the plane $P$ (Steiner's theorem).

3. A rigid line segment $AB$ describes in a plane $\Pi$ one closed motion of a connecting-rod: $B$ describes a closed counterclockwise circular motion with center $C$, while $A$ describes a (closed) rectilinear motion on a line passing through $C$. Apply the results of the previous example to determine the area of the closed curve in $\Pi$ described by a point $M$ rigidly connected to the line segment $AB$.

4. The end points $A$ and $B$ of a rigid line segment $AB$ describe one full turn on a closed convex curve $\Gamma$. A point $M$ on $AB$, where $AM = a$, $MB = b$, describes as a result of this motion a closed curve $\Gamma'$. Prove that the area between the curves $\Gamma$ and $\Gamma'$ is equal to $\pi ab$ (Holditch's theorem).

5. Prove that if we apply to each element $ds$ of a twisted, closed, and rigid curve $\Gamma$ a force of magnitude $ds/\rho$ in the direction of the principal normal vector (Chapter 2 p. 213). the curve $\Gamma$ remains in equilibrium; $1/\rho$ is the curvature of $\Gamma$ at $ds$ and is supposed to be finite and continuous at every point of $\Gamma$. (By the principles of the statics of a rigid body, we have to prove that

$$\int_\Gamma \frac{\mathbf{n}}{\rho}\, ds = 0, \qquad \int_\Gamma \frac{\mathbf{x} \times \mathbf{n}}{\rho}\, ds = 0.$$

where $\mathbf{n}$ denotes the unit principal normal vector of $\Gamma$ at $ds$, and $\mathbf{x}$ is the position vector of $ds$.)

6. Prove that a closed rigid surface $\Sigma$ remains in equilibrium under a uniform inward pressure on all its surface elements. (If by $\mathbf{n}'$ we denote the inward-drawn unit vector normal to the surface element $d\sigma$ and by $\mathbf{x}$ the position vector of $d\sigma$, the statement becomes equivalent to the vector equations

$$\iint_\Sigma \mathbf{n}'\, d\sigma = 0, \qquad \iint_\Sigma \mathbf{x} \times \mathbf{n}'\, d\sigma = 0.)$$

7. A rigid body of volume $V$ bounded by the surface $\Sigma$ is completely immersed in a fluid of specific gravity unity. Prove that the statical effect of the fluid pressure on the body is the same as that of a single force $\mathbf{f}$ of magnitude $V$, vertically upward, applied at the centroid $C$ of the volume $V$.

8. Let $p$ denote the distance from the center of the ellipsoid $\Sigma$

$$\frac{x^2}{a^2} + \frac{y^2}{b^2} + \frac{z^2}{c^2} = 1$$

to the tangent plane at the point $P(x, y, z)$ and $dS$ the element of area at this point. Prove the relations

(i) $$\iint_\Sigma p\, dS = 4\pi abc,$$

(ii)
$$\iint_\Sigma \frac{1}{p}\, dS = \frac{4\pi}{3abc}(b^2c^2 + c^2a^2 + a^2b^2).$$

9. An ordinary plane angle is measured by the length of the arc that its sides intercept on a unit circle with center at the vertex. This idea can be extended to a *solid angle* bounded by a conical surface with vertex $A$ as follows: The magnitude of the solid angle is by definition equal to the area that it intercepts on a unit sphere with center $A$. Thus, the measure of the solid angle of the domain $x \geq 0$, $y \geq 0$, $z \geq 0$ is $4\pi/8 = \pi/2$. Now let $\Gamma$ be a closed curve, $\Sigma$ a surface bounded by $\Gamma$, and $A$ a fixed point outside both $\Gamma$ and $\Sigma$. An element of area $dS$ at a point $M$ of $\Sigma$ defines an elementary cone with its vertex at $A$, and the solid angle of this cone is readily found by an elementary argument to be

$$\frac{\cos\theta}{r^2}\, dS,$$

where $r = AM$ and $\theta$ is the angle between the vector $\overrightarrow{MA}$ and the normal to $\Sigma$ at $M$. This elementary solid angle is positive or negative according to whether $\theta$ is acute or obtuse. Interpret the surface integral

$$\Omega = \iint_\Sigma \frac{\cos\theta}{r^2}\, dS$$

geometrically as a solid angle and show that

$$\Omega = \iint_\Sigma \frac{(a-x)\, dy\, dz + (b-y)\, dz\, dx + (c-z)\, dx\, dy}{[(a-x)^2 + (b-y)^2 + (c-z)^2]^{3/2}}$$

where $(a, b, c)$ and $(x, y, z)$ are the Cartesian coordiantes of $A$ and $M$, respectively.

10. Prove, first directly and then by interpretation of the integral as a solid angle, that

$$\int_{-\infty}^\infty \int_{-\infty}^\infty \frac{dx\, dy}{(x^2 + y^2 + 1)^{3/2}} = 2\pi.$$

11. Prove that the solid angle that the whole surface of the hyperboloid of one sheet $(x^2/a^2) + (y^2/b^2) - (z^2/c^2) = 1$ subtends at its center $(0, 0, 0)$ is

$$8c \int_0^{\pi/2} \sqrt{\frac{b^2\cos^2\varphi + a^2\sin^2\varphi}{a^2b^2 + b^2c^2\cos^2\varphi + a^2c^2\sin^2\varphi}}\, d\varphi.$$

12. Show that the value of the integral

$$\Omega = \iint_\Sigma \frac{(a-x)\, dy\, dz + (b-y)\, dz\, dx + (c-z)\, dx\, dy}{[(a-x)^2 + (b-y)^2 + (c-z)^2]^{3/2}}$$

is independent of the choice of the surface $\Sigma$, provided its boundary $\Gamma$ is kept fixed. By integrating over the outside of the surface, prove from this result that if $\Sigma$ is a closed surface, then $\Omega = 4\pi$ or $0$, according to whether $A(a, b, c)$ is within the volume bounded by $\Sigma$ or outside this volume.

13. Let the surface $\Sigma$ be bounded by the closed curve $\Gamma$ and consider the integral

$$\Omega(a,b,c) = \iint_\Sigma \frac{(a-x)\,dy\,dz + (b-y)\,dz\,dx + (c-z)\,dx\,dy}{r^3},$$

$$[r^2 = (a-x)^2 + (b-y)^2 + (c-z)^2],$$

as a function of $a$, $b$, $c$. Prove that the components of the gradient of $\Omega$ can be expressed as line integrals as follows:

$$\frac{\partial\Omega}{\partial a} = \int_\Gamma \frac{(z-c)\,dy - (y-b)\,dz}{r^3}, \qquad \frac{\partial\Omega}{\partial b} = \int_\Gamma \frac{(x-a)\,dz - (z-c)\,dx}{r^3},$$

$$\frac{\partial\Omega}{\partial c} = \int_\Gamma \frac{(y-b)\,dx - (x-a)\,dy}{r^3}.$$

These formulae, which have an important interpretation in electromagnetism, can be expressed by the following vector equation

$$\operatorname{grad}\Omega = -\int_\Gamma \frac{\mathbf{x}\times d\mathbf{x}}{|\mathbf{x}|^3},$$

where $\mathbf{x}$ is the vector with components $(x-a)$, $(y-b)$, $(z-c)$.

14. Verify that the expression

$$\frac{-4xy\,dx + 2(x^2 - y^2 - 1)\,dy}{(x^2 + y^2 - 1)^2 + 4y^2}$$

is the total differential of the angle that the segment $-1 \le x \le 1$, $y = 0$ subtends at the point $(x, y)$. Using this fact, prove the following result by a geometrical argument: Let $\Gamma$ be an oriented closed curve in the $x$, $y$-plane, not passing through either of the points $(-1, 0)$, $(1, 0)$. Let $p$ be the number of times $\Gamma$ crosses the line segment $-1 < x < 1$, $y = 0$ from the upper half-plane $y > 0$ to the lower half plane $y < 0$, and $n$ the number of times $\Gamma$ crosses this line segment from $y < 0$ to $y > 0$. Then,

$$\theta = \int_\Gamma \frac{-4xy\,dx + (x^2 - y^2 - 1)\,dy}{(x^2 + y^2 - 1) + 4y^2} = 2\pi(p - n).$$

Thus, if $\Gamma$ is the curve $r = 2\cos 2\theta$ $(0 \le \theta \le 2\pi)$, in polar coordinates, $\theta = 0$.

15. Consider the unit circle $C$

$$x' = \cos\varphi, \quad y' = \sin\varphi, \quad z' = 0 \qquad (0 \le \varphi \le 2\pi)$$

in the $x$, $y$-plane. Denote by $\Omega$ the solid angle which the circular disc $x^2 + y^2 \le 1$, $z = 0$, subtends at the point $P = (x, y, z)$. Now let $P$ describe an oriented closed curve $\Gamma$ that does not meet the circle $C$. Let $p$ be the number of times $\Gamma$ crosses the circular disc $x^2 + y^2 < 1$, $z = 0$, from the upper half-space $z > 0$ to the lower half-space $z < 0$, and $n$ the number of times $\Gamma$ crosses this disc from $z < 0$ to $z > 0$. If $P$ starts from a point $P_0$ on $\Gamma$ with $\Omega = \Omega_0$, then $P$, describing $\Gamma$ (while $\Omega$ varies continuously with $P$), will return to $P_0$ with a value $\Omega = \Omega_1$. Prove by a geometrical argument that

$$\Omega_1 - \Omega_0 = \int_\Gamma d\Omega = 4\pi(p - n).$$

Using the vector equation found above,

$$\operatorname{grad} \Omega = -\int_\sigma \frac{\overrightarrow{PP'} \times dP'}{|PP'|^3}$$

(Exercise 13), prove that

$$\int_C \int_\Gamma \frac{1}{|PP'|^3} \begin{vmatrix} x' - x & dx & dx' \\ y' - y & dy & dy' \\ z' - z & dz & dz' \end{vmatrix}$$

$$= \int_\Gamma \int_C \frac{(x'-z)(dy\,dz'-dz\,dy') + (y'-y)(dz\,dx'-dx\,dz') + (z'-z)(dx\,dy'-dy\,dz')}{[(x'-x)^2 + (y'-y)^2 + (z'-z)^2]^{3/2}}$$

$$= 4\pi(p - n).$$

[This repeated line integral, which is due to Gauss, gives the number of times $\Gamma$ is wound around $C$. It should be remarked that its vanishing is necessary if the two curves $\Gamma$ and $C$ (thought of as being two strings) are to be separable, but not sufficient, as is shown by the example in Fig. 5.13, where $p = n = 1$, yet $\Gamma$ and $C$ cannot be separated.]

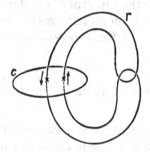

**Figure 5.13**

16. Let $\Gamma$ be a closed curve in space on which a definite sense of description of the curve has been assigned. Prove that there is a vector **a** with the following characteristic property: for any unit vector **n** the scalar product **a·n** is equal to the algebraic value of the area enclosed by the orthogonal projection of $\Gamma$ on the plane $\Pi$ orthogonal to **n**. (Note that **n** gives the orientation of $\Pi$, and $\Gamma$ gives the orientation of its projection on $\Pi$.) In particular, the projection of $\Gamma$ on any plane parallel to **a** has the algebraic area zero. (The vector **a** may be called the *area vector* of $\Gamma$.)

17. Let $f(x, y)$ be a continuous function with continuous first and second derivatives. Prove that if
$$f_{xx}f_{yy} - f_{xy}^2 \ne 0,$$
the transformation
$$u = f_x(x, y), \qquad v = f_y(x, y), \qquad w = -z + xf_x(x, y) + yf_y(x, y)$$
has a unique inverse, which is of the form
$$x = g_u(u, v), \qquad y = g_v(u, v), \qquad z = -w + ug_u(u, v) + vg_v(u, v).$$

18. Represent the gravitational vector field

$$X = \frac{x}{\sqrt{(x^2 + y^2 + z^2)^3}}, \qquad Y = \frac{y}{\sqrt{(x^2 + y^2 + z^2)^3}}$$

$$Z = \frac{z}{\sqrt{(x^2 + y^2 + z^2)^3}},$$

as a curl.

### 5.11  Integral Identities in Higher Dimensions

The formulae of Gauss and Stokes discussed in the previous sections all can be considered as extensions to more dimensions of the *fundamental theorem of calculus*

$$(76) \qquad \int_a^b f'(x) \, dx = f(b) - f(a).$$

That theorem expresses the integral of the derivative of a function of a single variable over an interval in terms of the values of the function at the boundary points of the interval. In a similar way, Gauss's theorem

$$(77) \quad \iiint_R (f_x + g_y + h_z) \, dx \, dy \, dz = \iint_S \left( f \frac{dx}{dn} + g \frac{dy}{dn} + h \frac{dz}{dn} \right) dS$$

($n$ = outward-drawn normal) expresses an integral over a set $R$ in terms of quantities taken on the boundary of $R$. In vector form, with $A = (f, g, h)$ the divergence theorem becomes

$$\iiint_R \operatorname{div} A \, dx \, dy \, dz = \iint_S A \cdot n \, dS.$$

Obviously, the expression div $A$ plays the role of the derivative $f'$ in the simple formula (76).

In three-dimensions we obtained in addition formulae expressing integrals of differential expressions over curves or surfaces in terms of boundary integrals. The curve integrals considered took the form

$$(78) \qquad \int_C A \cdot t \, ds,$$

($t$ = unit tangent vector of the curve $C$) and surface integrals the form

$$\iint_S A \cdot n \, dS$$

(**n** = unit normal vector to the surface $S$). There are bound to be restrictions on the vector **A** if integrals of these types are to be expressible in a form that only involves boundary points of $C$ or of $S$. The reason is that there are many curves or surfaces in three-space with the same boundary. An identity expressing an integral in terms of functions on the boundary alone implies that the integral does not depend on the particular curve or surface chosen and this can only be the case for vectors **A** of special types.

Thus, we found that if the line integral of **A·t** over a curve $C$ is to depend only on the end points $P$ and $Q$ of $C$, then the vector field $A(x, y, z)$ has to be *irrotational*; that is, curl **A** = 0. If this condition is satisfied in a simply connected set containing $C$, we can find a scalar $U = U(x, y, z)$ such that $\mathbf{A} = \text{grad } U = (U_x, U_y, U_z)$; in that case, we indeed have an integral identity of the desired type:

$$\int_C \mathbf{A} \cdot \mathbf{t} \, ds = \int_C dU = U(Q) - U(P).$$

Similarly, for the surface integral

$$\iint_C \mathbf{A} \cdot \mathbf{n} \, dS$$

to depend only on the boundary curve $C$ of $S$, the vector **A** has to satisfy the necessary condition[1] div **A** = 0. If the condition div **A** = 0 is satisfied, we can represent **A** in the form **A** = curl **B** (see p. 315) and express the integral of **A · n** over the surface $S$ in terms of an integral over $C$ by Stokes's theorem

(79)     $$\iint_S \mathbf{A} \cdot \mathbf{n} \, dS = \iint_S (\text{curl } \mathbf{B}) \cdot \mathbf{n} \, dS = \int_C \mathbf{B} \cdot \mathbf{t} \, ds.$$

From these examples one would expect that there exist more general formulae expressing appropriate combinations of derivatives of functions over an $m$-dimensional set in $M$-dimensional euclidean space as integrals of the functions over the $(m - 1)$-dimensional

---

[1]Assume that the double integral of **A · n** over any surface $S$ depends only on the boundary $C$ of $S$. Then the integral is the same for any two surfaces with the same boundary if we define the direction **n** consistently on the two surfaces (i.e., so that the normal vectors **n** go into each other if one surface is deformed smoothly into the other). In case the two surfaces together form the boundary $\sigma$ of a set $R$ in space, the integral of **A · N** over $\sigma$ is 0 if **N** denotes the unit normal of $\sigma$ pointing away from $R$. By the divergence theorem, it follows then that the integral of div **A** over $R$ vanishes. Since $R$ is arbitrary, we find by space differentiation that div **A** = 0.

boundary of the set. For $m = M$ Gauss's theorem (77) suggests an obvious generalization:

$$\iint_R \cdots \int (f^1_{x_1} + f^2_{x_2} + \cdots + f^M_{x_M}) \, dx_1 \cdots dx_M$$

$$= \int \cdots \int_S \left( f^1 \frac{dx_1}{dn} + \cdots + f^M \frac{dx_M}{dn} \right) dS.$$

Here $R$ is a set in $M$-space bounded by the $(M-1)$-dimensional hyper-surface $S$ with outward-drawn normal $\mathbf{n}$, and $f^1, f^2, \ldots, f^M$ are functions of $x_1, \ldots, x_M$. On the other hand, the formula of Stokes in the form (79) has no such obvious analogue. However, the calculus of *exterior,* or *alternating,* differential forms leads one immediately to conjecture the *general Stokes's formula*

(80)
$$\int \cdots \int_{S^*} d\omega = \int \cdots \int_{\partial S^*} \omega$$

for arbitrary differential forms $\omega$ of order $m - 1$ and arbitrary $m$-dimensional oriented surfaces $S^*$ with suitably oriented $(m-1)$-dimensional boundary $\partial S^*$. In the Appendix to this chapter we shall prove the general formula (80) without using any new ideas beyond those already arising in the rigorous proof of the special cases (77) and (79).

## Appendix: General Theory of Surfaces and of Surface Integrals

Rigorous proofs of the theorems of Gauss and Stokes and their extensions to higher dimensions require a more careful analysis of the notions of surface, of orientation of surfaces, and of integrals over surfaces. These are provided in the present appendix.

### A.1  Surfaces and Surface Integrals in Three Dimensions

#### a.  Elementary Surfaces

Elementary surfaces are essentially the analogues of the simple arcs defined in Volume I, p. 334. They form the building blocks making up surfaces of more complicated structure.

An elementary surface $\sigma$ in $x$, $y$, $z$-space is a set of points $P = (x, y, z)$ represented parametrically by three functions,

(1a) $$x = f(u, v), \quad y = g(u, v), \quad z = h(u, v)$$

where (1) the domain $U$ of the functions is an open bounded set in the $u$, $v$-plane; (2) $f$, $g$, $h$ are continuous and have continuous first derivatives in $U$; (3) the inequality

(1b) $$W = \sqrt{\begin{vmatrix} f_u & f_v \\ g_u & g_v \end{vmatrix}^2 + \begin{vmatrix} g_u & g_v \\ h_u & h_v \end{vmatrix}^2 + \begin{vmatrix} h_u & h_v \\ f_u & f_v \end{vmatrix}^2}$$
$$= \sqrt{(f_u g_v - f_v g_u)^2 + (g_u h_v - g_v h_u)^2 + (h_u f_v - h_v f_u)^2} > 0$$

is satisfied at all points $U$; and (4) the mapping of the set $U$ in the $u$, $v$-plane on the set $\sigma$ in $x$, $y$, $z$-space is 1–1 and the inverse mapping from $\sigma$ onto $U$ is also continuous.

The quantity $W$ represents the length of the vector with components

(2) $$A = g_u h_v - g_v h_u, \quad B = h_u f_v - h_v f_u, \quad C = f_u g_v - f_v g_u$$

that is the vector product of the two vectors

(3) $$(f_u, g_u, h_u) \quad \text{and} \quad (f_v, g_v, h_v).$$

The two vectors in (3) are tangential to the surface, while the vector $(A, B, C)$ is perpendicular to those two and, hence, normal to the surface. Equation (1b) guarantees that there are only two directions normal to the surface, namely that of the vector $(A, B, C)$ and of its opposite $(-A, -B, -C)$.

At each point of $\sigma$, at least one of the three quantities $A$, $B$, $C$ does not vanish. If, say, $C \neq 0$ at a point $P_0 = (x_0, y_0 \; z_0)$ corresponding to a parameter point $(u_0, v_0)$ in $U$, we can find for a sufficiently small positive $\varepsilon$ a number $\delta > 0$ such that each pair $(x, y)$ with

(4) $$\sqrt{(x - x_0)^2 + (y - y_0)^2} < \delta$$

is representable uniquely in the form

(5) $$x = f(u, v), \quad y = g(u, v)$$

with

(6) $$\sqrt{(u - u_0)^2 + (v - v_0)^2} < \varepsilon.$$

The values $u$, $v$ determined by $x$, $y$ are functions

(7) $$u = \phi(x, y), \qquad v = \psi(x, y),$$

which are continuous and have continuous first derivatives for $(x, y)$ satisfying (4). By the assumed continuous dependence of $(u, v)$ on $P$ we see that every point $P$ on the surface $\sigma$ that is sufficiently close to $P_0$ has parameters $(u, v)$ satisfying (6). If, moreover, the distance from $P$ to $P_0$ is $< \delta$, the coordinates $x$, $y$ of $P$ will satisfy (4). Thus, for all $P$ on $\sigma$ sufficiently close to $P_0$, we can express the parameter values $u$, $v$ in terms of $x$, $y$ by (7). On substituting these values in the equation $z = h(u, v)$, we then have a *nonparametric representation*

(8) $$z = h(\phi(x, y), \psi(x, y)) = H(x, y),$$

which applies to all points of the surface $\sigma$ that are sufficiently close to $P_0$. If the quantity $B$ does not vanish, we obtain similarly a local representation of the form $y = G(x, z)$ and in case $A \neq 0$ a representation of the form $x = F(y, z)$.

The same elementary surface $\sigma$ has many different parameter representations, all of which, however, are related in a simple fashion. Let

(9) $$\bar{x} = \bar{f}(\bar{u}, \bar{v}), \quad \bar{y} = \bar{g}(\bar{u}, \bar{v}), \quad \bar{z} = \bar{h}(\bar{u}, \bar{v}) \qquad \text{for} \qquad (\bar{u}, \bar{v}) \text{ in } \bar{U}$$

be a second parameter representation for $\sigma$ also satisfying all our four requirements. The bi-unique and bi-continuous correspondence between $U$ and $\sigma$ and between $\bar{U}$ and $\sigma$ establishes then a 1–1 and continuous mapping with continuous inverse of the set $\bar{U}$ onto the set $U$:

(10) $$u = \alpha(\bar{u}, \bar{v}), \quad v = \beta(\bar{u}, \bar{v}) \qquad \text{for} \qquad (\bar{u}, \bar{v}) \text{ in } \bar{U}.$$

If, here, for a certain $(\bar{u}_0, \bar{v}_0)$ in $\bar{U}$ the corresponding values $(u_0, v_0)$ are such that the quantity $C(u_0, v_0)$ is not zero, then the representation (7) applies for all $(u, v)$ near $(u_0, v_0)$, and hence, we find from (9) that

$$u = \alpha(\bar{u}, \bar{v}) = \phi(\bar{f}(\bar{u}, \bar{v}), \bar{g}(\bar{u}, \bar{v}))$$

$$v = \beta(\bar{u}, \bar{v}) = \psi(\bar{f}(\bar{u}, \bar{v}), \bar{g}(\bar{u}, \bar{v}))$$

for all $(\bar{u}, \bar{v})$ sufficiently close to $(\bar{u}_0, \bar{v}_0)$. Since $\phi$, $\psi$, $f$, $g$ all are functions with continuous first derivatives, it follows that the functions α, β describing the change of parameters (10) not only are continuous but have continuous first derivatives as well.

Putting

(11)
$$\Delta = \frac{d(u, v)}{d(\bar{u}, \bar{v})} = \frac{\partial \alpha}{\partial \bar{u}} \frac{\partial \beta}{\partial \bar{v}} - \frac{\partial \alpha}{\partial \bar{v}} \frac{\partial \beta}{\partial \bar{u}},$$

we find from the rules for the Jacobian of the product of two mappings [see (31b), p. 258] that

(12a)
$$\bar{C} = \frac{d(x, y)}{d(\bar{u}, \bar{v})} = \frac{d(x, y)}{d(u, v)} \cdot \frac{d(u, v)}{d(\bar{u}, \bar{v})} = C\Delta$$

and, similarly, that

(12b)
$$\bar{B} = B\Delta, \qquad \bar{A} = A\Delta.$$

In particular, we find that the Jacobian of the mapping (10) between the two parameter regions does not vanish, since by (12a, b)

(13) $\quad \bar{W} = \sqrt{\bar{A}^2 + \bar{B}^2 + \bar{C}^2} = \sqrt{\Delta^2(A^2 + B^2 + C^2)} = |\Delta|\, W$

and, by assumption, $\bar{W} \neq 0$.

Of course the same statements are valid for the expressions of $\bar{u}, \bar{v}$ in terms of $u, v$. The important fact is that *the relation between two parameter systems for the same elementary surface satisfy all of the assumptions made in the proofs of the transformation laws for areas and integrals.*

#### b. Integral of a Function over an Elementary Surface

There is nothing difficult in the notion of a *continuous function F defined in the points P of an elementary surface* σ. We just require that with every $P \in \sigma$ there is associated a value $F = F(P)$ in such a way that for a sequence of points $P_n$ on σ that converges to a point $P$ of σ, we have

$$\lim_{n \to \infty} F(P_n) = F(P).$$

In any particular parametric representation (1a), $F$ becomes a function of $u, v$ in the domain $U$ and continuity of $F$ on σ becomes equivalent[1] to continuity of $F$ as a function of $u$ and $v$.

---

[1]We make use here of the bi-continuous character of the relation between σ and $U$.

We restrict ourselves here to continuous functions $F$ on $\sigma$ that are zero outside some compact (i.e., closed and bounded) subset $s$ of $\sigma$. The corresponding parameter points $(u, v)$ form then a compact[1] subset $S$ of $U$. We then define the integral of $F$ over the elementary surface $\sigma$ by the formula

(14)
$$\iint_\sigma F \, dA = \iint FW \, du \, dv,$$

where $W$ is the expression given by (1b). Here $FW$ is continuous function of $u$, $v$, which we define as 0 for $(u, v)$ outside $S$; hence, $FW$ is integrable. One still has to show that the surface integral of $F$ over $\sigma$ defined by (14) does not depend on the particular parameter representation (1a). This follows immediately from the law of transformation (13) for $W$ and from the general formula (16b), p. 403, for transformation of double integrals under a change of variables from $u$, $v$ to $\bar{u}$, $\bar{v}$. Indeed,

$$\iint FW \, du \, dv = \iint FW \left| \frac{d(u, v)}{d(\bar{u}, \bar{v})} \right| d\bar{u} \, d\bar{v}$$

$$= \iint FW \, |\Delta| \, d\bar{u} \, d\bar{v} = \iint F\bar{W} \, d\bar{u} \, d\bar{v}$$

The independence of the integral of $FW$ from the particular parametric representation means that the *differential form* $W \, du \, dv = dA$ is invariant; it can be identified with the *element of area*.

It would be easy to extend the notion of integral over an elementary surface to more general functions, although we will not do so in the sequel. This involves the extension of the notion of Jordan-measurability to a set $s$ whose closure is contained in the elementary surface $\sigma$; we merely require that the corresponding set $S$ of points $(u, v)$ in the parameter plane be a Jordan-measurable set whose closure lies in $U$. It is seen immediately from the relations between different parameter representations that Jordan-measurability of $s$ does not depend on the particular representation.[2] The same holds for the area of $s$ that we can define as

---

[1]For $(u_n, v_n) \in S$ and $(u_n, v_n) \to (u, v)$ the corresponding points $P_n$ of $\sigma$ lie in $s$. Compactness of $s$ implies that a subsequence of the $P_n$ converges toward a point $P$ of $s$. By continuity convergence of $P_n$ to $P$ implies convergence of the $(u_n, v_n)$ to the corresponding parameter point in $S$. Thus, $(u, v) \in S$, which proves that $S$ is closed. It is bounded as a subset of the bounded set $U$.

[2]See p. 539

$$A(s) = \iint_S dA = \iint_S W \, du \, dv.$$

Of particular importance are the sets $s$ whose closure lies on $\sigma$ and that have area 0. They correspond to sets $S$ in the $u$, $v$-plane of area 0; this means that $S$ can be covered by a finite number of squares contained in $U$ of arbitrarily small total area.

### c. *Oriented Elementary Surfaces*

A particular parameter representation (1a) of the elementary surface $\sigma$ is said to define a particular *orientation* of $\sigma$ (the one that is positive with respect to the $u$, $v$-system). Two parameter sets $u$, $v$ and $\bar{u}$, $\bar{v}$ for the same elementary surface $\sigma$ are said to give $\sigma$ the same orientation if the Jacobian

$$\frac{d(\bar{u}, \bar{v})}{d(u, v)}$$

is positive throughout the parameter domains and to give the opposite orientations if the Jacobian is negative throughout the parameter domains. The combination of the elementary surface $\sigma$ with a particular orientation is called an *oriented elementary* surface $\sigma^*$.

By our assumptions, the Jacobian cannot vanish. Since it is also a *continuous* function of the parameters, we can be sure that it has constant sign when the parameter domain is a *connected* set. In that case there are only two possible orientations for an elementary surface $\sigma$ that may be distinguished as $\sigma^*$ and $-\sigma^*$. It is clear, however, that the number of possible orientations is larger for disconnected sets, where orientations of the parts of $\sigma$ corresponding to the different components of $U$ can be changed independently of each other.

Orientation of the elementary surface is intimately connected with picking a normal direction on $\sigma$ or with "distinguishing the sides" of $\sigma$. A particular parameter representation (1a) of $\sigma$ defines by formulae (2) at each point $P$ quantities $A$, $B$, $C$ that can be considered as the components of a vector perpendicular to $\sigma$ at $P$. This vector has the same direction as the *unit vector* with components

(15)
$$\xi = \frac{A}{W}, \qquad \eta = \frac{B}{W}, \qquad \zeta = \frac{C}{W}.$$

When we change parameters from $u$, $v$ to $\bar{u}$, $\bar{v}$ the quantities $A$, $B$, $C$ change and are replaced by the proportional quantities $\bar{A}$, $\bar{B}$, $\bar{C}$, according to the laws (11) and (12a). Here the factor of proportionality is just the quantity

$$\Delta = \frac{d(u, v)}{d(\bar{u}, \bar{v})}$$

Hence, *the unit normal* $(\xi, \eta, \zeta)$ *is the same for equal orientations of* $\sigma$ *and opposite for opposite orientations.* Equivalently, the orientation of $\sigma^*$ picks out at each point a certain *side* of $\sigma$, namely, that one to which the normal $(\xi, \eta, \zeta)$ points.[1]

The orientation of $\sigma^*$ can also assign a definite sense to every simple closed curve $C$ lying on $\sigma$ by ascribing to $C$ that sense that is positive on the closed curve $\gamma$ in the $u, v$-plane that corresponds to $C$ with respect to the finite region enclosed by $\gamma$.

Specification of an orientation for the elementary surface becomes mandatory when we consider instead of integrals of the form $\iint F\, dA$, where $F$ is a scalar, an integral of a differential form

(16) $$\omega = a\, dy\, dz + b\, dz\, dx + c\, dx\, dy,$$

where, say, $a, b, c$ are continuous functions on $\sigma$ vanishing outside a closed and bounded subset. Here the natural interpretation for the integral suggested by the substitution formulae is, of course,

$$\iint \omega = \iint \left[ a \frac{d(y, z)}{d(u, v)} + b \frac{d(z, x)}{d(u, v)} + c \frac{d(x, y)}{d(u, v)} \right] du\, dv$$

$$= \iint (aA + bB + cC)\, du\, dv$$

$$= \iint (a\xi + b\eta + c\zeta)\, W\, du\, dv = \iint (a\xi + b\eta + c\zeta)\, dA$$

where we have made use of the relations (15) and (14). Here $\xi, \eta, \zeta$ are the direction cosines of the normal determined by the choice of the parameters $u, v$; their sign depends on the orientation of our surface $\sigma$. Thus, we first define the integral of $\omega$ over one of the *oriented* surfaces $\sigma^*$ arising from $\sigma$. We put

(17) $$\iint_{\sigma^*} \omega = \iint \left[ a \frac{d(y, z)}{d(u, v)} + b \frac{d(z, x)}{d(u, v)} + c \frac{d(x, y)}{d(u, v)} \right] du\, dv$$

---

[1]This is the *positive* side of $\sigma^*$, which depends on the orientation of the $x, y, z$-coordinate system; see p. 580. In the notation used on p. 581, we have
$$\Omega(\sigma^*) = \Omega(u, v).$$

$$= \iint (a\xi + b\eta + c\zeta) \, dA,$$

where $u$, $v$ must be one of the parameter systems used to define the orientation of $\sigma^*$ or connected with such a system by a substitution with positive Jacobian and where $\xi$, $\eta$, $\zeta$ is the normal direction induced by the orientation of $\sigma^*$. If $-\sigma^*$ is the elementary surface with the opposite orientation, we have

(18) $$\iint_{-\sigma^*} \omega = - \iint_{\sigma^*} \omega.$$

### d. Simple Surfaces

Let $\sigma$ be an elementary surface with a parametric representation (1a) where the parameter point $(u, v)$ varies over the open set $U$. If $U'$ is any open subset of $U$, the points of $\sigma$ with $(u, v)$ restricted to $U'$ clearly form an elementary surface $\sigma'$ contained in $\sigma$. Indeed, all four of our conditions immediately apply to $\sigma'$, using the same parameters $u$, $v$. As an example, we note that the points of $\sigma$ of distance $< \varepsilon$ from a given point $(x_0, y_0, z_0)$ again form an elementary surface (if not empty), for those are the points whose parameter values $u$, $v$ satisfy

(19) $$[f(u, v) - x_0]^2 + [g(u, v) - y_0]^2 + [h(u, v) - z_0]^2 < \varepsilon^2,$$

and since $f$, $g$, $h$ are continuous functions in $U$, the set $U'$ of such points $(u, v)$ is open.

It is less obvious that *the most general elementary surface $\sigma'$ contained in the elementary surface $\sigma$ can be obtained by restricting the parameter domain of $\sigma$ to a suitable open set.*

For the proof, let the elementary surface $\sigma$ have the parametric representation (1a) for $(u, v) \in U$. Let $\sigma'$ be an elementary surface with the parametric representation (9) with $(\bar{u}, \bar{v})$ varying over the set $\bar{U}$. Let $\sigma'$ be a subset of $\sigma$. Then every $(\bar{u}, \bar{v}) \in \bar{U}$ determines a point $P \in \sigma$, which in turn determines a point $(u, v) \in \bar{U}$ whose coordinates are functions of $\bar{u}$, $\bar{v}$:

(20) $$u = \alpha(\bar{u}, \bar{v}), \qquad v = \beta(\bar{u}, \bar{v}) \qquad \text{for} \qquad (\bar{u}, \bar{v}) \in \bar{U}.$$

The set $\bar{U}$ is mapped by (20) onto a subset $U'$ of $U$. It is clear then that the set $\sigma'$ arises from $\sigma$ by restricting the parameter points $(u, v)$ to the subset $U'$ of $U$. It only remains to see that $U'$ is *open*. Let $P_0 =$

$(x_0, y_0, z_0)$ be a point of $\sigma'$ corresponding, respectively, to the parameter points $(\bar{u}_0, \bar{v}_0)$ in $\bar{U}$ and $(u_0, v_0)$ in $U'$. Let $C$ and $\bar{C}$ be both different from 0 at that point.[1] Then a neighborhood of $(\bar{u}_0, \bar{v}_0)$ is mapped by

$$x = \bar{f}(\bar{u}, \bar{v}), \qquad y = \bar{g}(\bar{u}, \bar{v})$$

onto a set in the $x$, $y$-plane that covers a neighborhood of $(x_0, y_0)$; the corresponding points $(u, v)$ obtained from (7) then cover a neighborhood of $(u_0, v_0)$, so that $U'$ is seen to be an open set.

We see in addition that the two surfaces $\sigma$ and $\sigma'$ agree in a sufficiently small neighborhood of $P_0$, since every $P$ on $\sigma$ sufficiently near $P_0$ has parameter values $(u, v)$ arbitrarily near $(u_0, v_0)$; thus, for $P$ sufficiently close to $P_0$, we have $(u, v) \in U'$, since $(u_0, v_0)$ is an interior point of $U'$, and hence, we see that $P \in \sigma'$. We have proved:

*If the elementary surface $\sigma'$ is contained in the elementary surface $\sigma$ and if $P_0$ is a point of $\sigma'$, then we can find a sufficiently small neighborhood of $P_0$ in which $\sigma$ and $\sigma'$ agree.*

Any orientation imposed on the elementary surface $\sigma$ immediately determines a unique orientation on any elementary surface $\sigma'$ contained in $\sigma$. We need only refer $\sigma'$ to the same parameter system that defines the orientation of $\sigma$ and take that system to fix the orientation of $\sigma'$.

We are now in a position to give precise meaning to the more general notion of a simple surface, as an object "patched together" from elementary surfaces:

*A set $\tau$ in $x$, $y$, $z$-space is called a simple surface if for every point $P_0$ on $\tau$ there exists an $\varepsilon > 0$ such that the points of $\tau$ that have distance less than $\varepsilon$ from $P_0$ form an elementary surface.*

Thus, for every $P_0 \in \tau$ there is an elementary surface $\sigma$ that agrees with $\tau$ near $P_0$ and is contained in $\tau$. We can show that the intersection of two elementary surfaces $\sigma'$ and $\sigma''$ contained in the simple surface $\tau$ is again an elementary surface (if not empty), for if $P_0$ is a common point of $\sigma'$ and $\sigma''$, we can find an $\varepsilon$-neighborhood $N_\varepsilon$ of $P_0$ such that $\sigma = N_\varepsilon \cap \tau$ is an elementary surface. Here $\sigma$ contains the two elementary surfaces $N_\varepsilon \cap \sigma'$ and $N_\varepsilon \cap \sigma''$. Consequently, $\sigma'$ and $\sigma''$ agree with $\sigma$, and thus with each other, at all points sufficiently near to $P_0$. If $\sigma'$ is referred to parameters $u$, $v$ with $u_0$, $v_0$ corresponding to $P_0$, all $(u, v)$ sufficiently close to $(u_0, v_0)$ will correspond to points

---

[1] We can assume that all three quantities $\bar{A}, \bar{B}, \bar{C}$ are $\neq 0$ at $P_0$, applying, if necessary, a suitable rotation to $x$, $y$, $z$-space. At least one of the quantities $A$, $B$, $C$ does not vanish at $P_0$; let it be $C$.

of $\sigma'$ that lie in $\sigma''$. Hence, the parameter points $(u, v)$ corresponding to points $(x, y, z)$ in $\sigma' \cap \sigma''$ form an open set. Thus, $\sigma' \cap \sigma''$ is an elementary surface.

We define an *oriented simple surface* analogously:

*The simple surface $\tau$ is oriented if $\tau$ is represented as the union of elementary surfaces each of which has been given an orientation, provided the orientations agree in the intersection of any two of the elementary surfaces. Two orientations of $\tau$ are considered identical if they lead to the same orientations at the points common to any two of the oriented elementary surfaces used in defining the orientations of $\tau$. Equivalently, two orientations are identical if they lead to the same choice of a normal direction at each point of $\tau$.*

A case of special importance arises when the simple surface $\tau$ is the boundary of a set $R$ in $x$, $y$, $z$-space. We assume here that $R$ is the closure of a bounded open set.[1] In that case, we can assign an orientation to $\tau$ for which the positive sense assigned by the orientation to each normal of $\tau$ is that of the "direction pointing away from $R$" or that of the "exterior normal." Indeed, for each point $P_0 = (x_0, y_0, z_0)$ on $\tau$, we can find a neighborhood in which $\tau$ agrees with an elementary surface. We can even choose the neighborhood so small that $\tau$ can be represented nonparametrically in that neighborhood, say, by an equation

$$(21) \qquad z = F(x, y) \qquad \text{valid for} \qquad (x - x_0)^2 + (y - y_0)^2 < \varepsilon^2$$

If two points $P$ and $P'$ in space can be joined by an arc that contains no point of the boundary $\tau$ of $R$, either both or neither lie in $R$. This is clearly the case for any two points satisfying either condition

$$(22a) \qquad F(x, y) < z < F(x, y) + \delta, \qquad (x - x_0)^2 + (y - y_0)^2 < \varepsilon^2$$

or

$$(22b) \qquad F(x, y) - \delta < z < F(x, y), \qquad (x - x_0)^2 + (y - y_0)^2 < \varepsilon^2,$$

provided $\delta$ is a sufficiently small positive number. Thus, each of the to sets (22a) and (22b) either is completely contained in $R$ or has no points in common with $R$. They cannot both be contained in $R$, for then the set (21) also would belong to $R$, since $R$ is closed; but then $P_0$ would not be a boundary point of $R$. Neither can both sets be free of points of $R$, since then $P_0$ could not be a limit of interior points of

---

[1] This means that $R$ is closed and bounded and that every boundary point of $R$ is the limit of interior points.

$R.$ Thus, exactly one of the sets (22a) and (22b) is contained in $R.$ If (22b) is the set contained in $R,$ we choose the parameters $u = x, v = y$ to assign an orientation to the elementary surface (21), writing

$$x = u, \quad y = v, \quad z = F(u, v).$$

The corresponding normal direction has directio ncosines [see (2) and (15)]

$$\xi = -\frac{F_u}{W}, \quad \eta = -\frac{F_v}{W}, \quad \zeta = \frac{1}{W}.$$

Since $\zeta > 0,$ the normal at any point of the surface *points* away from $R,$ in the sense that any point on the normal at a point of (21) that is sufficiently close to the surface will lie in the set (22a) and, hence, outside R. Similarly, if the set (22a) belongs to $R,$ we define the orientation of (21) by the parametric representation

$$x = v, \quad y = u, \quad z = F(u, v),$$

which leads to $\zeta = -1/W < 0$ and again singles out the normal direction away from $R.$

We have thus represented $\tau$ as a union of oriented simple surfaces, where, because of the geometric meaning of the orientation in relation to the set $R,$ orientations agree in overlapping simple surfaces. We call $\tau$ oriented positively with respect to $R$ [1].

### e.  Partitions of Unity and Integrals over Simple Surfaces

Given a simple surface $\tau,$ we wish to define

$$\iint_\tau F \, dA$$

under the assumption that $F$ is a continuous function on $\tau$ that vanishes outside some closed and bounded subset $s$ of $\tau.$ (In case the whole surface $\tau$ is closed and bounded, the definition will furnish the integral over $\tau$ of an *arbitrary* continuous function on $\tau.$) We make use of a device known as *partition of unity* to reduce our integrals to integrals over compact subsets of elementary surfaces that have been defined already.

---

[1] We assume here that $R$ has the orientation of the $x, y, z$-coordinate system.

A partition of unity consists of a finite number of functions $\chi_1(P)$, $\chi_2(P), \ldots, \chi_N(P)$ defined and continuous in the points $P$ of the set $s$ with the properties:

1. $\chi_i(P) \geqq 0$ for all $P \in s$ and $i = 1, \ldots, N$;

2. $\chi_1(P) + \chi_2(P) + \cdots + \chi_N(P) = 1$ for all $P \in s$

3. for each $i = 1, \ldots, N$ there exists an elementary surface $\sigma_i$ contained in $\tau$ such that $\chi_i(P) = 0$ for $P$ in $s$ outside a certain compact subset of $\sigma_i$.

(It is, of course, property 2 that accounts for the name *partition of unity*).

Assume that we have such a partition of unity for $s$. We can write for $P \in s$

$$(23a) \quad F(P) = F(P)\,\chi_1(P) + F(P)\,\chi_2(P) + \cdots + F(P)\,\chi_N(P).$$

Here each term is defined and continuous for $P$ in $s$. However, since $F(P)$ is assumed to be defined and continuous on the whole of $\tau$ and to vanish outside the set $s$, we can extend each term $F(P)\,\chi_i(P)$ over the whole of $\tau$ as a continuous function just by defining $F\,\chi_i$ as zero for points of $\tau$ not in $s$.

We then define the integral of $F$ over $\tau$ by the formula

$$(23b) \qquad \iint_\tau F\,dA = \sum_{i=1}^N \iint_{\sigma_i} F\,\chi_i\,dA$$

Here the integrals on the right have a meaning since $F\,\chi_i$ is continuous on the elementary surface $\sigma_i$ and vanishes outside a compact subset of $\sigma_i$.

To complete the definition, we have to show that the expression (23b) for the integral of $F$ over $\tau$ does not depend on the *particular* partition of unity used. Assume that we have a second partition consisting of functions $\chi_1'(P), \chi_2'(P), \ldots, \chi_m'(P)$ vanishing, respectively, outside compact subsets of elementary surfaces $\sigma_1', \ldots, \sigma_m'$. For each $i = 1, \ldots, N$ and $k = 1, \ldots, m$ the set

$$\sigma_i \cap \sigma_{k'}$$

is again an elementary surface (if not empty), since both $\sigma_i$ and $\sigma_{k'}$ lie on $\tau$. Moreover, the function $F\,\chi_i\,\chi_{k'}$ vanishes outside a compact subset of that surface. Hence, formula (23b) yields

$$\iint_\tau F \, dA = \sum_i \iint_{\sigma_i} F \, \chi_i \, dA$$

$$= \sum_{i,k} \iint_{\sigma_i} F \, \chi_i \, \chi_k' \, dA$$

$$= \sum_{i,k} \iint_{\sigma_i \cap \sigma_k} F \, \chi_i \, \chi_k' \, dA$$

$$= \sum_{i,k} \iint_{\sigma_k} F \, \chi_i \, \chi_k' \, dA$$

$$= \sum_k \iint_{\sigma_{k'}} F \, \chi_k' \, dA,$$

which shows that a different partition leads to the same value for the integral.

It remains to exhibit an actual partition of unity. By definition, we have for every point $Q$ of the simple surface $\tau$ a number $\varepsilon_Q > 0$ such that the points of $\tau$ within distance $\varepsilon_Q$ from $Q$ form an elementary surface $\sigma_Q$. We associate with $Q$ the function of $P$ defined by

(24a) $$\psi_Q(P) = \begin{cases} \varepsilon_Q - 2\overline{PQ} \text{ for } \overline{PQ} < \dfrac{1}{2}\,\varepsilon_Q \\[2mm] 0 \text{ for } \overline{PQ} \geqq \dfrac{1}{2}\,\varepsilon_Q. \end{cases}$$

Here $\overline{PQ}$ denotes the distance between the two points $P$ and $Q$. The function $\psi_Q(P)$ is defined and continuous for all $P$ in space and, hence, in particular, is continuous on $\sigma_Q$. The number $\varepsilon_Q$ can be chosen so small that the set of points $P$ on $\sigma_Q$ for which $\overline{PQ} \leqq \frac{1}{2}\,\varepsilon_Q$ is closed.[1] These points then form a compact subset of $\sigma_Q$ outside of which the function $\psi_Q(P)$ vanishes.

---

[1] The reason is that all points $P$ in the closure of an elementary surface $\sigma$ that are sufficiently near to a given point $Q$ of $\sigma$ have to belong to the set $\sigma$ itself: Let $\sigma$ correspond to the open set $U$ in the parameter plane, with $Q$ corresponding to a point $q$. Let $P_n$ be a sequence of points on $\sigma$ with images $p_n$ in $U$, and let $P_n \to P$. For $P_n$ sufficiently close to $Q$ the $p_n$ lie in a closed disc about $q$ contained in $U$. A subsequence of the $p_n$ converges to a point $p$ of $U$. The point on $\sigma$ corresponding to $p$ is just $P$. Now by definition of $\tau$ there exists a positive $\delta_Q$ such that the points $P$ of $\tau$ with $\overline{PQ} < \varepsilon_Q$ form an elementary surface $\sigma$. There exists then a positive $\varepsilon_Q \leqq \delta_Q$ (depending on the choice of $\delta_Q$) such that the points $P$ of the closure of $\sigma$ for which $\overline{PQ} \leqq \frac{1}{2}\,\varepsilon_Q$ belong to $\sigma$. Let $\sigma_Q \subset \sigma$ denote the set of points $P$ of $\tau$ with $\overline{PQ} < \varepsilon_Q$. Then the closure of the set of points $P$ of $\sigma_Q$ with $\overline{PQ} \leqq \frac{1}{2}\,\varepsilon_Q$ belongs to $\sigma$, and hence also to $\sigma_Q$ since $\frac{1}{2}\,\varepsilon_Q < \varepsilon_Q$.

We take now for each $Q$ on $\tau$ the open ball of radius $\frac{1}{2}\varepsilon_Q$ in which the function $\psi_Q$ is positive. By the Heine-Borel theorem a finite number of these balls, say the ones with centers $Q_1, \ldots, Q_N$, already covers the closed and bounded set $s$. We then define the partition functions $\chi_i$ for $i = 1, \ldots, N$ by

(24b)
$$\chi_i(P) = \frac{\psi_{Q_i}(P)}{\psi_{Q_1}(P) + \cdots + \psi_{Q_n}(P)}$$

Here the denominator is different from zero for each $P$ in $s$, so that $\chi_i(P)$ is defined and continuous in $s$. It is clear that in $s$ the $\chi_i(P)$ are nonnegative and have sum 1. Moreover, $\chi_i(P) = 0$ outside a compact subset of the elementary surface $\sigma_{Q_i}$. Thus, the $\chi_i(P)$ form a partition of unity.

Having defined the integral of a function $F$ over a simple surface, we can immediately obtain the integral of a differential form

(25a)
$$\omega = a \, dy \, dz + b \, dz \, dx + c \, dx \, dy$$

over an *oriented simple surface* $\tau^*$, assuming the coefficients $a$, $b$, $c$ to vanish outside a compact subset $s$ of $\tau^*$. We simply take

(25b)
$$\iint_{\tau^*} \omega = \iint_\tau (a\xi + b\eta + c\zeta) \, dA,$$

where $\tau$ is the unoriented surface and $\xi$, $\eta$, $\zeta$ are the direction cosines of the normal singled out by the orientation of $\tau^*$ with respect to the coordinate axes.

## A.2 The Divergence Theorem

### a. *Statement of the Theorem and Its Invariance*

In several variables the role of the fundamental theorem of calculus, which connects the operations of differentiation and integration, is played by the *Gauss divergence theorem*. Under suitable assumptions, for a set $R$ in $x$, $y$, $z$-space with boundary surface $\tau$ the theorem takes the form

(26)
$$\iiint_R (a_x + b_y + c_z) \, dx \, dy \, dz = \iint_\tau (a\xi + b\eta + c\zeta) \, dA,$$

where $\xi$, $\eta$, $\zeta$ denote the direction cosines of the *exterior* normal (i.e., of the normal pointing away from $R$) in the points of $\tau$.

We shall prove the theorem here under the assumptions that $R$ is the closure of an open bounded set in $x, y, z$-space and that the boundary of $R$ is a simple surface. The functions $a(x, y, z)$, $b(x, y, z)$, $c(x, y, z)$ shall be continuous in $R$ and have continuous and bounded first derivatives in the interior points of $R$.

An important feature of formula (26) is its *invariance* under rigid motions of space. This fact is more easily verified if subscripts rather than different letters are used to distinguish variables. We replace the quantities $x, y, z$ by $x_1, x_2, x_3$ and $a, b, c$ by $a_1, a_2, a_3$, and $\xi, \eta, \zeta$ by $\xi_1, \xi_2, \xi_3$. Formula (26) becomes

(27a)
$$\iiint_R \sum_i \frac{\partial a_i}{\partial x_i} \, dx_1 \, dx_2 \, dx_3 = \iint_\tau \sum_i a_i \, \xi_i \, dA,$$

where $i = 1, 2, 3$. Of course, the analogous formula with $i$ ranging from 1 to $n$ holds in $n$ dimensions.

A rigid motion is given by a linear transformation from $x$- to $y$-variables of the form

(27b)
$$x_i = \sum_k c_{ik} \, y_k + d_i$$

where the $c_{ik}$ and $d_i$ are constants and the $c_{ik}$ satisfy the *orthogonality relations* [see (47) p. 156]

(27c)
$$\sum_i c_{ij} \, c_{ik} = \begin{cases} 0 & \text{for } j \neq k \\ 1 & \text{for } j = k. \end{cases}$$

The same law of transformation, but with the "inhomogeneous" terms $d_i$ omitted, applies to vectors, since their components are just differences of the coordinates of their end points. Thus, we associate with the $a_i$ the components $b_k$ of the same vector in the new system determined by

$$a_i = \sum_k c_{ik} \, b_k$$

This law of transformation also applies to the direction cosines of the normal on the boundary, which are just the components of the exterior unit normal. The new direction cosines $\eta_k$ are connected with the $\xi_i$ by the formulae

$$\xi_i = \sum_k c_{ik} \, \eta_k.$$

Then, obviously,

$$\sum_i \frac{\partial a_i}{\partial x_i} = \sum_{i,k} c_{ik} \frac{\partial b_k}{\partial x_i} = \sum_{i,k} \frac{\partial x_i}{\partial y_k} \frac{\partial b_k}{\partial x_i} = \sum_k \frac{\partial b_k}{\partial y_k},$$

where we have made use of the *chain rule of differentiation* (see p. p. 208–209). Similarly, using (27c)

$$\sum_i a_i \xi_i = \sum_{i,j,k} c_{ik} b_k c_{ij} \eta_j = \sum_k b_k \eta_k$$

Hence, (27a) implies that

$$\iiint \sum_k \frac{\partial b_k}{\partial y_k} \, dy_1 \, dy_2 \, dy_3 = \iint \sum_k b_k \eta_k \, dA$$

and, thus, represents a relation that is invariant under rigid motions of space.[1]

### b. Proof of the Theorem

The proof of the general formula (26) is again simplified considerably by the use of *partitions of unity*. This device permits us for a given region $R$ with boundary $\tau$ to reduce the formula for general $a$, $b$, $c$ to the case where $a$, $b$, $c$ are zero except in the neighborhood of a point. We shall prove the following:

*If every point $Q$ in $R$ has a neighborhood of radius $\varepsilon_Q$ such that (26) holds for all $a$, $b$, $c$ vanishing outside that neighborhood,[2] then the formula holds for general $a$, $b$, $c$.*

For the proof of this assertion, we use the auxiliary functions $\psi_Q(P)$ defined by

$$\psi_Q(P) = \begin{cases} (\varepsilon_Q{}^2 - 4\overline{PQ}{}^2)^2 & \text{for } \overline{PQ} < \frac{1}{2}\,\varepsilon_Q \\[2mm] 0 & \text{for } \overline{PQ} \geqq \frac{1}{2}\,\varepsilon_Q \end{cases}$$

---

[1]The invariance of the volume element follows because the Jacobian of the transformation (27b), that is, the determinant of the $c_{ik}$, has the value $\pm 1$ (see p. 175), while that of the surface element $dA = W \, du \, dv$ follows by transforming the expression (1b) for $W$.

[2]We consider only functions $a$, $b$, $c$ satisfying the assumptions stated: They are continuous in $R$ and have continuous derivatives in the interior points of $R$.

that are continuous and have continuous first derivatives for all $P$. Since $R$ is closed and bounded, we can pick a finite number of points $Q$, say $Q_1, Q_2, \ldots, Q_N$, such that the corresponding balls $\overline{PQ_i} < \frac{1}{2} \varepsilon_{Q_i}$ cover all of $R$. We again introduce functions

$$\chi_i(P) = \frac{\psi_{Q_i}(P)}{\psi_{Q_1}(P) + \cdots + \psi_{Q_n}(P)}$$

that are defined and have continuous first derivatives in all points $P$ of $R$ and, besides, satisfy the conditions for a partition of unity

(a)                     $\chi_i(P) \geqq 0 \quad \text{in } R$

(b)                     $\sum_i \chi_i(P) = 1$

(c)                     $\chi_i(P) = 0 \quad \text{for} \quad \overline{PQ_i} > \frac{1}{2}\varepsilon_{Q_i}$

The function $a$ can then be decomposed into

$$a = \sum_i a\,\chi_i$$

where the individual terms $a\,\chi_i$ are again continuous in $R$ and have continuous first derivatives in the interior points of $R$. Similarly, $b$ and $c$ can be decomposed. Then, since formula (26) applies to the individual terms, it obviously applies to the whole expression.

Hence, we only have to prove (26) for functions $a$, $b$, $c$ vanishing outside an arbitrarily small neighborhood of a point $Q$. We distinguish the cases of $Q$ in the interior of $R$ and $Q$ on the boundary surface $\tau$.

For a point $Q$ interior to $R$, we choose $\varepsilon_Q$ so small that the ball of radius $2\varepsilon_Q$ and center $Q$ lies in $R$. For $a$, $b$, $c$ vanishing outside the ball of radius $\varepsilon_Q$, the surface integral vanishes and we only have to prove that

(28)                     $\iiint (a_x + b_y + c_z)\,dx\,dy\,dz = 0$

Here $a$, $b$, $c$ are defined and have continuous derivatives in the whole space if we put $a = b = c = 0$ outside $R$. The first derivatives of $a$, $b$, $c$ are integrable over every parallel to the coordinate axes. Applying formula (29), p. 531 for the reduction of a triple integral to single integrals we find, for example,

$$\iiint c_z \, dx \, dy \, dz = \iint h(x, y) \, dx \, dy$$

where

$$h(x, y) = \int c_z(x, y, z) \, dz = 0.$$

In this way (28) is established.

Now consider the case where $Q$ is a boundary point of $R$. We can assume that the normal of the surface $\tau$ at $Q$ is not parallel to any of the three coordinate planes; this can always be brought about by a suitable rigid motion of space, which does not change the formula to be proved. In a neighborhood of $Q$ of sufficiently small radius $\varepsilon_Q$, no normal will be parallel to a coordinate plane; that is, none of the direction cosines $\xi$, $\eta$, $\zeta$ will vanish. If the neighborhood is sufficiently small, the portion of $\tau$ contained in it can be represented nonparametrically, expressing any one of the three variables $x$, $y$, $z$ as a function of the other two. For example, we can represent $\tau$ by an equation

$$z = F(x, y)$$

The set $R$ in that neighborhood will be characterized either by $z \leq F(x, y)$ or by $z \geq F(x, y)$; (see p. 633). We assume, with no loss of generality, that $R$ is characterized locally by $z \leq F(x, y)$; the exterior normal of $\tau$ then has the direction cosines $\xi$, $\eta$, $\zeta$ where $\zeta > 0$. For $a, b, c$ vanishing outside the neighborhood, and using $u = x$ and $v = y$ as surface parameters, we have

(29) $$\iint_\tau c\zeta \, dA = \iint c \, dx \, dy,$$

in agreement with our orientation. On the other hand, continuing $c$ as 0, where not defined,[1]

$$\iiint_R c_z \, dx \, dy \, dz = \iiint_{z \leq F(x,y)} c_z \, dx \, dy \, dz = \iint h(x, y) \, dx \, dy,$$

---

[1] The corresponding function $c_z$ is then bounded and continous except in the set of points $(x, y, z)$ near $Q$ for which $z = F(x, y)$. This latter set has Jordan measure zero. Hence $c_z (x, y, z)$ is Riemann integrable as a function of $x, y, z$, and also as a function of $z$ alone for fixed $x, y$. (See footnote 2 on p. 407). Thus fromula (29), p. 531 applies.

where

$$h(x, y) = \int_{-\infty}^{F(x,y)} c_z\,(x, y, z)\,dz = c(x, y, F(x, y)).$$

Only points near $Q$ contribute to the integrals, so that the function $F(x, y)$ also has to be defined only for $(x, y, z)$ near $Q$. Comparison with (29) establishes that

$$\iint_\tau c\zeta\,dA = \iiint_R c_z\,dx\,dy\,dz.$$

Similarly, with $y$, $z$ or $x$, $z$ as parameters, it also follows that

$$\iint_\tau a\xi\,dA = \iiint_R a_x\,dx\,dy\,dz, \qquad \int_\tau b\eta\,dA = \iiint_R b_y\,dx\,dy\,dz.$$

This completes the proof of the divergence theorem (26).

## A.3   Stokes's Theorem

We consider a simple surface $\tau$, which need not be closed. Given a subset $\sigma$ of $\tau$ we define the *relative interior* of $\sigma$ (that is "relative" to the surface $\tau$) as the set of points $P$ of $\tau$ with the property that in some suitable neighborhood of $P$ all points of $\tau$ belong to $\sigma$. Similarly, the *relative boundary* of $\sigma$ consists of the points $P$ of $\tau$ for which every neighborhood contains points of $\tau$ belonging to $\sigma$ as well as points of $\tau$ not belonging to $\sigma$. The set $\sigma$ is *relatively open* if each of its points is a relatively interior point.

We now consider a closed and bounded subset $s$ of $\tau$ that shall consist of a relatively open set $\sigma$ and of its relative boundary. This relative boundary shall be a simple closed curve $C$, given parametrically in the form

(30)                $x = \alpha(t), \quad y = \beta(t), \quad z = \gamma(t),$

where $\alpha$, $\beta$, $\gamma$ are functions of period $p$ with continuous first derivatives, for which $\alpha'^2 + \beta'^2 + \gamma'^2 > 0$ for all $t$. We assume that the surface $\tau$ is oriented and that $\xi$, $\eta$, $\zeta$ are the direction cosines of the positive normal on the oriented surface $\tau^*$. We can then assign a special orientation to the curve $C$ determined by the orientation of $\tau$ and by the "side" of $C$ on which $\sigma$ lies and, thus, make $C$ into an oriented curve $C^*$. This "positive" orientation of $C$ with respect to $\tau^*$ can be defined in two equivalent ways. In $x$, $y$, $z$-space the tangent vector of $C$ corresponding to the direction of increasing $t$ points in the direction given by the vector $(\alpha'(t),\ \beta'(t),\ \gamma'(t)\ )$. The exterior product of this tangent vector and of the surface normal $(\xi,\ \eta,\ \zeta)$ is the vector with components

(31)  $$\beta'\zeta - \gamma'\eta, \quad \gamma'\xi - \alpha'\zeta, \quad \alpha'\eta - \beta'\xi.$$

Its direction, which is perpendicular to that of the tangent of $C$ and tangential to the surface, gives a distinguished normal direction for $C$ relative to the surface. The orientation assigned to $C$ shall now be that of increasing $t$ if the vector (31) points away from $s$ and that of decreasing $t$ if it points into $s$.

A different way of arriving at the same orientation uses the parameter representation for $\tau$ in the neighborhood of the point $P$:

(32)  $$x = f(u, v), \quad y = g(u, v), \quad z = h(u, v)$$

where we assume that the parameters $u$, $v$ are those defining the orientation of $\tau$ near $P$, that is, that the vector $(A,\ B,\ C)$ defined by (2), p. 625 points in the direction of the distinguished normal of $\tau$ [1]. The curve $C$ near $P$ will be mapped onto an arc $\gamma$ in the $u$, $v$-plane; the set $s$ near $P$ will be mapped into a set $\rho$ in the $u$, $v$-plane. We can define the orientation of $C$ as that corresponding to the positive orientation of $\gamma$ with respect to the set $\rho$, in the sense imparted by the orientation. We could also say that the orientation of $\gamma$ is that of increasing $t$ if the vector with components $dv/dt$ and $-du/dt$ points away from $\rho$.

Given now three functions $a(x, y, z)$, $b(x, y, z)$, $c(x, y, z)$, which are defined and have continuous first derivatives in a neighborhood of the set $s$, *Stokes's theorem* is represented by the formula

(33)  $$\iint_S [(c_y - b_z)\,\xi + (a_z - c_x)\eta + (b_x - a_y)\,\zeta]\,dA$$

$$= \int_{C^*} (a\,dx + b\,dy + c\,dz).$$

The proof of the theorem follows a pattern that should be familiar to the reader by now. By using a suitable partition of unity, we can restrict ourselves to the case where the functions $a$, $b$, $c$ vanish out-

---

[1] The parametric representation (32) of $\tau$ is only *local* (i.e., valid near the point $P$).

side an arbitrarily small neighborhood of a point $Q$ of $s$. Near this
point the surface $\tau$ has a parametric representation of the form (32) for
which the normal vector with components $A$, $B$, $C$ given by (2), p.000
has the direction fixed by the orientation of $\tau^*$. We can write

$$\iint_{S^*} [(c_y - b_z)\,\xi + (a_z - c_x)\eta + (b_x - a_y)\zeta]\,dA$$

$$= \iint_\rho [(c_y - b_z)A + (a_z - c_x)B + (b_x - a_y)\,C]\,du\,dv$$

$$= \iint_\rho (\lambda_u + \mu_v)\,du\,dv,$$

where

$$\lambda = ax_v + by_v + cz_v, \qquad -\mu = ax_u + by_u + cz_u,$$

as is easily verified algebraically by substituting the expressions
(2), p. 625 for $A$, $B$, $C$ and using the chain rule of differentiation

$$a_u = a_x f_u + a_y g_u + a_z h_u,$$

and so on.[1]

If $Q$ is now a point in the relative interior of $s$, then the functions
$\lambda(u, v)$ and $\mu(u, v)$ vanish near the boundary $\gamma$ of $\rho$, and from the
*divergence theorem* for two dimensions, we find

$$\iint_\rho (\lambda_u + \mu_v)\,du\,dv = 0.$$

On the other hand, if $Q$ is on the relative boundary of $s$ the corresponding point in the $u$, $v$-plane lies on $\gamma$ and $\lambda$, $\mu$ vanish outside a small
neighborhood of that point. In this case again, the two-dimensional
divergence theorem yields

$$\iint_\rho (\lambda_u + \mu_v)\,du\,dv = \int_\gamma (\lambda p + \mu q)\,d\gamma,$$

where $d\gamma$ is the element of length and $p$, $q$ are the direction cosines
of the normal pointing away from $\rho$ on the curve $\gamma$. Describing $\gamma$ in
the positive sense with respect to $\rho$, we have

---

[1]Formula (63b), p. 321 is another version of this identity with $L = a\,dx + b\,dy + c\,dz$,
$\lambda = L/dv$, $\mu = L/du$.

$$\int_\gamma (\lambda p + \mu q) \, d\gamma = \int_{\gamma^*} (\lambda \, dv - \mu du)$$

$$= \int_{\gamma^*} (ax_u + by_u + cz_u) \, du + (ax_v + by_v + cz_v) \, dv$$

$$= \int_{C^*} (a \, dx + b \, dy + c \, dz),$$

which was to be proved.

## A.4 Surfaces and Surface Integrals in Euclidean Spaces of Higher Dimensions

### a. Elementary Surfaces

Let $E_M$ be M-dimensional euclidean space referred to Cartesian coordinates $x_1, \ldots, x_M$. We first define $m$-dimensional elementary surfaces" in $E_M$ as sets of points that can be represented "nicely" with the help of $m$ parameters. We say a set $S$ in $E_M$ is an $m$-dimensional elementary surface if we can find $M$ functions $f^1(u_1, \ldots, u_m)$, $f^2(u_1, \ldots, u_m), \ldots, f^M(u_1, \ldots, u_m)$ defined in an open set $U$ of $u_1, u_2, \ldots, u_m$-space with the following properties:

1. The equations

$$x_1 = f^1(u_1, \ldots, u_m), \ldots, x_M = f^M(u_1, \ldots, u_m)$$

define a 1–1 continuous mapping of $U$ onto $S$ whose inverse is also continuous.

2. The functions $f^i(u_1, \ldots, u_m)$ have continuous first derivatives in $U$.

3. For any point $(u_1, \ldots, u_m)$ in $U$ and for $i = 1, \ldots, m$, let $\mathbf{A}^i = \mathbf{A}^i (u_1, \ldots, u_m)$ be defined as the vector in $E_M$ with components $(f_{u_i}^1, f_{u_i}^2, \ldots, f_{u_i}^M)$. We require that the $m$ vectors $\mathbf{A}^i$ be independent, that is, that

(34) $$W = \sqrt{\Gamma(\mathbf{A}^1, \mathbf{A}^2, \ldots, \mathbf{A}^m)} > 0,$$

where $\Gamma$ is the *Gram determinant* defined by (81a), p. 194.

One proves, as on p. 626, that if we represent $S$ in the same manner with the help of some other parameters $v_1, \ldots, v_m$, there is a 1–1 continuously differentiable relation between corresponding parameter points $(u_1, \ldots, u_m)$ and $(v_1, \ldots, v_m)$ with a nonvanishing Jacobian:

$$(35) \qquad \frac{d\,(u_1,\,\ldots,\,u_m)}{d\,(v_1,\,\ldots,\,v_m)} \neq 0.$$

If $F(x_1, \ldots, x_M)$ is a function defined and continuous on the elementary surface $S$ which has *compact support* on $S$ (that is, $F$ vanishes outside a closed and bounded subset of $S$), we define[1] the integral of $F$ over $S$ by

$$(36) \qquad \underset{S}{\iint} \cdots \int F\,dS = \iint \cdots \int_U FW\,du_1 \cdots du_m.$$

The integral defined in this manner does not depend[2] on the particular parametric representation used for $S$.

At a point $P_0$ of $S$ we form the corresponding vectors $\mathbf{A}^i$, give them initial point $P_0$, and denote their final points by $P_i$, so that $\mathbf{A}^i = \overrightarrow{P_0P_i}$. The $m+1$ points $P_0, P_1, \ldots, P_m$ lie in an $m$-dimensional plane $p_0$, the *tangent plane* of $S$ at $P_0$. If $p_0$ is endowed with an orientation (see p. 200), converting it into the oriented tangent plane $p_0{}^*$ we have

$$(37a) \qquad \Omega(p_0{}^*) = \varepsilon(p_0)\,\Omega(\mathbf{A}^1, \ldots, \mathbf{A}^m),$$

where $\varepsilon(p_0)$ has either the value $+1$ or $-1$. We call the surface $S$ oriented if at every point $P$ of $S$ we orient the tangent plane $p^* = p^*(P)$ so that the orientation depends *continuously* on $P$; that is, for

$$\Omega(p^*) = \Omega(\mathbf{B}^1, \ldots, \mathbf{B}^m)$$

with suitable vectors $\mathbf{B}^1, \ldots, \mathbf{B}^m$ in $p^*$, we require that[3]

$$[\mathbf{B}^1(P), \ldots, \mathbf{B}^m(P); \mathbf{B}^1(P_0), \ldots, \mathbf{B}^m(P_0)] > 0$$

---

[1]The cube with edges of length $h$ parallel to the coordinate axes in $u_1, \ldots, u_m$-space is mapped up to terms of higher order onto a parallelepiped in $x_1, \ldots, x_M$-space spanned by the vectors $h\mathbf{A}^1, \ldots, h\mathbf{A}^m$ and, hence, of $m$-dimensional volume

$$\sqrt{\Gamma(h\mathbf{A}^1, \ldots, h\mathbf{A}^M)} = h^m W.$$

This makes it plausible that $dS$ should be identified with the element of volume in $u_1, \ldots, u_m$-space multiplied by the factor $W$.

[2]To prove this, we observe that under changes of parameters, $W$ is multiplied by the absolute value of the Jacobian of the parameter transformation, for such a transformation results in a linear substitution for the vectors $\mathbf{A}^i$ that changes the volume $W$ of the parallelepiped spanned by the vectors only by a factor equal to the determinant of the substitution (see p. 202).

[3]The symbol in brackets stands for the determinant defined by (85a), p. 198.

for all points $P$ on $S$ sufficiently close to a point $P_0$. Since the vectors $A^i$ vary continuously with the point $P$ of contact, the orientation of $p^*$ varies continuously with the point of contact $P$ if the factor $\varepsilon(P)$ defined by (37a) varies continuously with $P$ on $S$. Since $\varepsilon$ can only have the values $+1$ or $-1$, it follows, as on p. 579, that *for a connected elementary surface there are only two possible orientations.* In any case, the oriented surface $S^*$ determines an orientation of the set $U$ in the parameter space $u_1, \ldots, u_m$, namely, the one given by

$$(37b) \qquad \Omega(U) = \varepsilon(P)\, \Omega(u_1, \ldots, u_m)$$

[see (40n, o, p), p. 580–1]. Here, under a change of parameters from $u_1, \ldots, u_m$ to $v_1, \ldots, v_m$ the quantity $\varepsilon$ is just multiplied by the sign of the Jacobian (35).

### b. *Integral of a Differential form over an Oriented Elementary Surface*

After these preliminaries we are ready to define the integral of an $m$th-order differential form $\omega$ over an $m$-dimensional oriented elementary surface $S^*$. The form $\omega$ is some linear combination of ordered products of $m$ of the differentials $dx_1, \ldots, dx_M$ at a time, say,

$$\omega = a\; dx_1\, dx_2 \cdots dx_m + b\; dx_2\, dx_3 \cdots dx_{m+1} + c\; dx_1\, dx_3 \cdots dx_m + \cdots,$$

where the coefficients $a(x_1, \ldots, x_M),\ b(x_1, \ldots, x_M),\ \ldots$ are assumed to be continuous and to have compact support on $S^*$.[1] Let $S^*$ be represented parametrically with the help of parameters $u_1, \ldots, u_m$ that vary over the set $U^*$, oriented in accordance with the orientation of $S^*$. We then define

$$\int_{S^*} \cdots \int \omega = \int_{U^*} \cdots \int \frac{\omega}{du_1 \cdots du_m}\, du_1 \cdots du_m$$

$$= \int_{U^*} \cdots \int \left[ a\, \frac{d(x_1, x_2, \ldots, x_m)}{d(u_1, u_2, \ldots, u_m)} + b\, \frac{d(x_2, x_3, \ldots, x_{m+1})}{d(u_1, u_2, \ldots, u_m)} + \cdots \right] du_1 \cdots du_m.$$

---

[1] That is, $a, b, c, \ldots$ vanish outside some closed and bounded subset of $S^*$.

Our notation[1] has been arranged in such a way that the value of the integral does not depend on the particular parameter representation used for $S^*$.

### c. Simple m-Dimensional Surfaces

By "patching together" elementary surfaces, we can obtain *simple* surfaces just as in three-space. A set $\tau$ in $M$-dimensional euclidean space is called an $m$-dimensional simple surface if each point $P_0$ of $\tau$ has a neighborhood intersecting $\tau$ in an elementary $m$-dimensional surface. If each of the elementary surfaces occurring in the characterization of a simple surface is oriented and if the orientations of two of these elementary surfaces agree, whenever they overlap we say that the simple surface $\tau$ has been oriented.

At each point of an $m$-dimensional oriented simple surface $\tau^*$ we can choose m vectors $\mathbf{A}^1(P), \ldots, \mathbf{A}^m(P)$ such that

$$\Omega(\tau^*) = \Omega[\mathbf{A}^1(P), \ldots, \mathbf{A}^m(P)]$$

and

$$[\mathbf{A}^1(P), \ldots, \mathbf{A}^m(P); \mathbf{A}^1(Q), \ldots, \mathbf{A}^m(Q)] > 0$$

for $Q$ sufficiently close to $P$.

For subsets $s$ of an $m$-dimensional simple surface $\tau$ we can define the *relative boundary*[2] of $s$, that is, the boundary of $s$ relative to the surface $\tau$. The relative boundary of $s$ consists of those points of $s$ for which each neighborhood contains points of $s$ and points of $\tau$ not belonging to $s$. The *relative closure*[3] of $s$ consists of $s$ and of relative boundary points of $s$. The set $s$ is called *relatively open* if it has no

---

[1]Here, for a continuous integrand $F(u_1, \ldots, u_m)$, the integral of $F$ over an oriented set $U^*$ with orientation

$$\Omega(U^*) = \varepsilon\Omega(u_1, \ldots, u_m)$$

($\varepsilon = \pm 1$ and continuous) is defined by

$$\iint \cdots_{U^*} \int F \, du_1 \cdots du_m = \iint \cdots_{U} \int F\varepsilon \, du_1 \cdots du_m$$

where the integral on the right side has the ordinary meaning that gives positive values for positive integrands.

[2]This notion is needed when we want to discuss, say, the boundary curve of a two-dimensional surface $s$ in spaces of dimensions $M > 2$. The ("absolute") boundary of the surface $s$ taken with respect to the whole space always contains the whole surface $s$.

[3]The relative closure of $s$ also is the set of all points of $\tau$ that are limits of sequences formed from points of $s$.

points in common with its relative boundary and called *relatively closed* if it contains its relative boundary.

Of particular interest is the case where $s$ is a subset of the $m$-dimensional simple surface $\tau$ whose relative boundary itself is an $(m-1)$-dimensional simple surface $\partial s$. We assume furthermore that $s$ is the relative closure of a relative open set. In the neighborhood of a point $P$ of $\partial s$ we can always represent $\partial s$ and $\tau$ "nonparametrically"; that is, we can use some of the Cartesian coordinates $x_1, \ldots, x_M$ in space as independent variables; after a suitable renumbering of coordinates we then have for $\tau$ near $P$ the parametric representation

$$x_i = f_i(x_1, \ldots, x_m) \qquad (i = m+1, \ldots, M),$$

and on $\partial s$ we have an additional condition

$$x_1 = g(x_2, \ldots, x_m)$$

with continuously differentiable functions $f_i$ and $g$. Moreover, the points of $s$ are characterized near $P$ by either the inequality

$$g(x_2, \ldots, x_m) \leqq x_1$$

or by

$$g(x_2, \ldots, x_m) \geqq x_1.$$

If we deal with an oriented set $s^*$, we can assign a unique orientation to the relative boundary $\partial s$. Let there be given $m-1$ independent vectors $\mathbf{A}^2, \ldots, \mathbf{A}^m$ at a point $P$ of $\partial s$ that are tangential to $\partial s$ and an additional vector $\mathbf{A}^1$ that is tangential to $\tau$ but not to $\partial s$ at $P$ and that points away from $s^*$. We then have

$$(38) \qquad \Omega(s^*) = \varepsilon\Omega(\mathbf{A}^1, \ldots, \mathbf{A}^{m-1}, \mathbf{A}^m)$$

where $\varepsilon$ has either the value $+1$ or $-1$. The boundary $\partial s^*$ is then called *oriented positively* with respect to $s^*$ if

$$(39) \qquad \Omega(\partial s^*) = \varepsilon\Omega(\mathbf{A}^2, \ldots, \mathbf{A}^m).$$

In particular, let $m = M$ and $\tau$ be the whole $M$-dimensional space. Let $s$ be the closure of an open[1] set and let the boundary of $s$ be an

---

[1] We can omit here the word *relative*.

$(m - 1)$-dimensional simple surface $\partial s$. Assume that in a neighborhood of a point $P$ the surface $\partial s$ has the nonparametric representation

$$x_1 = g(x_2, \ldots, x_m).$$

We can define a quantity $\delta = \pm 1$ so that

(40a) $$[x_1 - g(x_2, \ldots, x_m)]\delta \leqq 0$$

for points $(x_1, \ldots, x_m)$ in $s$ near $P$. We choose for $\mathbf{A}^2, \ldots, \mathbf{A}^m$ the vectors

$$\mathbf{A}^2 = (g_{x_2}, 1, 0, \ldots, 0, 0), \ldots, \mathbf{A}^m = (g_{x_m}, 0, \ldots, 0, 1)$$

tangential to $\partial s$, and for $\mathbf{A}^1$ the vector

$$\mathbf{A}_1 = (\delta, 0, \ldots, 0)$$

that points away from $s$. Then in $x_1, \ldots, x_m$-coordinates

$$\det(\mathbf{A}^1, \ldots, \mathbf{A}^{m-1}, \mathbf{A}^m) = \delta,$$

so that [see (83a, b), p. 197]

$$\Omega(\mathbf{A}^1, \ldots, \mathbf{A}^{m-1}, \mathbf{A}^m) = \delta\Omega(x_1, \ldots, x_m).$$

For the oriented set $s^*$ let $\varepsilon = \pm 1$ be defined near $P$ by (38). Then,

(40b) $$\Omega(s^*) = \varepsilon\delta\Omega(x_1, \ldots, x_m),$$

while for the boundary $\partial s^*$ oriented positively with respect to $s^*$, relation (39) holds. Consequently, if $x_2, \ldots, x_m$ are considered as parameters for the surface $\partial s^*$ near $P$ then the orientation of $x_2, \ldots, x_m$-space determined by $\partial s^*$ is

(40c) $$\varepsilon\Omega(x_2, \ldots, x_m)$$

[see (37b), p. 647]. *Thus, for a set $s^*$ oriented positively with respect to $x_1, \ldots, x_m$-coordinates ($\varepsilon\delta = 1$), the positively oriented boundary has the orientation of the $x_2, \ldots, x_m$-system where $s$ lies "below" the boundary, and the opposite one where $s$ lies "above" the boundary (compare p. 634).*

## A.5   Integrals over Simple Surfaces, Gauss's Divergence Theorem and the General Stokes Formula in Higher Dimensions

We define integrals over simple surfaces by means of *partitions of unity* exactly as on p. 635. In particular, if $\tau^*$ is an $m$-dimensional oriented simple surface and $\omega$ an $m$th-order differential form the integral

$$\int \cdots \int_{\tau^*} \omega$$

is defined provided the coefficients of $\omega$ are continuous and vanish outside a bounded and closed[1] subset of $\tau^*$.

Now let $\tau$ be an $m$-dimensional simple surface in $M$-space and $s^*$ an oriented bounded and closed subset of $\tau$. We assume that $s^*$ is the closure of a relatively open set and that the relative boundary of $s^*$, oriented positively with respect to $s^*$, is an $(m - 1)$-dimensional oriented simple surface $\partial s^*$. Let $\omega$ be a differential form of order $m - 1$ with coefficients that have continuous first derivatives. *Stokes's general theorem* asserts that

(41) $$\int \cdots \int_{\partial s^*} \omega = \int \cdots \int_{s^*} d\omega.$$

We shall first treat the special case where $m = M$, which is *Gauss's divergence theorem* in $m$ dimensions. In this case, we take $\tau$ as the whole space, $s^*$ as an oriented set that is the closure of an open set bounded by an $(m - 1)$-dimensional simple surface $\partial s^*$ oriented positively with respect to $s^*$. The form $\omega$ of degree $m - 1$ can be written as

$$a_1 \, dx_2 \, dx_3 \cdots dx_m + a_2 \, dx_3 \, dx_4 \cdots dx_m \, dx_1 + \cdots$$
$$+ a_m \, dx_1 \, dx_2 \cdots dx_{m-1},$$

where the $a_i$ are functions of $x_1, \ldots, x_m$. Then,

(42a) $$d\omega = da_1 \, dx_2 \, dx_3 \cdots dx_m + da_2 \, dx_3 \, dx_4 \cdots dx_m \, dx_1 +$$
$$\cdots + da_m \, dx_1 \, dx_2 \cdots dx_{m-1}$$

---

[1]Not just relatively closed.

$$= \frac{\partial a_1}{\partial x_1} dx_1\, dx_2 \cdots dx_m + \frac{\partial a_2}{\partial x_2} dx_2\, dx_3 \cdots dx_m\, dx_1 + \cdots$$

$$+ \frac{\partial a_m}{\partial x_m} dx_m\, dx_1 \cdots dx_{m-1}$$

$$= K\, dx_1 \cdots dx_m,$$

where

$$(42b) \quad K = \frac{\partial a_1}{\partial x_1} + (-1)^{m-1} \frac{\partial a_2}{\partial x_2} + \frac{\partial a_3}{\partial x_3} + (-1)^{m-1} \frac{\partial a_4}{\partial x_4} + \cdots$$

$$+ (-1)^{m-1} \frac{\partial a_m}{\partial x_m}.$$

The proof of formula (41) for this case proceeds exactly as in the special case $m = 3$ discussed on pp. 639–642, and there is no point in recapitulating the individual steps. The only item to be checked is the *sign* in the final formula. The proof finally reduces to the case where $a_2, \ldots, a_m$ vanish identically and $a_1$ vanishes outside a neighborhood of a point $P$ of the surface $\sigma^*$. Here near $P$ the surface is given by an equation

$$x_1 = g(x_2, \ldots, x_m)$$

and $s^*$ is given by the inequality

$$[x_1 - g(x_2, \ldots, x_m)]\delta \leqq 0,$$

where $\delta = \pm 1$. Let the number $\varepsilon = \pm 1$ be defined at $P$ by

$$\Omega(s^*) = \varepsilon\delta\Omega(x_1, \ldots, x_m)$$

[see (40b)]. Then, by (42a, b),

$$\int \cdots \int_{s^*} d\omega = \varepsilon\delta \int \cdots \int \frac{\partial a_1}{\partial x_1} dx_1 \cdots dx_m = \varepsilon \int \cdots \int_{x_1=g} a_1\, dx_2 \cdots dx_m.$$

On the other hand [see (40b) and (40c)], we also have

$$\int \cdots \int_{\partial s^*} \omega = \varepsilon \int \cdots \int_{x_1=g} a_1\, dx_2 \cdots dx_m.$$

This completes the proof of the divergence theorem.

The general Stokes formula for arbitrary $m < M$ is an immediate consequence. Using partitions of unity, it is again sufficient to establish it for differential forms that vanish outside a neighborhood of a point $P$ of the simple surface $\tau$. In that neighborhood $\tau$ is identical with an elementary surface. Introducing local parameters $u_1, \ldots,$ $u_m$ to describe $\tau$, the identity (41) goes over into the corresponding identity in $m$-dimensional parameter space, where now everything is reduced to Gauss's divergence theorem discussed above. In this way, the general Stokes theorem is established.

This kind of argument makes it pretty clear that the fact that our $m$-dimensional surface $\tau$ is embedded in a euclidean space of dimension $M$ is rather irrelevant. All that counts are the local parametric representations mapping $\tau$ onto a set in euclidean $m$-space. This suggests that similar formulae will hold on more general $m$-dimensional abstract manifolds that near every point can be described by parameters. However, in order to avoid topological considerations beyond the scope of this book, we have restricted ourselves to *simple* surfaces in euclidean spaces.

# CHAPTER
# 6

# *Differential Equations*

We have already discussed special cases of differential equations in Volume I, Chapter 9. We cannot attempt to develop the general theory in detail within the scope of this book. In this chapter, however, starting with further examples from mechanics, we shall give at least a sketch of some of the principles of the subject, making use of the calculus of functions of several variables.

## 6.1 The Differential Equations for the Motion of a Particle in Three Dimensions

### a. The Equations of Motion

In Volume I (Chapter 4, pp. 397–423), we discussed the motion of a particle constrained to move in the $x$, $y$-plane. We now drop this restriction and consider a mass $m$ that we suppose concentrated at a point with coordinates $(x, y, z)$. The position vector from the origin to the particle has components $x$, $y$, $z$ and we denote it by $\mathbf{R}$. A motion of the particle will then be represented mathematically if we can express $(x, y, z)$ or $\mathbf{R}$ as a function of the time $t$. If, as before, we denote differentiation with respect to the time $t$ by a dot, then the vector $\dot{\mathbf{R}} = (\dot{x}, \dot{y}, \dot{z})$ of length

$$(1) \qquad v = \sqrt{\dot{x}^2 + \dot{y}^2 + \dot{z}^2}$$

represents the *velocity*, and the vector $\ddot{\mathbf{R}} = (\ddot{x}, \ddot{y}, \ddot{z})$, the *acceleration* of the particle.

The fundamental tool for determining the motion is *Newton's second law*[1], according to which the product of the acceleration vector

---

[1]"Mutationem motus proportionalem esse vi motrici impressae, et fieri secundum

$\ddot{\mathbf{R}}$ and the mass $m$ is equal to the force vector $\mathbf{F} = (x, y, z)$ acting on the particle:

(2a) $$m\ddot{\mathbf{R}} = \mathbf{F},$$

or, in components,

(2b) $$m\ddot{x} = X, \quad m\ddot{y} = Y, \quad m\ddot{z} = Z.$$

These relations[1] can be used to find the motion, provided we are given sufficient information about the force $\mathbf{F}$.

One example is the constant field of force representing gravity near the surface of the earth. If we take gravity as acting in the direction of the negative $z$-axis, we know the force to be represented by the vector

(3) $$\mathbf{F} = (0, 0, -mg) = -mg(\operatorname{grad} z),$$

where $g$ is the constant acceleration due to gravity (see Volume I, p. 399).

Another example is the field of force produced by a mass $\mu$ concentrated at the origin of the coordinate system and attracting according to Newton's law of gravitation (see Volume I, p. 413). If $r = \sqrt{x^2 + y^2 + z^2} = |\mathbf{R}|$ is the distance of the particle $(x, y, z)$ with mass $m$ from the origin, the field of force is given by the expression

(4a) $$\mathbf{F} = \mu m \gamma \left( \operatorname{grad} \frac{1}{r} \right),$$

where $\gamma$ is the universal gravitational constant. In this case, Newton's law of motion (2a) states that

(4b) $$\ddot{\mathbf{R}} = \mu \gamma \operatorname{grad} \frac{1}{r}$$

or, in components,

$$\ddot{x} = -\mu\gamma \frac{x}{r^3}, \quad \ddot{y} = -\mu\gamma \frac{y}{r^3}, \quad \ddot{z} = -\mu\gamma \frac{z}{r^3}.$$

---

lineam rectam qua vis illa imprimitur" (i.e.,"Change of motion is proportional to the force applied and takes place in the direction of the straight line in which the force acts").

[1]The vector $m\dot{\mathbf{R}}$ is called the *momentum*, so that Newton's law states that "force equals the rate of change of momentum".

In general, if $\mathbf{F}$ is a given field of force with components $X(x, y, z)$, $Y(x, y, z)$, $Z(x, y, z)$, which are known functions of position, the equations of motion

$$m\ddot{x} = X(x, y, z), \qquad m\ddot{y} = Y(x, y, z), \qquad m\ddot{z} = Z(x, y, z)$$

form a system of three *differential equations* for the three unknown functions $x(t)$, $y(t)$, $z(t)$. The fundamental problem of the mechanics of a particle is to determine the path of the particle from the differential equations, when at the beginning of the motion, say at the time $t = 0$, the *position* of the particle [i.e., the coordinates $x_0 = x(0)$, $y_0 = y(0)$, $z_0 = \dot{z}(0)$] and the *initial velocity* [i.e., the quantities $\dot{x}_0 = \dot{x}(0)$, $\dot{y}_0 = \dot{y}(0)$, $\dot{z}_0 = \dot{z}(0)$] are given. The problem of finding three functions that satisfy these initial conditions and also satisfy the three differential equations for all values of $t$ is known as the problem of the *solution* or *integration*[1] of the system of differential equations.

### b. The Principle of Conservation of Energy

The equations of motion (2a) for a particle have an important consequence obtained by forming the scalar product with the velocity vector $\dot{\mathbf{R}}$:

(6a) $$m\dot{\mathbf{R}} \cdot \ddot{\mathbf{R}} = \mathbf{F} \cdot \dot{\mathbf{R}} = X\dot{x} + Y\dot{y} + Z\dot{z}.$$

Here the left-hand side can be written as

(6b) $$\frac{d}{dt}\left(\frac{1}{2}\, m\dot{\mathbf{R}} \cdot \dot{\mathbf{R}}\right) = \frac{d}{dt}\,\frac{1}{2}\, mv^2,$$

that is, as the time derivative of the *kinetic energy* $\frac{1}{2}mv^2$ (*energy of motion*) of the particle. Integrating equation (6a) with respect to $t$ from $t_0$ to $t_1$, we find that the change in kinetic energy of the particle during the time interval from $t_0$ to $t_1$ is given by

(6c) $$\frac{1}{2}\, mv_1^2 - \frac{1}{2}\, mv_0^2 = \int_{t_0}^{t_1}\left(X\frac{dx}{dt} + Y\frac{dy}{dt} + Z\frac{dz}{dt}\right)dt$$

$$= \int (X\, dx + Y\, dy + Z\, dz),$$

where the line integral is extended over the path described by the particle during the time from $t_0$ to $t_1$. The integral

---

[1] The word is used here because the solution of differential equations may be regarded as a generalization of the process of ordinary integration.

$$\int X\,dx + Y\,dy + Z\,dz$$

taken over an oriented arc is called the *work done by the force* $\mathbf{F} = (X, Y, Z)$ in moving along this arc.[1] Hence, (6c) can be stated as the *equation of energy: The gain in kinetic energy is equal to the work done by the force during the motion.*

In the important case where the field of force can be represented as the gradient of a function, say

(7a) $$\mathbf{F} = \operatorname{grad} \phi,$$

the integral of the differential form

$$X\,dx + Y\,dy + Z\,dz = d\phi$$

is independent of the path and depends only on the initial and final points of the path (see p. 95). Following Helmholtz, a field of force of the type (7a) is called *conservative.*[2] We introduce the *potential energy U (energy of position)* of the conservative force field by $U = -\phi$. The equations of motion then have the form

$$m\ddot{\mathbf{R}} = -\operatorname{grad} U$$

or, in components,

(7b) $$m\ddot{x} = -U_x, \qquad m\ddot{y} = -U_y, \qquad m\ddot{z} = -U_z.$$

The potential energy as a function of position $(x, y, z)$ is determined by the force field only within an arbitrary additive constant. For the work done by the conservative forces during the motion we find

$$\int X\,dx + Y\,dy + Z\,dz = -\int dU = U_0 - U_1$$

---

[1] See Volume I, p. 420. Introducing the arc length $s$ as parameter, the line integral takes the form

$$\int \mathbf{F} \cdot \frac{d\mathbf{R}}{ds}\,ds$$

and thus is equal to the limit of the sums of the component of force in the direction of motion multiplied with the distances.

[2] "Conservative" by virtue of the theorem of the conservation of energy, which we shall deduce shortly.

where $U_0$ and $U_1$ are the respective values of the potential energy for the positions of the particle at the times $t_0$ and $t_1$. Comparison with (6c) shows that

$$\frac{1}{2} mv_1^2 + U_1 = \frac{1}{2} mv_0^2 + U_0.$$

Hence, the quantity $\frac{1}{2}mv^2 + U$ has the same value at any times $t_0$ and $t_1$ during the motion. Without going into the physical explanation of these concepts, we have arrived at a form of the *law of conservation of energy* for a particle in a conservative field of force:

*The total energy—that is, the sum of the kinetic energy $\frac{1}{2}mv^2$ and of the potential energy U—remains constant during the motion.*

In the examples in the next sections we show how this theorem can be used in the actual solution of the equations of motion.

We notice that both the force fields defined by equations (3) and (4a) are conservative. The equations of motion under the uniform gravitational field (3) reduce to

(8a) $$\ddot{x} = 0, \quad \ddot{y} = 0, \quad \ddot{z} = -g.$$

Their general solution trivially is given by

(8b) $$x = a_1 t + a_2, \qquad y = b_1 t + b_2, \qquad z = -\frac{1}{2} g t^2 + c_1 t + c_2.$$

Here, obviously, the constants $(a_2, b_2, c_2)$ give the initial position, and the constants $(a_1, b_1, c_1)$, the initial velocity of the particle at the time $t = 0$. The trajectory of a particle given parametrically in terms of the time $t$ by equations (8b) is a parabola with axis parallel to the $z$-axis. Since the force field is $-mg$ grad $z$, the potential energy is $U = mgz + $ constant. Changes in $U$ are proportional to changes in elevation $z$. The law of conservation of energy thus takes the form

(8c) $$\frac{1}{2}mv^2 + mgz = \text{constant} = \frac{1}{2} mv_0^2 + mgz_0$$

$$= \frac{1}{2}m(a_1^2 + b_1^2 + c_1^2) + mgc_2.$$

The velocity $v$ is therefore least at the highest point of the trajectory.

Instead of a freely falling particle, we can consider a particle moving under the influence of the gravitational field $\mathbf{F} = -mg$ grad $z$, where the particle is constrained to stay on a surface $z = f(x, y)$

by a *reaction force* perpendicular to the surface.[1] Since the reaction force has no component in the direction of motion, and hence does no work, the work done during the motion is that done by the conservative gravitational field. We arrive thus at the same equation of energy

$$(9) \qquad \frac{1}{2} mv^2 + mgz = \text{constant},$$

as for the freely falling body, the only difference being that $z = f(x, y)$ is now a prescribed function of the coordinates $x$, $y$.

### c. Equilibrium. Stability

The equations of motion

$$(10a) \qquad m\ddot{\mathbf{R}} = -\text{grad } U$$

of a particle in a conservative force field enable us to discuss motions near a position of equilibrium. We say that the particle is *in equilibrium under the influence of the field of force* if it remains at rest. In order that this may be the case, its velocity and its acceleration must both be 0 throughout the interval of time under consideration. The equations of motion (10a) therefore yield

$$(10b) \qquad \text{grad } U = 0$$

or

$$(10c) \qquad U_x = U_y = U_z = 0$$

as the necessary conditions for equilibrium. Thus, *a position of equilibrium* $(x_0, y_0, z_0)$ *necessarily is a critical point of the potential energy* $U$. Conversely, every critical point $(x_0, y_0, z_0)$ of $U$ is a possible position of rest, since obviously the constant vector

$$\mathbf{R} = (x_0, \ y_0, \ z_0)$$

satisfies the equations (10a).

Of great practical importance is the notion of *stability* of equilibrium. We mean by *stability* that if we slightly disturb the state of

---

[1]An example is furnished by the spherical pendulum where a mass is constrained to move on a sphere. Compare with the motions on a curve discussed in Volume I, pp. 405 ff.

equilibrium, the whole resulting motion will differ only slightly from the state of rest.[1] More precisely, let $r_1$ and $v_1$ be any positive numbers. We can find corresponding to $r_1$ and $v_1$ two positive numbers $r_0$, $v_0$ so small that if the particle is moved a distance not more than $r_0$ from its position of equilibrium and started off with a velocity not greater than $v_0$, then in its whole subsequent motion it will never reach a distance greater than $r_1$ from the point of equilibrium and a velocity greater than $v_1$.

It is particulary interesting that *the equilibrium is stable at a point at which the potential energy $U$ has a strict relative minimum.*[2] It is remarkable that we can prove this statement about stability without actually solving the equations of motion. For simplicity, we assume that the position of equilibrium under consideration is the origin, which we can always bring about by a translation. Moreover, since the potential energy is only determined within a constant, we can assume that $U(0, 0, 0) = 0$. Since $U$ has a strict relative minimum at the origin, we can find a positive number $r < r_1$ such that $U > 0$ everywhere on the surface of the sphere of radius $r$ about the origin and in its interior, except at the origin. The minimum value of $U$ on the surface of the sphere is then a positive number $a$. Since $U$ is continuous, we can find an $r_0 < r$ such that $U(x, y, z) < \frac{1}{2}a$ and $U(x, y, z) < \frac{1}{4}mv_1^2$ in the solid sphere of radius $r_0$ about the origin. Let, moreover, the positive number $v_0$ be so small that $\frac{1}{2}mv_0^2 < \frac{1}{2}a$ and $\frac{1}{2}mv_0^2 < \frac{1}{4}mv_1^2$. Then, for an initial position of the particle of distance less than $r_0$ from the origin and an initial velocity less than $v_0$, we have initially for the total energy the inequalities

(11a)
$$\frac{1}{2}mv^2 + U(x,\ y,\ z) \le \frac{1}{2}mv_0^2 + \frac{1}{2}a < a$$

(11b)
$$\frac{1}{2}mv^2 + U(x,\ y,\ z) < \frac{1}{4}mv_1^2 + \frac{1}{4}mv_1^2 = \frac{1}{2}mv_1^2.$$

---

[1]The notion can be illustrated best by the analogous two-dimensional problem of a particle moving under gravity but constrained to stay on a surface $z = f(x, y)$. Here the positions of equilibrium are the critical points of the potential energy $mgz = mgf(x, y)$, that is, the highest or lowest points or saddle points of the surface $z = f(x, y)$. The equilibrium is stable for a particle resting, say, under the influence of gravity at the lowest point of a spherical bowl, which is concave upward. On the other hand, a particle resting at the highest point of a spherical bowl that is concave downward is in *unstable* equilibrium; the slightest disturbance results in a large change of position. Since the small disturbances can always be assumed to be present in practice, unstable equilibrium is not maintained and unlikely to be observed.

[2]At a *strict* minimum point the value of $U$ is lower than at all other points of a sufficiently small neighborhood. See page 325–6 for the definitions.

Since the energy is constant throughout the motion, we see from (11a) that at all subsequent times

$$\frac{1}{2}\,mv^2 + U(x, y, z) < a,$$

and consequently,

$$U(x, y, z) < a.$$

Since initially the particle is inside the sphere of radius $r$ and since $U \geq a$ on that sphere, the particle can never reach the surface of the sphere. This shows that the distance of the particle from the origin never exceeds the value $r < r_1$. Since also $U \geq 0$ inside the sphere of radius $r$, it follows from (11b) that

$$\frac{1}{2}\,mv^2 < \frac{1}{2}\,mv_1^2$$

and, consequently, that the velocity of the particle never exceeds the value $v_1$, as was to be proved.

### d. Small Oscillations About a Position of Equilibrium

The motion of a particle about a position of stable equilibrium, corresponding to a minimum of the potential energy, can be approximated in a simple way. For the sake of brevity, we restrict ourselves to a motion in the $x$, $y$-plane and assume that there is no force acting in the direction of the $z$-axis. We also assume that the potential $U(x, y)$ has a minimum at the origin and that $U(0, 0) = 0$. Moreover, at the minimum point, $U = U_0 = 0$. We imagine $U$ expanded by Taylor's theorem in the form

$$U = \frac{1}{2}\,(ax^2 + 2bxy + cy^2) + \cdots.$$

The function $U$ will have a strict relative minimum at the origin if the quadractic form

(12a) $$Q(x, y) = \frac{1}{2}\,(ax^2 + 2bxy + cy^2)$$

is *positive definite*,[1] that is, that

---

[1]See page 347. The positive definite character of $Q$ is sufficient, but not necessary, for a strict relative minimum. However, it is necessary that $Q$ be neither indefinite nor negative definite.

(12b)                       $$a > 0, \quad ac - b^2 > 0.$$

We assume that conditions (12b) are satisfied and that *in a sufficiently small neighborhood of the position of equilibrium at the origin the potential energy U can be replaced with sufficient accuracy by the quadratic form Q.* [1] With these assumptions the equations of motion take the form

$$m\ddot{\mathbf{R}} = -\,\text{grad}\; Q$$

or

(12c)[2]          $$m\ddot{x} = -ax - by, \quad m\ddot{y} = -bx - cy.$$

The equations (12c) can be integrated completely if we first rotate the x- and y-axes through a suitably chosen angle $\phi$ so that the new coordinate axes coincide with the *principal axes* of the ellipses $Q =$ constant. We make the orthogonal substitution

---

[1]No serious attempt at justifying this "plausible" assumption can be made here.
[2]We again can interpret these equations as approximating the equations of motion under gravity of a particle constrained to move on a surface $z = f(x, y)$ near a minimum point of that surface. The precise equations of motion here have the form

$$\ddot{x} = -\lambda f_x, \quad \ddot{y} = -\lambda f_y, \quad \ddot{z} = -g + \lambda,$$

taking into account that the forces acting on a particle consist of the gravitational force $(0, 0, -mg)$ and a *reaction force* $(-\lambda f_x, -\lambda f_y, \lambda)$ perpendicular to the surface and containing an indeterminate multiplier $\lambda$. We can eliminate $\lambda$ by observing that

$$\ddot{z} = \frac{d^2 f}{dt^2} = f_x\ddot{x} + f_y\ddot{y} + f_{xx}\dot{x}^2 + 2f_{xy}\dot{x}\dot{y} + f_{yy}\dot{y}^2$$

and find the equations

$$\ddot{x} = -\lambda f_x, \quad \ddot{y} = -\lambda f_y$$

with

$$\lambda = \frac{g + f_{xx}\dot{x}^2 + 2f_{xy}\dot{x}\dot{y} + f_{yy}\dot{y}^2}{1 + f_x^2 + f_y^2}$$

for the two unknown functions $x, y$. If $f$ has a minimum at the origin and is approximated there by the quadratic

(13a)                       $$f = \frac{1}{2}\,(\alpha x^2 + 2\beta xy + \gamma y^2),$$

we find near the origin, neglecting all nonlinear terms, the differential equations
(13b)          $$\ddot{x} = -g(\alpha x + \beta y), \quad \ddot{y} = -g(\beta x + \gamma y),$$
which are of the form (12c). If, for example, the surface is the sphere

$$z = L - \sqrt{L^2 - x^2 - y^2}$$

("spherical pendulum of length $L$"), we find

(13c)                       $$\ddot{x} = -\frac{g}{L}\,x, \quad \ddot{y} = -\frac{g}{L}\,y.$$

$$x = \xi \cos \phi - \eta \sin \phi, \qquad y = \xi \sin \phi + \eta \cos \phi,$$

where $\phi$ is determined from the condition that

$$Q = \frac{1}{2}(ax^2 + 2bxy + cy^2) = \frac{1}{2}(\alpha\xi^2 + \gamma\eta^2)$$

with suitable positive constants $\alpha$, $\gamma$ [1]. In the new rectangular coordinates $\xi$, $\eta$ the equations of motion (12c) transform into

(14a) $$m\ddot{\xi} = -\alpha\xi, \qquad m\ddot{\eta} = -\gamma\eta.$$

As in Volume I (p. 404), both these equations can be integrated completely. We obtain

(14b) $$\xi = A_1 \sin\sqrt{\frac{\alpha}{m}}(t - c_1), \qquad \eta = A_2 \sin\sqrt{\frac{\gamma}{m}}(t - c_2),$$

where $c_1$, $c_2$, $A_1$, $A_2$ are constants of integration that enable us to make the motion satisfy any arbitrarily assigned initial conditions.[2]

The form of the solution shows that the motion about a position of stable equilibrium results from the superposition of simple harmonic oscillations in the two *principal directions,* the $\xi$-direction and the $\eta$-direction, the frequencies of these oscillations being given by $\sqrt{\alpha/m}$ and $\sqrt{\gamma/m}$. [3] A general discussion of these oscillations, which we shall not carry out here, shows that the resultant motion may take a great variety of forms.

To give a few examples of these compound oscillations, we first consider the motion represented by the equations

$$\xi = \sin(t + c), \quad \eta = \sin(t - c)$$

By eliminating the time $t$, we obtain the equation

---

[1]One finds immediately that $\phi$ is determined from the equation

$$\tan 2\phi = \frac{2b}{a - c}.$$

The positivity of $\alpha$, $\gamma$ follows from the positive definiteness of $Q$.

[2]It is of interest to observe that in cases of *unstable* equilibrium, one or both of the constants $\alpha$, $\gamma$ might be negative. In that case, the trigonometric functions occuring in (14b) would have to be replaced by hyperbolic ones and the coordinates $\xi$, $\eta$ do not both stay bounded for all $t$.

[3]In the case (13c) of the spherical pendulum, the two frequencies have the same value $\sqrt{g/L}$.

$$(\xi + \eta)^2 \sin^2 c + (\xi - \eta)^2 \cos^2 c = 4 \sin^2 c \cos^2 c,$$

which represents an ellipse. The two components of the oscillation have the same frequency 1 and the same amplitude 1, but a difference of phase $2c$. If this difference of phase successively takes all values between 0 and $\pi/2$, the corresponding ellipse passes from the degenerate straight-line case $\xi - \eta = 0$ to the circle $\xi^2 + \eta^2 = 1$, and the oscillation passes from the so-called linear oscillation to the circular (cf. Figs. 6.1–6.3).

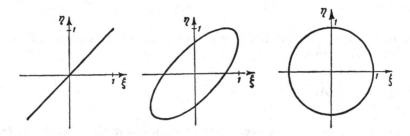

**Figures 6.1–6.3**   Oscillation diagrams.

If, as a second example, we consider the motion represented by the equations

$$\xi = \sin t, \qquad \eta = \sin 2(t - c),$$

where the frequencies are no longer equal, we obtain oscillation diagrams decidedly more complicated. In Figs. 6.4–6.6 these curves are given for the phase differences $c = 0$, $c = \pi/8$, and $c = \pi/4$, respectively. In the first two cases, the particle moves continuously on a closed curve, but in the last case, it swings backward and forward

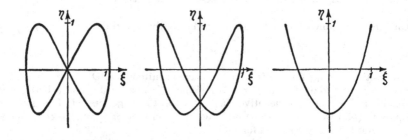

**Figures 6.4–6.6**   Oscillation diagrams.

on an arc of the parabola $\eta = 2\xi^2 - 1$. The curves obtained by the superposition of different simple harmonic oscillations in directions at right angles to one another are given the general name of *Lissajous figures*.

### e. Planetary Motion[1]

In the examples discussed above, the differential equations of the motion can immediately (or after a simple transformation) be written in such a way that each of the coordinates occurs in one differential equation only and can be determined by elementary integration. We shall now consider the most important case of a motion in which the equations of motion are no longer separable in this simple way, so that their integration involves a somewhat more difficult calculation. The problem in question is the *deduction of Kepler's laws of planetary motion from Newton's law of attraction*. We suppose that at the origin of the coordinate system there is a body of mass $\mu$ (e.g., the sun) whose gravitational field of force per unit mass is given by the vector

$$\gamma\mu \ \mathrm{grad} \ \frac{1}{r}.$$

What is the motion of a particle of mass $m$ (a planet) under the influence of this field of force? The equations of motion are (see p. 655)

$$(15) \qquad \ddot{x} = -\gamma\mu\, \frac{x}{r^3}, \qquad \ddot{y} = -\gamma\mu\, \frac{y}{r^3}, \qquad \ddot{z} = -\gamma\mu\, \frac{z}{r^3}.$$

In order to integrate them, we first state the theorem of conservation of energy (see p. 658) for the motion in the form

$$\frac{1}{2}\, m\, (\dot{x}^2 + \dot{y}^2 + \dot{z}^2) - \frac{\gamma\mu m}{r} = C,$$

where $C$ is constant throughout the motion and is determined by the initial conditions.

From the equations of motion (15) we can deduce other equations in which only the components of the velocity, not the acceleration, are present. If we multiply the first equation of motion by $y$, the second by $x$, and then subtract, we obtain

$$\ddot{x}y - x\ddot{y} = 0 \qquad \text{or} \qquad \frac{d}{dt}\, (\dot{x}y - \dot{y}x) = 0,$$

---

[1]The special case of circular motion has been discussed in Volume I (pp. 413 ff.).

whence, by integration, we have

$$x\dot{y} - y\dot{x} = c_1.$$

Similarly, from the remaining equation of motion we obtain[1]

$$y\dot{z} - z\dot{y} = c_2, \quad z\dot{x} - x\dot{z} = c_3.$$

These equations enable us to simplify our problem very considerably in a way that is highly plausible from the intuitive point of view. Without loss of generality, we can choose the coordinate system in such a way that at the beginning of the motion, that is, at $t = 0$, the particle lies in the $x$, $y$-plane and its velocity vector at that time also lies in that plane. Then $z(0) = 0$, and $\dot{z}(0) = 0$; and by substituting these values in the above equations and remembering that the right-hand sides are constants, we obtain

(16a)     $$x\dot{y} - y\dot{x} = c_1 = h,$$

(16b)     $$y\dot{z} - z\dot{y} = 0,$$

(16c)     $$z\dot{x} - x\dot{z} = 0.$$

From these equations we conclude in the first place that the whole motion takes place in the plane $z = 0$. Since we naturally exclude the possibility of an initial collision between the sun and planet, we assume that initially the three coordinates $(x, y, z)$ do not vanish

---

[1] We can also arrive at these three equations using vector notation if we form the vector product of both sides of the equation of motion and the position vector $\mathbf{R}$. Since the force vector is in the same direction as the position vector, we obtain zero on the right, while the expression $\mathbf{R} \times \ddot{\mathbf{R}}$ on the left is the derivative of the vector $\mathbf{R} \times \dot{\mathbf{R}}$ with respect to the time. It therefore follows that this vector $\mathbf{R} \times \dot{\mathbf{R}} = \mathbf{C}$ has a value constant in time; this is exactly what is stated by the coordinate equations above.

As we see, this equation does not depend on our special problem but holds in general for every motion in which the force has the same direction as the position vector.

The vector $\mathbf{R} \times \dot{\mathbf{R}}$ is called the *moment of velocity* and the vector $m\mathbf{R} \times \dot{\mathbf{R}}$ the *moment of momentum* of the motion. From the geometrical meaning of the vector product we easily obtain the following intuitive interpretation of the relation just given (cf. the subsequent discussions in the text). If we project the moving particle on to the coordinate planes and in each coordinate plane consider the area that the radius vector from the origin to the point of projection sweeps over in time $t$, this area is proportional to the time *(theorem of areas)*.

simultaneously, so that at the time $t = 0$ at which $z(0) = 0$, we have, say, $x(0) \neq 0$. Now, from (16c), it follows that

$$\frac{d}{dt}\left(\frac{z}{x}\right) = -\frac{z\dot{x} - \dot{z}x}{x^2} = 0.$$

Therefore, $z = ax$, where $a$ is a constant. If we put $t = 0$ here, then from the equations $z(0) = 0$ and $x(0) \neq 0$, it follows that $a = 0$, so that $z$ is always 0.

We therefore reduce our problem to integration of the two differential equations

(17a)
$$\frac{1}{2}\, m(\dot{x}^2 + \dot{y}^2) - \frac{\gamma\mu m}{r} = C,$$

(17b)
$$x\dot{y} - y\dot{x} = h.$$

We next use the equations $x = r\cos\theta$, $y = r\sin\theta$ to transform the rectangular coordinates $(x, y)$ into the polar coordinates $(r, \theta)$, which are now to be determined as functions of $t$. Since

$$\dot{x}^2 + \dot{y}^2 = \dot{r}^2 + r^2\dot{\theta}^2, \qquad x\dot{y} - y\dot{x} = r^2\dot{\theta},$$

we have the two differential equations

(17c)
$$\frac{1}{2}\, m\,(\dot{r}^2 + r^2\dot{\theta}^2) - \frac{\gamma\mu m}{r} = C,$$

(17d)
$$r^2\dot{\theta} = h$$

for the polar coordinates $r$, $\theta$. The first of these equations is the theorem of the *conservation of energy*, while the second expresses *Kepler's law of areas*. In fact (cf. Volume I, pp. 371–372) the expression $\frac{1}{2}r^2\dot{\theta}$ is the derivative with respect to the time of the area swept out in time $t$ by the radius vector from the origin to the particle. This is found to be constant, or, as Kepler expressed it, *the radius vector describes equal areas in equal times.*

If the *area constant $h$* is zero, $\dot{\theta}$ must vanish; that is, $\theta$ must remain constant, so that the motion must take place on a straight line through the origin. We exclude this special case and expressly assume that $h \neq 0$.

In order to find the geometrical form of the orbit, we shall no longer describe it parametrically in terms of the time[1] but consider the angle $\theta$ as a function of $r$ or $r$ as a function of $\theta$, and from our two equations we calculate the derivative $dr/d\theta$ as a function of $r$.

If we substitute the value $\dot\theta = h/r^2$ from the area equation in the energy equation and recall the equation

$$\dot r = \frac{dr}{dt} = \frac{dr}{d\theta}\,\dot\theta,$$

we at once obtain the differential equation of the orbit in the form

$$\frac{m}{2}\left\{\frac{h^2}{r^4}\left(\frac{dr}{d\theta}\right)^2 + \frac{h^2}{r^2}\right\} - \frac{\gamma\mu m}{r} = C$$

or

(17e)
$$\left(\frac{dr}{d\theta}\right)^2 = r^4\left(\frac{2C}{mh^2} + \frac{2\gamma\mu}{h^2}\frac{1}{r} - \frac{1}{r^2}\right).$$

To simplify the later calculations, we make the substitution

$$r = \frac{1}{u}$$

and introduce the following abbreviations:

$$\frac{1}{p} = \frac{\gamma\mu}{h^2}, \qquad \varepsilon^2 = 1 + \frac{2Ch^2}{m\gamma^2\mu^2}.$$

The differential equation (17e) then becomes

$$\left(\frac{du}{d\theta}\right)^2 = \frac{\varepsilon^2}{p^2} - \left(u - \frac{1}{p}\right)^2,$$

and this can be integrated immediately. We have

$$\theta - \theta_0 = \int \frac{du}{\sqrt{(\varepsilon^2/p^2 - (u - 1/p)^2)}},$$

---

[1] The course of the motion as a function of the time can be determined subsequently by means of the equation

$$\int_{\theta_0}^{\theta} r^2\,d\theta = h(t - t_0),$$

in which we suppose that $r$ is known as a function of $\theta$ (cf. p. 670).

or, if for the moment we introduce $u - 1/p = v$ as a new variable,

$$\theta - \theta_0 = \int \frac{dv}{\sqrt{(\varepsilon^2/p^2) - v^2}}.$$

For the integral [by Volume I, p. 270, formula (24)] we obtain the value arc sin $(vp/\varepsilon)$ and thus find the equation of the orbit in the form

$$\frac{1}{r} - \frac{1}{p} = v = \frac{\varepsilon}{p} \sin(\theta - \theta_0).$$

The angle $\theta_0$ can be chosen arbitrarily, since it is immaterial from which fixed line the angle $\theta$ is measured. If we take $\theta_0 = \pi/2$—that is, if we let $v = 0$ correspond to the value $\theta = \pi/2$—we finally obtain the equation of the orbit in the form

$$r = \frac{p}{1 - \varepsilon \cos \theta}.$$

This is the familiar equation in polar coordinates of a conic having one focus at the origin.[1]

Our result therefore gives Kepler's law:

*The planets move in conics with the sun at one focus.*

It is interesting to relate the constants of integration

$$p = \frac{h^2}{\gamma\mu}, \qquad \varepsilon^2 = 1 + \frac{2Ch^2}{m\gamma^2\mu^2}$$

to the initial motion. The quantity $p$ is known as the semi-latus rectum or parameter of the conic; in the case of the ellipse and the hyperbola it is connected with the semiaxes $a$ and $b$ by the simple relation

$$p = \frac{b^2}{a}.$$

The square of the eccentricity, $\varepsilon^2$, determines the character of the conic; it is an ellipse, a parabola, or a hyperbola, according to whether $\varepsilon^2$ is less than, equal to, or greater than 1.

From the relation

----

[1]This is seen easily by transforming the equation to rectangular coordinates:

$$(x - \varepsilon a)^2 + \frac{y^2}{1 - \varepsilon^2} = a^2 \qquad\qquad \left(a = \frac{p}{1 - \varepsilon^2}\right).$$

$$\varepsilon^2 = 1 + \frac{2Ch^2}{m\gamma^2\mu^2}$$

we see at once that the three different possiblities can also be stated in terms of the energy constant $C$; the orbit is an ellipse, a parabola, or a hyperbola, according to whether $C$ is less than, equal to, or greater than zero.

If we suppose that at time $t = 0$ the particle is at the point $\mathbf{R}_0$ in the field of force and is moving with initial velocity $\dot{\mathbf{R}}_0$, then the relation

$$C = \frac{1}{2}mv_0{}^2 - \frac{\gamma\mu m}{r_0}$$

gives the suprising fact that the character of the orbit—ellipse, parabola, or hyperbola—does not depend on the direction of the initial velocity at all, but only on its absolute value $v_0$.

Kepler's third law is a simple consequence of the other two:

*For a planet in elliptic orbit the square of the period bears a constant ratio to the cube of the major semiaxis, the ratio depending on the field of force only and not on the particular planet.*

If we denote the period $T$ and the major semiaxis by $a$, we should then have

$$\frac{T^2}{a^3} = \text{constant,}$$

where the constant on the right is independent of the particular problem and depends only on the magnitude of the attracting mass and on the gravitational constant.

To prove this we use the theorem of areas (17d) in the integrated form

$$\int_{\theta_0}^{\theta} r^2 \, d\theta = h(t - t_0),$$

which defines the motion as a function of the time. If we take the integral over the interval from $0$ to $2\pi$, we obtain on the left twice the area of the orbital ellipse, and that, by previous results, is $2\pi ab$; on the right the time difference $t = t_0$ is replaced by the period $T$. Therefore,

$$2\pi ab = hT \qquad \text{or} \qquad 4\pi^2 a^2 b^2 = h^2 T^2.$$

We already know that $h^2$ is connected with the $a$ and $b$ of the orbit by the relation $h^2/\gamma\mu = p = b^2/a$. If we replace $h^2$ in the above equations by $(b^2/a)\,\gamma\mu$, it follows at once that

$$\frac{T^2}{a^3} = \frac{4\pi^2}{\gamma\mu},$$

which exactly expresses Kepler's third law.

### Exercises 6.1e

1. Treat in detail the motion of an orbiting body in a straight line trajectory [$h = 0$ in equation (17d)].
2. Prove that as $t \to \infty$ the velocity $v$ of a planet tends to 0 if its orbit is a parabola and to a positive limit if it is a hyperbola.
3. Prove that a body attracted toward a center 0 by a force of magnitude $mr$ moves on an ellipse with center 0.
4. Prove that the orbit of a body repelled by a force of magnitude $f(r)$, where $f$ is a given function, from a center 0 is given in polar cordinates $(r, \theta)$ by

$$\theta = \int^r \frac{dr}{r^2\sqrt{2c/h^2 + 2\int^r r\,f(r)\,dr/h^3 - 1/r^2}}.$$

5. Prove that the equation of the orbit of a body repelled with a force $\mu/r^3$ from a center 0 is

$$\frac{1}{r} = \begin{cases} \dfrac{2c}{h^2 k}\cos(k\theta + \varepsilon) & \text{for } \mu < h^2 \\[2mm] \dfrac{2c}{h^2 k}\cosh(k\theta + \varepsilon) & \text{for } \mu > h^2 \end{cases}$$

if

$$k = \sqrt{\left|1 - \frac{\mu}{h^2}\right|}$$

and $\varepsilon$ is a constant of integration.
6. A planet is moving on an ellipse, and $\omega = \omega(t)$ denotes the angle $P'\,MP_s$, where $P'$ is the point on the auxiliary circle corresponding to $P$, the position of the planet at that time $t$; $P_s$ its position at the time $t_s$ when it is nearest to the sun $S$; and $M$ the center of the ellipse. Prove that $\omega$ and $t$ are connected by Kepler's equation

$$h(t - t_s) = ab(\omega - \varepsilon\sin\omega).$$

7. Prove that in a central field of force the attraction $p$ per unit mass is given by

$$p = \frac{h^2}{q^3}\frac{dq}{dr},$$

where $q$ is the distance of the tangent of the orbit from the pole and $h$ the area constant (p. 667). Hence prove that the cardioid $r = a(1 + \cos \theta)$ can be described under an attraction to the pole equal to $\mu r^{-4}$ per unit mass.

8. A particle of unit mass moves under the action of two forces, of which the first is always toward the origin and is equal to $\lambda^2$ times the distance of the particle from that point, while the second is always at right angles to the path of the particle and is equal to $2\mu$ times its velocity. Prove that if the particle is projected from the origin along the axis of $x$ with velocity $u$, its coordinates at any subsequent time $t$ are

$$x = \frac{u}{\sqrt{\lambda^2 + \mu^2}} \sin (\sqrt{\lambda^2 + \mu^2}\, t) \cos \mu t,$$

$$y = \frac{u}{\sqrt{\lambda^2 + \mu^2}} \sin (\sqrt{\lambda^2 + \mu^2}\, t) \sin \mu t.$$

9. Let there be $n$ fixed particles in a plane, all attracting with a central force of magnitude $1/r$. Prove that there are not more than $n - 1$ positions of equilibrium for a particle in the field.

Calculate these positions for the case of four attracting particles with coordinates $(a, b)$, $(a, -b)$, $(-a, b)$, $(-a, -b)$, where $a > b > 0$.

### f. Boundary Value Problems. The Loaded Cable and the Loaded Beam.

In the problems of mechanics and the other examples previously discussed, we selected from the whole family of functions satisfying the differential equation a particular one by means of so-called *initial conditions*; that is, we chose the constants of integration in such a way that the solution and, in certain cases, some of its derivatives assume preassigned values at a definite point. In many applications we are concerned neither with finding the general solution nor with solving definite initial-value problems but with solving a so-called *boundary value problem*. In a boundary value problem we seek a solution that satisfies preassigned conditions at *several* points and satisfies the differential equation in the intervals between those points. Here we shall discuss a few typical examples without going into the general theory of such boundary value problems.

*Example 1—The Differential Equation of a Loaded Cable*

In a vertical $x, y$-plane—in which the $y$-axis is vertical—we suppose that a cable with (constant) horizontal component of tension $S$ is stretched from the origin to the point $x = a$, $y = b$, (cf. Fig. 6.7). The cable is acted on by a load whose density per unit length of horizontal projection is given by a sectionally continuous function $p(x)$. Then the sag $y(x)$ of the cable, that is, the $y$-coordinate, is given by the differential equation

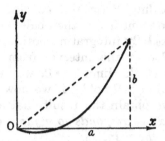

**Figure 6.7** Loaded cable.

$$(18) \qquad y''(x) = g(x) \qquad\qquad g(x) = \frac{p}{S}.$$

The shape of the cable will then be given by that solution $y(x)$ of the differential equation that satisfies the conditions $y(0) = 0$, $y(a) = b$. The solution of this boundary value problem can be written down at once, since the general solution of the homogeneous equation $y'' = 0$ is the linear function $c_0 + c_1 x$, and the solution of the nonhomogeneous equation that, with its first derivative, vanishes at the origin is given by the integral $\int_0^x g(\xi)(x - \xi)\, d\xi$ [see (42), p. 78]. In the general solution

$$y(x) = c_0 + c_1 x + \int_0^x g(\xi)(x - \xi)\, d\xi$$

the condition $y(0) = 0$ at once gives $c_0 = 0$, and then the condition $y(a) = b$ determines $c$, through the quation

$$b = c_1 a + \int_0^a g(\xi)(a - \xi)\, d\xi$$

In practice, we must often deal with a more complicated form of this boundary value problem in which the cable is subject not only to the continuously distributed load but also to concentrated loads, that is, loads that are concentrated at a definite point of the cable, say, at the point $x = x_0$. Such concentrated loads we shall consider as ideal limiting cases arising as $\varepsilon \to 0$ from a loading $p(x)$ that acts only in the interval $x_0 - \varepsilon$ to $x_0 + \varepsilon$ and for which

$$\int_{x_0 - \varepsilon}^{x_0 + \varepsilon} p(x)\, dx = P,$$

In this, the total loading $P$ remains constant during the passage to the limit $\varepsilon \to 0$; the number $P$ is then called the concentrated load acting at the point $x_0$.[1] By integrating both sides of the differential equation $y'' = p(x)/S$ over the interval from $x - \varepsilon$ to $x + \varepsilon$ before making the passage to the limit $\varepsilon \to 0$, we see that the equation $y'(x_0 + \varepsilon) - y'(x_0 - \varepsilon) = P/S$ holds. If we now perform the passage to the limit $\varepsilon \to 0$, we obtain the result that a *concentrated load P acting at the point $x_0$ corresponds to a jump of the derivative $y'(x)$ by an amount P/S at the point $x_0$.*

The following example shows how the presence of a concentrated load modifies the boundary value problem. We suppose that the cable is stretched between the points $x = 0$, $y = 0$ and $x = 1$, $y = 1$ and that the only load is a concentrated load of magnitude $P$ acting at the midpoint $x = \frac{1}{2}$. This physical problem corresponds to the following mathematical problem: to find a continuous function $y(x)$ that satisfies the differential equation $y'' = 0$ everywhere in the interval $0 \leq x \leq 1$ except at the point $x_0 = \frac{1}{2}$; that takes the values $y(0) = 0$, $y(1) = 1$ on the boundary; and whose derivative has a jump of the amount $P/S$ at the point $x_0$. In order to find this solution, we express it in the following way:

$$y(x) = ax + b \qquad\qquad (0 \leq x \leq \tfrac{1}{2})$$

and

$$y(x) = c(1 - x) + d \qquad\qquad (\tfrac{1}{2} \leq x \leq 1).$$

The condition $y(0) = 0$, $y(1) = 1$ gives $b = 0$, $d = 1$. From the condition that both parts of the function shall give the same value at the point $x = \frac{1}{2}$, we find that

$$\frac{1}{2}a = \frac{1}{2}c + 1.$$

---

[1]One often thinks of the concentrated load as described purely formally by a distributed load

$$p(x) = P\,\delta(x - x_0),$$

where $\delta(x)$ stands for a *generalized* function (the so-called *Dirac function*) for which

$$\delta(x) = 0 \quad \text{for} \quad x \neq 0 \quad \text{and} \quad \int_{-\infty}^{\infty} \delta(x)\, dx = 1,$$

with no value assigned to $\delta(0)$. No finite value of $\delta(0)$ would be compatible with the other conditions imposed.

Finally, the requirement that the derivative $y$ shall increase by the amount $P/S$ on passing the point $\frac{1}{2}$ gives the condition

$$-c - a = \frac{P}{S}.$$

These conditions yield

$$a = 1 - \frac{P}{2S}, \qquad b = 0, \qquad c = -1 - \frac{P}{2S}, \qquad d = 1,$$

and our solution has been found. Moreover, no other solution with the same properties exists.

### Example 2—The Loaded Beam[1]

The treatment of a loaded beam is very similar (cf. Fig. 6.8). Let us suppose that in its position of rest the beam coincides with the

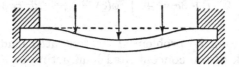

**Figure 6.8** Loaded beam.

$x$-axis between the abscissas $x = 0$ and $x = a$. Then it is found that the sag (*vertical displacement*) $y(x)$ due to a force acting vertically in the $y$-direction is given by the linear differential equation of the fourth order

(19a) $$y'''' = \varphi(x),$$

where the right-hand side $\varphi(x)$ is $p(x)/EI$, $p(x)$ being the density of loading, $E$ the modulus of elasticity of the material of the beam ($E$ is the stress divided by the elongation), and $I$ the moment of inertia of the cross section of the beam about a horizontal line through the center of mass of the cross section.

The general solution of this differential equation can at once be written [(42), p. 78] in the form

$$y(x) = c_0 + c_1 x + c_2 x^2 + c_3 x^3 + \int_0^x \varphi(\xi) \frac{(x - \xi)^3}{3!} d\xi,$$

---

[1]For the theory of loaded beams, cf. v. Karman and Biot, *Mathematical Methods in Engineering*.

where $c_0$, $c_1$, $c_2$, $c_3$ are arbitrary constants of integration. The real problem, however, is not that of finding this general solution but of finding a particular solution, that is, of determining the constants of integration in such a way that certain definite boundary conditions are satisfied. If for example, the beam is *clamped* at the ends, the boundary conditions

$$y(0) = 0, \qquad y(a) = 0, \qquad y'(0) = 0, \qquad y'(a) = 0$$

hold. It then follows at once that $c_0 = c_1 = 0$, and the constants $c_2$ and $c_3$ are to be determined from the equations

$$c_2 a^2 + c_3 a^3 + \int_0^a \varphi(\xi) \frac{(a - \xi)^3}{3!} \, d\xi = 0,$$

$$2c_2 a + 3c_3 a^2 + \int_0^a \varphi(\xi) \frac{(a - \xi)^2}{2!} \, d\xi = 0.$$

For beams, too, the problem of concentrated loads is important. We again think of the concentrated load acting at the point $x = x_0$ as arising from a loading $p(x)$, distributed continuously over the interval $x_0 - \varepsilon$, to $x_0 + \varepsilon$, for which $\int_{x_0-\varepsilon}^{x_0+\varepsilon} p(\xi) \, d\xi = P$; we again let $\varepsilon$ approach zero and at the same time let $p(x)$ increase in such a way that the value of $P$ remains constant during the passage to the limit $\varepsilon \to 0$. $P$ is then the value of the concentrated load at $x = x_0$. Just as in the example above, we integrate both sides of the differential equation (19a) over the interval from $x - \varepsilon$ to $x + \varepsilon$ and then pass to the limit as $\varepsilon \to 0$. It is found that the third derivative of the solution $y(x)$ must have a jump at the point $x = x_0$, amounting to

(19b) $$y''' (x_0 + 0) - y''' (x_0 - 0) = \frac{P}{EI} .$$

Here $y(x_0 + 0)$ means the limit of $y(x_0 + h)$ as $h$ tends to 0 through positive values, $y(x_0 - 0)$ being the corresponding limit from the left.

Thus, the following mathematical problem arises: we attempt to find a solution of $y'''' = 0$ that, together with its first and second derivatives, is continuous, for which $y(0) = y(1) = y'(0) = y'(1) = 0$, and whose third derivative has a jump of the amount $P/EI$ at the point $x = x_0$ and elsewhere is continuous.

If the beam is *fixed* at a point $x = x_0$ (cf. Fig. 6.9)—that is, if at this point the sag has the fixed preassigned value $y = 0$—we can think of

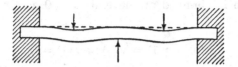

**Figure 6.9**   Sag of beam supported in the middle.

this constraint as being achieved by means of a concentrated load acting at that point. By the mechanical principle that action is equal to reaction, the value of this concentrated load will be equal to the force that the fixed beam exerts on its support. The magnitude $P$ of this force is then given at once by the formula [see (19b)]

$$P = EI \{ y''' (x_0 + 0) - y''' (x_0 - 0) \},$$

where $y(x)$ satisfies the differential equation $y'''' = p/EI$ everywhere in the interval $0 \leq x \leq 1$ except at the point $x = x_0$ and in addition also satisfies the conditions $y(0) = y(1) = y'(0) = y'(1) = 0, y(x_0) = 0$, and $y$, $y'$, and $y''$ are also continuous at $x = x_0$.

In order to illustrate these ideas, we consider a beam that extends from the point $x = 0$ to the point $x = 1$, is clamped at its end points $x = 0$ and $x = 1$, carries a uniform load of density $p(x) = 1$, and is supported at the point $x = \frac{1}{2}$ (cf. Fig. 6.9). For the sake of simplicity we assume that $EI = 1$, so that the beam satisfies the differential equation

$$y'''' = 1$$

everywhere, except at the point $x = \frac{1}{2}$.

As the formula shows, the general solution of the differential equation is a polynomial of the fourth degree in $x$, the coefficient of $x^4$ being $1/4!$. The solution will be expressed by a polynomial of this type in each of the two half-intervals. For the first half-interval we write the polynomial in the form

$$y = b_0 + b_1 x + b_2 x^2 + b_3 x^3 + \frac{1}{4!} x^4,$$

in the second half-interval, in the form

$$y = c_0 + c_1(x - 1) + c_2(x - 1)^2 + c_3(x - 1)^3 + \frac{1}{4!} (x - 1)^4.$$

Since the beam is clamped at the ends $x = 0$ and $x = 1$, it follows that

$$y(0) = y(1) = y'(0) = y'(1) = 0,$$

whence we obtain $b_0 = b_1 = c_0 = c_1 = 0$. In addition, $y(x)$, $y'(x)$, $y''(x)$ must be continuous at the point $x = \frac{1}{2}$; that is, the values of $y(\frac{1}{2})$, $y'(\frac{1}{2})$, $y''(\frac{1}{2})$ calculated from the two polynomials must be the same, and the value of $y(\frac{1}{2})$ must be 0. This gives

$$\frac{1}{4} b_2 + \frac{1}{8} b_3 + \frac{1}{384} = \frac{1}{4} c_2 - \frac{1}{8} c_3 + \frac{1}{384} = 0,$$

$$b_2 + \frac{3}{4} b_3 + \frac{1}{48} = - c_2 + \frac{3}{4} c_3 - \frac{1}{48},$$

$$2b_2 + 3b_3 = 2c_2 - 3c_3.$$

From this we obtain the following values for $b_2$, $b_3$, $c_2$, $c_3$:

$$b_2 = c_2 = \frac{1}{96}; \; b_3 = -c_3 = -\frac{1}{24},$$

and the force that must act on the beam at the point $x = \frac{1}{2}$ in order that no sag may occur at that point is given by

$$y'''\left(\frac{1}{2} + 0\right) - y'''\left(\frac{1}{2} - 0\right) = \left(6c_3 - \frac{1}{2}\right) - \left(6b_3 + \frac{1}{2}\right) = -\frac{1}{2}.$$

## 6.2   The General Linear Differential Equation of the First Order

### a. Separation of Variables

A differential equation is said to be *of the first order* if it involves, besides $x$ and $y(x)$, the first derivative of the function $y(x)$ but no higher derivative. The most general equation of this type is

(20a) $$F(x, y, y') = 0,$$

where $F$ is a given function of its three arguments $x$, $y$, $y'$. We can assume that in a certain region of the $x$, $y$-plane the differential equation (20a) can be solved uniquely for $y'$ and thus expressed in the form

(20b) $$y' = f(x, y).$$

Explicit formulae for the general solution of a differential equation (20b) can only be found in special cases.[1] The simplest situation arises when the function $f(x, y)$ is the quotient of a function of $x$ alone and of a function of $y$ alone, that is, when the differential equation has the form

(21a)
$$y' = \frac{\alpha(x)}{\beta(y)}.$$

In this case we can "separate" the variables $x, y$, writing the equation symbolically in the form

(21b)
$$\beta(y) \, dy = \alpha(x) \, dx.$$

We now introduce the two indefinite integrals

(21c)
$$A(x) = \int \alpha(x) \, dx, \qquad B(y) = \int \beta(y) \, dy$$

obtained by ordinary quadratures. Then by (21a)

$$\frac{dB(y)}{dx} = \frac{dB(y)}{dy} \frac{dy}{dx} = \beta(y) \, y' = \alpha(x) = \frac{dA(x)}{dx}.$$

It follows that for every solution of (21a)

(21d)
$$B(y) - A(x) = c,$$

where $c$ is a constant (depending on the solution).[2] Equation (21d) may now be solved for $y$, assigning any value to $c$, and the required solution of (21a) is thus obtained by quadratures.

As a matter of fact, we already have used this method of separation of variables in a variety of problems leading to differential equations (see Volume I, p. 406; Volume II, p. 668). Another type of differential equation that can be reduced to the form (21a) is the so-called *homogeneous* equation

(21e)
$$y' = f\left(\frac{y}{x}\right).$$

---

[1]We shall, however, discuss on p. 704 a general approximation scheme giving the solution of (20b) in all cases, where the function $f$ has continuous first derivatives.
[2]Instead of using the chain rule in the derivation of (21d), we could also argue that by (21b, c)

$$d(B - A) = dB - dA = \beta \, dy - \alpha \, dx = 0$$

and, hence, that $B - A$ is constant.

Introducing the new unknown function $z = y/x$, we arrive at a differential equation

$$z' = \frac{xy' - y}{x^2} = \frac{f(z) - z}{x},$$

which is separable. The general solution is then found from the relation

(21f) $$\int \frac{dz}{f(z) - z} = \int \frac{dx}{x} + c = c + \log|x|,$$

where $c$ is a constant. We use this equation to express $z$ as a function of $x$ and put $y = xz$ to obtain the required solution.

As an example, consider the equation

$$y' = \frac{y^2}{x^2}$$

corresponding to $f(z) = z^2$. Here relation (21f) becomes

$$\int \frac{dz}{z^2 - z} = \log \frac{z - 1}{z} = c + \log|x|.$$

Hence,

$$y = \frac{x}{1 - kx},$$

where $k = \pm e^c$ is a constant.

### b. The Linear First-Order Equation

A differential equation is called *linear* if it represents a linear relation between the unknown function $y$ and its derivatives with coefficients that are given functions of $x$. Thus, the general first-order linear differential equation has the form

(22a) $$y' + a(x)\, y = b(x)$$

where $a(x)$ and $b(x)$ are given.

We first suppose that $b = 0$. Then the differential equation is separable and can be written as

$$\frac{dy}{y} = -a(x)\, dx.$$

Hence,

$$\log |y| = -\int a(x)\, dx + \text{constant}.$$

If we denote by $A(x)$ any indefinite integral of the function $a(x)$, that is, any function with derivative $a(x)$, we find that

(22b) $$y = ce^{-A(x)}$$

where $c$ is an arbitrary constant of integration. This formula gives a solution, even when $c = 0$, namely, $y = 0$.

If $b(x)$ is not zero we seek a solution of the form

(22c) $$y = u(x)e^{-A(x)}$$

where $A$ is defined as before and $u(x)$ must be suitably determined.[1] One finds by substitution into (22a) that

$$y' + ay = u'e^{-A} - uA'e^{-A} + aue^{-A} = u'e^{-A} = b.$$

Hence, the unknown function $u$ must have the derivative

$$u' = b(x)\, e^{A(x)}.$$

Thus,

$$u = c + \int b(x)\, e^{A(x)}\, dx,$$

where $c$ is a constant. We find for the solution $y$ of (22a) the expression

(22d) $$y = e^{-A(x)} \left( c + \int b(x)\, e^{A(x)}\, dx \right),$$

where $c$ is any constant and

(22e) $$A(x) = \int a(x)\, dx.$$

Since every function $y$ can be written in the form (22c) with a suitable function $u$, we see that formula (22d) represents the *most general*

---

[1]This device of replacing the constant $c$ in (22b) by the variable $u$ is known as *variation of parameters*.

*solution* of (22a). Thus, the general solution is formed from known functions merely by exponentiation and the ordinary process of integration. The solution really contains only *one* arbitrary constant, since any different choice of the constants of integration in $A(x)$ or in the indefinite integral occuring in (22d) can be compensated for by a suitable change in $c$.

For example, in the case of the differential equation

$$y' + xy = -x$$

we have

$$A(x) = \int x \, dx = \frac{1}{2} x^2$$

$$\int b(x)e^{A(x)} \, dx = -\int xe^{x^2/2} \, dx = -e^{x^2/2}$$

and, hence, obtain the solution

$$y = e^{-x^2/2} \left(c - e^{-x^2/2}\right) = -1 + ce^{-x^2/2}.$$

### Exercises 6.2

1. Integrate the following equations by separation of the variables:
   (a) $(1 + y^2)x \, dx + (1 + x^2) \, dy = 0$
   (b) $ye^{2x} \, dx - (1 + e^{2x}) \, dy = 0$.
2. Solve the follwing homogenous equations:
   (a) $y^2 \, dx + x(x - y) \, dy = 0$
   (b) $xy \, dx + (x^2 + y^2) \, dy = 0$
   (c) $x^2 - y^2 + 2xyy' = 0$
   (d) $(x + y) \, dx + (y - x) \, dy = 0$
   (e) $(x^2 + xy)y' = x\sqrt{x^2 - y^2} + xy + y^2$.
3. Show that a differential equation of the form
   $$y' = \phi\left[\frac{ax + by + c}{a_1x + b_1y + c_1}\right] \qquad (a, a_1, \ldots \text{constant})$$
   can be reduced to a homogeneous equation as follows. If $ab_1 - a_1b \neq 0$, we take a new unknown function and a new independent variable
   $$\eta = ax + by + c, \qquad \xi = a_1x + b_1y + c_1.$$
   If $ab_1 - a_1b = 0$, we need only change the unknown function by putting

$$\eta = ax + by$$

to reduce the equation to a new equation in which the variables are separated.

4. Apply the method of the previous exercise to

    (a) $(2x + 4y + 3)y' = 2y + x + 1$

    (b) $(3y - 7x + 3)y' = 3y - 7x + 7$.

5. Integrate the following linear differential equations of the first order:

    (a) $y' + y \cos x = \cos x \sin x$

    (b) $y' - \dfrac{ny}{x + 1} = e^x(x + 1)^n$

    (c) $x(x - 1)y' + (1 - 2x)y + x^2 = 0$

    (d) $y' - \dfrac{2}{x}y = x^4$

    (e) $(1 + x^2)y' + xy = \dfrac{1}{1 + x^2}$.

6. Integrate the equation

$$y' + y^2 = \frac{1}{x^2}.$$

7. A *Bernoulli equation* has the form

$$y' + f(x)y = g(x)y^n.$$

Show that such an equation is made separable by the substitution

$$y = v \exp\left\{-\int f(x)\, dx\right\} = vF(x).$$

8. Integrate the equation

$$xy' + y(1 - xy) = 0.$$

9. By any method available, solve

$$y' + y \sin x + y^n \sin 2x = 0.$$

## 6.3   Linear Differential Equations of Higher Order

### a. Principle of Superposition. General Solutions

Many of the examples previously discussed belong to the general class of linear differential equations. A differential equation in the unknown function $u(x)$ is said to be linear of the $n$th order if it has the form

(23) $$u^{(n)}(x) + a_1 u^{(n-1)}(x) + \cdots + a_n u(x) = \phi(x),$$

where $a_1$, $a_2$, $a_3$, . . ., $a_n$ are given functions of the independent
variable $x$, as is also the right-hand side $\phi(x)$. We denote the expression on the left side by $L[u]$ (where $L$ stands for "linear differential operator").

If $\phi(x)$ is identically zero in the interval under consideration, we
call the equation *homogeneous;* otherwise, we call it *nonhomogeneous.*
We see at once (as in the special case of the linear differential
equation of the second order with constant coefficients, discussed
in Volume I, p. 640) that the following *principle of superposition*
holds:

*If $u_1$, $u_2$ are any two solutions of the homogeneous equation, every
linear combination of them, $u = c_1u_1 + c_2u_2$, where the coefficients $c_1$,
$c_2$ are constants, is also a solution.*

If we know a single solution $v(x)$ of the nonhomogeneous equation
$L[u] = \phi(x)$, we can obtain all other such solutions by adding to
$v(x)$ any solution of the homogeneous equation.

For $n = 2$ and constant coefficients $a_1$, $a_2$ we proved in Volume
I (p. 636) that every solution of the homogeneous equation can be
expressed in terms of two suitably chosen solutions $u_1$, $u_2$ in the form
$c_1u_1 + c_2u_2$. An analogous theorem holds for any homogeneous
differential equation with arbitrary continuous coefficients.

To begin with, we explain what we mean by saying that functions
are linearly dependent or linearly independent, by means of the
following definition: $n$ functions $\phi_1(x)$, $\phi_2(x)$, . . ., $\phi_n(x)$ are *linearly
dependent* if $n$ constants $c_1$, . . ., $c_n$ that do not all vanish exist,
such that the equation

$$c_1\phi_1(x) + c_2\phi_2(x) + \cdots + c_n\phi_n(x) = 0$$

holds identically, that is, for all values of $x$ in the interval under
consideration. If, say, $c_n \neq 0$, then $\phi_n(x)$ may be expressed in the form

$$\phi_n(x) = a_1\phi_1(x) + \cdots a_{n-1}\phi_{n-1}(x),$$

and $\phi_n$ is said to be *linearly dependent on the other functions.* If no
linear relation of the form

$$c_1\phi_1(x) + c_2\phi_2(x) + \cdots + c_n\phi_n(x) = 0$$

exists, the $n$ functions $\phi_i(x)$ are said to be *linearly independent.*[1]

---

[1]Linear dependence of functions $\phi(x)$ is defined in exactly the same way as dependence of vectors (see p. 137). As a matter of fact, it often is convenient to visualize
a function $\phi(x)$ defined in an interval $I$ of the $x$-axis as a "vector $\phi$ with infinitely
many components," one component of value $\phi(x)$ corresponding to each $x$ in $I$.

## Example 1

The functions $1, x, x^2, \ldots, x^{n-1}$ are linearly independent. Otherwise, constants $c_0, c_1, \ldots, c_{n-1}$ would have to exist such that the polynomial

$$c_0 + c_1 x + \cdots + c_{n-1} x^{n-1}$$

vanishes for all values of $x$ in a certain interval. This, however, is impossible unless all the coefficients of the polynomial are zero.

## Example 2

The functions $e^{a_i x}$ are linearly independent, provided $a_1 < a_2 < \cdots < a_n$.

PROOF. We assume that this statement has been proved true for $(n-1)$ such exponential functions. Then if

$$c_1 e^{a_1 x} + c_2 e^{a_2 x} + \cdots + c_n e^{a_n x} = 0$$

is an identity in $x$, we divide by $e^{a_n x}$ and, putting $a_i - a_n = b_i$, obtain

$$c_1 e^{b_1 x} + c_2 e^{b_2 x} + \cdots + c_{n-1} e^{b_{n-1} x} + c_n = 0.$$

If we differentiate this equation with respect to $x$, the constant $c_n$ disappears and we have an equation that implies that the $(n-1)$ functions $e^{b_1 x}, e^{b_2 x}, \ldots, e^{b_{n-1} x}$ are linearly dependent, from which it follows that $e^{a_1 x}, e^{a_2 x}, \ldots, e^{a_{n-1} x}$ are linearly dependent, contrary to our original assumption. Hence, there cannot be a linear relation between the $n$ original functions either.

## Example 3

The functions $\sin x, \sin 2x, \sin 3x, \ldots, \sin nx$ are linearly independent in the interval $0 \leq x \leq \pi$. We leave the reader to prove this in Exercise 1, p. 690, using the fact that

$$\int_{-\pi}^{+\pi} \sin mx \sin nx \, dx = \begin{cases} 0 \text{ if } m \neq n, \\ \pi \text{ if } m = n, \end{cases}$$

(cf. Volume I, p. 274).

If we assume that the functions $\phi_i(x)$ have continuous derivatives up to, and including, the $n$th order, we have the following theorem:

*The necessary and sufficient condition that the system of functions $\phi_i(x)$ shall be linearly dependent is that the equation*

$$(24) \quad W = \begin{vmatrix} \phi_1(x) & \phi_2(x) & \cdots \phi_n(x) \\ \phi_1'(x) & \phi_2'(x) & \cdots \phi_n'(x) \\ \vdots & \vdots & \vdots \\ \phi_1^{(n-1)}(x) & \phi_2^{(n-1)}(x) & \cdots \phi_n^{(n-1)}(x) \end{vmatrix} = 0$$

*shall be an identity in* $x$. The function $W$ is called the *Wronskian* of the system of functions.[1]

That the condition is *necessary* follows immediately: if we assume that

$$\sum c_i \phi_i(x) = 0,$$

successive differentiation gives the further equations

$$\sum c_i \phi_i'(x) = 0, \cdots,$$
$$\sum c_i \phi_i^{(n-1)}(x) = 0.$$

These, however, form a homogeneous system of $n$ equations, which are satisfied by the $n$ coefficients $c_1, \ldots, c_n$; hence, $W$, the determinant of the system of equations, must vanish.

That the condition is sufficient, that is, that if $W = 0$ the functions are linearly dependent, may be proved as follows: From the vanishing of $W$ we may deduce that the system of equations

$$c_1\phi_1 + \cdots + c_n\phi_n = 0$$
$$c_1\phi_1' + \cdots + c_n\phi_n' = 0$$
$$\vdots \qquad \vdots \qquad \vdots$$
$$c_1\phi_1^{(n-1)} + \cdots + c_n\phi_n^{(n-1)} = 0$$

possesses a solution $c_1, c_2, \ldots, c_n$ that is not trivial (see p. 150) where $c_i$ may still be a function of $x$. Here we may assume without loss of generality that $c_n = 1$. Further, we may assume that $V$, the Wronskian of the $(n-1)$ functions $\phi_1, \phi_2, \ldots, \phi_{n-1}$ is not zero, for we may suppose that our theorem has already been proved for $(n-1)$ functions; then $V = 0$ implies the existence of a linear relation

---

[1] In this proof and the following one a knowledge of the elements of the theory of determinants is assumed. Notice that each column of the Wronskian determinant is the vector formed from a function $\phi$ and its derivatives of orders 1, 2, ..., $n-1$. Thus, vanishing of the Wronskian for a system of functions means that the corresponding vectors are dependent (see p. 175).

between $\phi_1$, $\phi_2$, . . ., $\phi_{n-1}$ and, hence, between $\phi_1$, $\phi_2$, $\phi_3$, . . ., $\phi_n$. By differentiating[1] the first equation with respect to $x$ and combining the result with the second, we obtain

$$c_1'\phi_1 + c_2'\phi_2 + \cdots + c_{n-1}'\phi_{n-1} = 0;$$

similarly, by differentiating the second equation and combining the result with the third, we obtain

$$c_1'\phi_1' + c_2'\phi_2' + \cdots + c_{n-1}'\phi_{n-1}' = 0,$$

and so on, up to

$$c_1'\phi_1^{(n-2)} + c_2'\phi_2^{(n-2)} + \cdots + c_{n-1}'\phi_{n-1}^{(n-2)} = 0.$$

Since $V$, the determinant of these equations, is assumed not to vanish, it follows that $c_1'$, $c_2'$, . . ., $c_{n-1}'$ are zero; that is, $c_1$, $c_2$, . . ., $c_{n-1}$ are constants. Hence, the equation

$$\sum_l^n c_i\phi_i(x) = 0$$

does express a linear relation, as was asserted.

We now state the fundamental theorem on linear differential equations:

*Every homogeneous linear differential equation*

$$(25) \qquad L[u] = a_0(x)\, u^{(n)}(x) + a_1(x)\, u^{n-1}(x) + \cdots a_n(x)\, u(x) = 0$$

*possesses systems of $n$ linearly independent solutions $u_1$, $u_2$, . . ., $u_n$. By superposing these fundamental solutions every other solution $u$ may be expressed[2] as a linear expression with constant coefficients $c_1$, . . ., $c_n$:*

---

[1] It is easy to see that the coefficients $c_i$ are continuously differentiable functions of $x$, for if the determinant $V$ is not zero, they can be expressed rationally in terms of the functions $\phi_i$ and their derivatives.

[2] Two different systems of fundamental solutions $u_1$, . . ., $u_n$; $v_1$, . . ., $v_n$ can be transformed into one another by a linear transformation

$$v_i = \sum_{k=1}^n c_{ik}\, u_k,$$

where the coefficients $c_{ik}$ are constants and form a matrix whose determinant does not vanish.

$$u = \sum_{i=1}^{n} c_i u_i.$$

In particular, a system of fundamental solutions can be determined by the following conditions. At a prescribed point, say $x = \xi$, $u_1$ is to have the value 1 and all the derivatives of $u_1$ up to the $(n - 1)$-th order are to vanish; $u_i$, where $i > 1$, and all the derivatives of $u_i$ up to the $(n - 1)$-th order, except the $i$-th, are to vanish, while the $i$-th derivative is to have the value 1.

The existence of a system of fundamental solutions will follow from the existence theorem proved on p. 702. It follows from Wronski's condition (24), which we have just proved, that a linear relation must exist between any further solution $u$ and $u_1, \ldots, u_n$, for the equations

$$\sum_{l=0}^{n} a_l u^{(n-l)} = 0$$

$$\sum_{l=0}^{n} a_l u_i^{(n-l)} = 0 \qquad\qquad (i = 1, \ldots, n)$$

imply that the Wronskian of the $(n + 1)$ functions $u, u_1, u_2, \ldots, u_n$ must vanish, so that $u, u_1, u_2, \ldots, u_n$ are linearly dependent. Since $u_1, \ldots, u_n$ are independent, $u$ depends linearly on $u_1, \ldots, u_n$.

### b. Homogeneous Differential Equations of the Second Order

We shall consider differential equations of the second order in more detail, as they have very important applications.

Let the differential equation be

(26) $$L[u] = au'' + bu' + cu = 0.$$

If $u_1(x)$, $u_2(x)$ form a system of fundamental solutions, $W = u_1 u_2' - u_2 u_1'$ is its Wronskian, and $W' = u_1 u_2'' - u_2 u_1''$.
Since

$$L[u_1] = 0 \qquad \text{and} \qquad L[u_2] = 0,$$

it follows that

$$u_1 L[u_2] - u_2 L[u_1] = aW' + bW = 0.$$

This is a first-order linear equation for $W$. Its general solution by formula (22b), p. 681 is given by

(27) $$W = ce^{-\int (b/a)\, dx} ,$$

where $c$ is a constant. This formula is used a great deal in the further development of the theory of differential equations of the second order.

Another property worth mentioning is that a linear homogeneous differential equation of the second order can always be transformed into an equation of the first order, known as *Riccati's differential equation*. Riccati's equation is of the form

$$v' + pv^2 + qv + r = 0,$$

where $v$ is a function of $x$. The linear equation (26) is transformed into Riccati's equation by putting $u' = uz$, so that $u'' = u'z + uz' = uz^2 + uz'$, and we have

$$az' + az^2 + bz + c = 0.$$

A third remark: if we know *one* solution $v(x)$ of our linear homogeneous differential equation of the second order, the problem can be reduced to that of solving a differential equation of the first order and can be carried out by quadratures. Specifically, if we assume that $L[v] = 0$ and put $u = zv$, where $z(x)$ is the new function that we are seeking, we obtain the differential equation

$$az''v + 2az'v' + bz'v + zL[v] = avz'' + (2av' + bv)\, z' = 0$$

for $z$. This, however, is a linear homogeneous differential equation for the unknown function $z' = w$; its solution is given by formula (22d) on p. 681. From $w$ we then obtain the factor $z$ and, hence, the solution $u$ by a further quadrature.[1]

For example, the linear equation of the second order

$$y'' - 2\frac{y'}{x} + 2\frac{y}{x^2} = 0$$

is equivalent to Riccati's equation

$$z' + z^2 - \frac{2}{x} z + \frac{2}{x^2} = 0,$$

---

[1]The same result is obtained by observing that the Wronskian $W$ formed from $v$ and any other solution $u$ is given by (27). But, for known $W$ and $v$ the equation $W = vu' - v'u$ represents a linear first-order equation for $u$ that can be solved by quadratures.

where $z = y'/y$. The original equation has $y = x$ as a particular solution; hence, it may be reduced to the equation of the first order

$$v''x = 0,$$

where $v = y/x$. That is, $v = ax + b$. Hence, the general integral of the original equation is given by

$$y = ax^2 + bx.$$

We mention that exactly the same method can be used to reduce a linear differential equation of the $n$th order to one of the $(n - 1)$-st order, when one solution of the first equation is known.

## Exercises 6.3b

1. Prove that the functions $\sin x$, $\sin 2x$, $\sin 3x$, . . . are linearly independent in the interval $0 \leq x \leq \pi$. Hint: Any two of these functions are orthogonal over the interval; namely, if $m \neq n$

$$\int_0^\pi \sin mx \sin nx \, dx = 0$$

(cf. Volume I, p. 274).

2. Prove that if $a_1, \ldots, a_k$ are different numbers and $P_1(x), \ldots, P_k(x)$ are arbitrary polynomials (not identically zero), then the functions

$$\phi_1(x) = P_1(x)e^{a_1 x}, \ldots, \phi_k(x) = P_k(x)e^{a_k x}$$

are linearly independent.

3. Show that the so-called Bernoulli equation (cf. Exercise 7 in Section 6.2)

$$y' + a(x)y = b(x)y^n \qquad\qquad (n \neq 1)$$

reduces to a linear differential equation for the new unknown function $z = y^{1-n}$. Use this to solve the equations

(a) $xy' + y = y^2 \log x$

(b) $xy^2(xy' + y) = a^2$

(c) $(1 - x^2)y' - xy = axy^2$.

4. Show that Riccati's differential equation

$$y' = P(x)y^2 + Q(x)y + R(x) = 0$$

can be transformed into a linear differential equation if we know a particular integral $y_1 = y_1(x)$. [Introduce the new unknown function $u = 1/(y - y_1)$].
    Use this to solve the equation

$$y' - x^2y^2 + x^4 - 1 = 0$$

that possesses the particular integral $y_1 = x$.

5. Find the integrals that are common to the two differential equations

    (a) $y' = y^2 + 2x - x^4$

    (b) $y' = -y^2 - y + 2x + x^2 + x^4$

6. Integrate the differential equation

$$y' = y^2 + 2x - x^4$$

in terms of definite integrals, using the particular integral found in Exercise 5. Draw a rough graph of the integral curves of the equation throughout the $x, y$-plane.

7. Let $y_1, y_2, y_3, y_4$ be four solutions of Riccati's equation (cf. Exercise 4). Prove that the expression

$$\frac{\dfrac{(y_1 - y_3)}{(y_1 - y_4)}}{\dfrac{(y_2 - y_3)}{(y_2 - y_4)}}$$

is a constant.

8. Show that if two solutions, $y_1(x)$ and $y_2(x)$, of Riccati's equation are known, then the general solution is given by

$$y - y_1 = c(y - y_2) \exp \left[ \int P(y_2 - y_1) \, dx \right],$$

where $c$ is an arbitrary constant.

   Hence find the general solution of

$$y' - y \tan x = y^2 \cos x - \frac{1}{\cos x},$$

which has solutions of the form $a \cos^n x$.

9. Prove that the equations

    (a) $(1 - x)\, y'' + xy' - y = 0$

    (b) $2x(2x - 1)y'' - (4x^2 + 1)y' + y(2x + 1) = 0$

have a common solution. Find it and hence, integrate both equations completely.

10. The tangent at a point $P$ of a curve cuts the axis of $y$ at a point $T$ below the origin $O$ and the curve is such that $OP = n \cdot OT$. Prove that its polar equation is of the form

$$r = a \frac{(1 + \sin \theta)^n}{\cos^{n+1} \theta}.$$

### c. The Nonhomogeneous Differential Equation. Method of Variation of Parameters

To solve the nonhomogeneous differential equation

(28a) $$L[u] = a_0 u^{(n)} + \cdots + a_n u = \phi(x)$$

in general, it is sufficient, by what we have said on p. 684, to find a single solution. This may be done as follows: By proper choice of the constants $c_1, c_2, \ldots, c_n$, we first determine a solution of the homogeneous equation $L[u] = 0$ in such a way that the equations

(28b)    $$u(\xi) = 0, \; u'(\xi) = 0, \ldots, u^{(n-2)}(\xi) = 0, \; u^{(n-1)}(\xi) = 1$$

are satisfied. This solution, which depends on the parameter $\xi$, we denote by $u(x, \xi)$. The function $u(x, \xi)$ is a continuous function of $\xi$ for fixed values of $x$, and so are its first $n$ derivatives with respect to $x$. As an example, for the differential equation $u'' + k^2 u = 0$ the solution $u(x, \xi)$ that fulfills the conditions (28b) has the form $[\sin k(x - \xi)]/k$.

We now assert that the formula

(28c)    $$v(x) = \int_0^x \phi(\xi) \, u(x, \xi) \, d\xi$$

gives a solution of $L[v] = \phi$ that, together with its first $n - 1$ derivatives, vanishes at the point $x = 0$. To verify this statement,[1] we differentiate the function $v(x)$ repeatedly with respect to $x$ by the rule for the differentiation of an integral with respect to a parameter [*cf.* (41) p. 77] and recall the relations following from (28b):

$$u(x, x) = 0, \; u'(x, x) = 0, \ldots, u^{(n-2)}(x, x) = 0, \; u^{(n-1)}(x, x) = 1$$

where, for example, $u'(x, x) = \partial u(x, \xi)/\partial x$ for $\xi = x$.

We thus obtain

$$v'(x) = \phi(\xi) \, u(x, \xi)|_{\xi=x} + \int_0^x \phi(\xi) \, u'(x, \xi) \, d\xi = \int_0^x \phi(\xi) \, u'(x, \xi) \, d\xi,$$

$$v''(x) = \phi(\xi) u'(x, \xi)|_{\xi=x} + \int_0^x \phi(\xi) \, u''(x, \xi) \, d\xi = \int_0^x \phi(\xi) \, u''(x, \xi) \, d\xi,$$

$$\cdot \quad \cdot \quad \cdot \quad \cdot \quad \cdot \quad \cdot \quad \cdot \quad \cdot \quad \cdot \quad \cdot \quad \cdot \quad \cdot$$

$$v^{(n-1)}(x) = \phi(\xi) \, u^{(n-2)}(x, \xi)|_{\xi=x} + \int_0^x \phi(\xi) \, u^{(n-1)}(x, \xi) \, d\xi$$

---

[1] It is possible to give a physical interpretation for this process. If $x = t$ denotes the time and $u$ the coordinate of a point moving on a straight line subject to a force $\phi(x)$, the effect of this force may be thought of as arising from the superposition of the small effects of small impulses. The above solution $u(x, \xi)$ then corresponds to an impulse of amount 1 at time $\xi$, and our solution gives the effect of impulses of amount $\phi(\xi)$ during the time between 0 and $x$.

$$= \int_0^x \phi(\xi) \, u^{(n-1)}(x, \xi) \, d\xi,$$

$$v^{(n)}(x) = \phi(\xi) \, u^{(n-1)}(x, \xi)|_{\xi=x} + \int_0^x \phi(\xi) \, u^{(n)}(x, \xi) \, d\xi$$

$$= \phi(x) + \int_0^x \phi(\xi) \, u^{(n)}(x, \xi) \, d\xi.$$

Since $L[u(x, \xi)] = 0$, this establishes the equation $L[v] = \phi(x)$ and shows that the initial conditions $v(0) = 0, v'(0) = 0, \ldots, v^{(n-1)}(0) = 0$ are satisfied.

The same solution can also be obtained by the following apparently different method, which generalizes the procedure used on p. 681 for a first-order equation. We seek a solution $u$ of the nonhomogeneous equation in the form of a linear combination of independent solutions $u_i$ of the homogeneous equation

(28d) 
$$u = \sum \gamma_i(x) \, u_i(x),$$

where now we allow the coefficients $\gamma_i$ to be functions of $x$. On these functions, we impose the following conditions:

$$\gamma_1' u_1 + \gamma_2' u_2 + \cdots + \gamma_n' u_n = 0$$
$$\gamma_1' u_1' + \gamma_2' u_2' + \cdots + \gamma_n' u_n' = 0$$
$$\cdots \cdots \cdots \cdots \cdots$$
$$\gamma_1' u_1^{(n-2)} + \gamma_2' u_2^{(n-2)} + \cdots + \gamma_n' u_n^{(n-2)} = 0.$$

From these it follows that the derivatives of $u$ are given by the following formulae:

$$u' = \sum \gamma_i u_i'$$
$$u'' = \sum \gamma_i u_i''$$
$$\cdots \cdots \cdots \cdots$$
$$u^{(n-1)} = \sum \gamma_i u_i^{(n-1)}$$
$$u^{(n)} = \sum \gamma_i' u_i^{(n-1)} + \sum \gamma_i u_i^{(n)}.$$

Substituting these expressions in the differential equation and remembering that $L[u] = \phi$, we have

$$\sum \gamma_i' u_i^{(n-1)} = \phi(x).$$

For the coefficients $\gamma_i'$ we obtain a linear system of equations, with determinant $W$, the Wronskian of the system of fundamental solutions $u_i$, which therefore does not vanish. Thus, the coefficients $\gamma_i'$ are determined, and hence, by quadratures, so are the coefficients $\gamma_i$. As the whole argument can be reversed, a solution of the equation has actually been found, which, in fact, is the general solution, by virtue of the integration constants concealed in the coefficients $\gamma_i$.

We leave it to the reader to show that the two methods are really identical, by expressing $u(x, \xi)$, the solution of the homogeneous equation defined above, in the form

$$u(x, \xi) = \sum a_i(\xi) u_i(x).$$

The latter method is known as *variation of parameters,* because it exhibits the solution as a linear combination of functions with variable coefficients, whereas in the case of the homogeneous equation these coefficients were constants.

*Example 1*

We consider the equation

$$u'' - 2\frac{u'}{x} + 2\frac{u}{x^2} = xe^x.$$

By p. 690, a system of independent solutions of the corresponding homogeneous equation

$$u'' - 2\frac{u'}{x} + 2\frac{u}{x^2} = 0$$

is given by $u_1 = x$, $u_2 = x^2$. Hence, if we seek solutions of the form

$$u = \gamma_1 x + \gamma_2 x^2,$$

we have the conditions

$$\gamma_1' x + \gamma_2' x^2 = 0,$$
$$\gamma_1' + 2\gamma_2' x = xe^x$$

for $\gamma_1$ and $\gamma_2$. That is,

$$\gamma_1' = -xe^x, \qquad \gamma_2' = e^x.$$

Hence, the general solution of the original nonhomogeneous equation is

$$u = xe^x + c_1 x + c_2 x^2.$$

## Example 2

As an application we give a method for dealing with forced vibrations, for which the right side of the differential equation need no longer be periodic, as in the cases considered in Volume I, Chapter 9, p. 641, but may instead be an arbitrary continuous function $f(t)$. For the sake of simplicity we restrict ourselves to the frictionless case and take $m = 1$ (or, what amounts to the same thing, divide through by $m$). Accordingly, we write the differential equation in the form

(28e) $$\ddot{x}(t) + k^2 x(t) = \phi(t),$$

where the quantity $k^2$ and $\phi$ are what we called $k$ and $f$ before.

According to (28c), the function

$$F(t) = \frac{1}{k} \int_0^t \phi(\lambda) \sin k(t - \lambda)\, d\lambda$$

is a solution of the differential equation (28e) and satisfies the initial conditions

$$F(0) = 0, \qquad F'(0) = 0.$$

For the general solution of the differential equation we thus obtain, just as before, the function

$$x(t) = \frac{1}{k} \int_0^t \phi(\lambda) \sin k(t - \lambda)\, d\lambda + c_1 \sin kt + c_2 \cos kt,$$

where $c_1$ and $c_2$ are arbitrary constants of integration.

In particular, if the function on the right side of the differential equation is a periodic function of the form $\sin \omega t$ or $\cos \omega t$, a simple calculation again yields the results of Volume I, Chapter 9, p. 642.

### Exercises 6.3c

1. Integrate the following equations:
   (a) $y''' - y = 0$.
   (b) $y''' - 4y'' + 5y' - 2y = 0$.
   (c) $y''' - 3y'' + 3y' - y = 0$
   (d) $y'''' - 3y'' + 2y = 0$
   (e) $x^2 y'' + xy' - y = 0$.

2. Prove that the linear homogeneous equation
$$L(y) = y^{(n)} + c_1 y^{(n-1)} + \cdots + c_{n-1} y' + c_n = 0$$
with *constant* coefficients $c$ has a system of fundamental solutions of the form $x^{\mu} e^{a_k x}$, where the $a_k$' s are the roots of the polynomial
$$f(z) = z^n + c_1 z^{n-1} + \cdots + c_n.$$

3. Let
$$a_0 y + a_1 y' + \cdots + a_n y^{(n)} = P(x)$$
be a linear nonhomogeneous differential equation of the $n$th order with constant coefficients, and let $P(x)$ be a polynomial. Let $a_0 \neq 0$ and consider the formal identity
$$\frac{1}{a_0 + a_1 t + \cdots + a_n t^n} = b_0 + b_1 t + b_2 t^2 + \cdots.$$
Prove that
$$y = b_0 P(x) + b_1 P'(x) + b_2 P''(x) + \cdots$$
is a particular integral of the differential equation.
   If $a_0 = 0$, but $a_1 \neq 0$, then the expansion
$$\frac{1}{a_1 t + a_2 t^2 + \cdots + a_n t^n} = b t^{-1} + b_0 + b_1 t + b_2 t^2 + \cdots$$
is possible. Prove that now
$$y = b \int P(x)\, dx + b_0 P(x) + b_1 P'(x) + b_2 P''(x) + \cdots$$
is a particular integral of the differential equation.

4. Apply the method of Exercise 3 to find particular integrals of
   (a) $y'' + y = 3x^2 - 5x$
   (b) $y'' + y' = (1 + x)^2$

5. A particular integral of the equation
$$a_0 y + a_1 y' + \cdots + a_n y^{(n)} = e^{kx} P(x),$$
where $k$, $a_0$, $a_1$, ... are real constants and $P(x)$ is a polynomial, can be found by introducing a new unknown function $z = z(x)$ given by
$$y = z e^{kx}$$
and applying the method of Exercise 3 to the equation in $z$.
   Use this method to find particular integrals of
   (a) $y'' + 4y' + 3y = 3e^x$
   (b) $y'' - 2y' + y = xe^x$.

6. Integrate the equation
$$y'' - 5y' + 6y = e^x(x^2 - 3)$$
completely.

7. (a) If $u$, $v$ are two independent solutions of the equation

$$f(x)y''' - f'(x)y'' + \phi(x)y' + \lambda(x)y = 0,$$

prove that the complete solution is $Au + Bv + Cw$, where

$$w = u \int \frac{vf(x)\ dx}{(uv' - u'v)^2} - v \int \frac{uf(x)dx}{(uv' - u'v)^2}.$$

and $A$, $B$, $C$ are arbitrary constants.

(b) Solve the equation

$$x^2(x^2 + 5)y''' - x(7x^2 + 25)y'' + (22x^2 + 40)y' - 30xy = 0$$

that has solutions of the form $x^n$.

## 6.4   General Differential Equations of the First Order

### a. *Geometrical Interpretation*

We begin by considering a differential equation of the first order

$$(29) \qquad\qquad F(x, y, y') = 0,$$

where we assume that the function $F$ is a continuously differentiable function of its three arguments $x, y, y'$. Geometrically at a point in the plane with rectangular coordinates $(x, y)$, the equation is a condition on the direction of the tangent to any curve $y(x)$ passing through this point that satisfies the differential equation. We assume that in a certain region $R$ of a plane, say in a rectangle, the differential equation $F(x, y, y') = 0$ can be solved uniquely for $y'$ and, thus, can be expressed in the form

$$(30) \qquad\qquad y' = f(x, y),$$

where the function $f(x, y)$ is continuously differentiable in $x$ and $y$. Then to each point $(x, y)$ of $R$ equation (30) assigns a *direction of advance*. The differential equation is therefore represented geometrically by a *field of directions;* and the problem of solving the differential equation geometrically consists in the finding of those curves that belong to this field of directions, that is, those whose tangents at every point have the direction preassigned by the equation $y' = f(x, y)$. We call these curves the *integral curves of the differential equation*.

It is now intuitively plausible that through each point $(x, y)$ of $R$ there passes a single integral curve of the differential equation $y' = f(x, y)$. These facts are stated more precisely in the following fundamental existence theorem:

*If in the differential equation* $y' = f(x, y)$ *the function f is continuous and has a continuous derivative with respect to y in a region R, then through each point* $(x_0, y_0)$ *of R there passes one, and only one, integral curve; that is, there exists in a neighborhood of* $x_0$ *one, and only one, solution* $y(x)$ *of the differential equation for which* $y(x_0) = y_0$.

We shall return to the proof of this theorem on p. 702 Here we confine ourselves to the consideration of some examples.

For the differential equation

$$(31a) \qquad\qquad y' = -\frac{x}{y},$$

that we consider in the region $y < 0$, say, the field at a point $(x, y)$ is readily seen to have a direction perpendicular to the vector from the origin to the point $(x, y)$. From this we infer by geometry that the circular arcs about the origin must be the integral curves of the differential equation. This result is very easily verified analytically, for by the method of separation of variables (p. 679), it follows that

$$x^2 + y^2 = \text{constant} = c,$$

which shows that these circles are the solutions of the differential equation.

At each point, the field of directions of the differential equation

$$(31b) \qquad\qquad y' = \frac{y}{x}$$

obviously has the direction of the line joining that point to the origin. Thus, the lines through the origin belong to this field of directions and are therefore integral curves. As a matter of fact, we see at once that the function $y = cx$ satisfies the differential equation for any arbitrary constant $c$.[1]

In the same way, we can verify analytically that the differential equation

$$y' = \frac{x}{y} \qquad\qquad\qquad (y \neq 0)$$

and

$$y' = -\frac{y}{x} \qquad\qquad\qquad (x \neq 0)$$

---

[1] At the origin the field of directions is no longer uniquely defined; this is connected with the fact that an infinite number of integral curves pass through this *singular point* of the differential equation.

are satisfied by the respective families of hyperbolas

$$y^2 = c + x^2$$

$$y = \frac{c}{x},$$

where $c$ is the parameter specifying the particular curve of the family.

Our fundamental theorem shows that, in general, differential equations of the first order are satisfied by a *one-parameter family* of functions. Functions of $x$ in such a family depend not only on $x$ but also on a parameter $c$, for example, on $c = y_0 = y(0)$; as we say, the solutions depend on an arbitrary *constant of integration*. Ordinary integration of a function $f(x)$ is merely the special case of the solution of the differential equation in which $f(x, y)$ does not involve $y$. The direction of the field at a point is then determined by the $x$-coordinate alone, and we see at once that the integral curves are obtained from one another by translation in the direction of the $y$-axis. Analytically, this corresponds to the familiar fact that the indefinite integral $y$, that is, the solution of the differential equation $y' = f(x)$, involves an arbitrary additive constant $c$.

The geometrical interpretation of the differential equation suggests an approximate graphical *construction* of the integral curves, in much the same way as in the special case of the indefinite integration of a function of $x$ (Volume I, p. 483). We have only to think of the integral curve as replaced by a polygon in which each side has the direction assigned by the field of directions for its initial point (or for any other one of its points). Such a polygon can be constructed by starting from an arbitrary point in R. The smaller we take the length of the sides of the polygon, the greater the accuracy with which the sides of the polygon will agree with the field of directions of the differential equation, not only at their initial points but throughout their whole length. Without going into the proof, we here state the fact that, by successively diminishing the length of the sides, a polygon constructed in this way may actually be made to approach closer and closer to the integral curve through the initial point.

### b. The Differential Equation of a Family of Curves. Singular Solutions. Orthogonal Trajectories

The existence theorem shows that every differential equation has a family of integral curves. This suggests that we ask the reverse question. Does every one-parameter family of curves $\phi(x, y, c) = 0$ or $y = g(c, x)$ have a corresponding differential equation

$$F(x, y, y') = 0$$

that is satisfied by all the curves of the family? If so, how can we find this differential equation? Here the essential point is that $c$, the parameter of the family of curves, does not occur in the differential equation, so that the differential equation is in a sense a representation of the family of curves *not* involving a parameter. In fact, it is easy to find such a differential equation. Differentiating with respect to $x$, in

(32a)                         $$\phi(x, y, c) = 0$$

we have

(32b)                         $$\phi_x + \phi_y y' = 0.$$

If we eliminate the parameter $c$ between this equation and the equation $\phi = 0$, the result is the desired differential equation. This elimination is always possible for a region of the plane in which the equation $\phi = 0$ can be solved for the parameter $c$ in terms of $x$ and $y$. We then have only to substitute the expression $c = c(x, y)$ thus found in the expressions for $\phi_x$ and $\phi_y$ in order to obtain a differential equation for the family of curves.

As a first example, we consider the family of concentric circles $x^2 + y^2 - c^2 = 0$, from which, by differentiation with respect to $x$, we obtain the differential equation

(32c)                         $$x + yy' = 0,$$

in agreement with (31a), p. 698.

Another example is the family $(x - c)^2 + y^2 = 1$ of circles with unit radius and center on the $x$-axis. By differentiation with respect to $x$, we obtain

$$(x - c) + yy' = 0,$$

and on eliminating $c$, we obtain the differential equation

$$y^2(1 + y'^2) = 1.$$

The family $y = (x - c)^2$ of parabolas touching the $x$-axis likewise leads by way of the equation $y' = 2(x - c)$ to the required differential equation

$$y'^2 = 4y.$$

In the last two examples we see that the corresponding differential equations are satisfied not only by the curves of the family but, in the first case, also by the lines $y = 1$ and $y = -1$ and, in the second case, also by the $x$-axis, $y = 0$. These facts, which can at once be verified analytically, also follow without calculation from the geometrical meaning of the differential equation. For these lines are the envelopes of the corresponding families of curves, and since the envelopes at each point touch a curve of the family, they must at that point have the direction prescribed by the field of directions. Therefore, every envelope of a family of integral curves must itself satisfy the differential equation. Solutions of the differential equation that are found by forming the envelope of a one-parameter family of integral curves are called *singular solutions*.

Let $R$ be a region that is simply covered by a one-parameter family of curves $\Phi(x, y) = c = $ constant. If to each point $P$ of $R$ we assign the direction of the tangent of the curve passing through $P$, we obtain a field of directions defined by the differential equation $y' = -\Phi_x/\Phi_y$ [see (32b)]. If, on the other hand, to each point $P$ we assign the direction of the normal to the curve passing through it, the resulting field of directions is defined by the differential equation

$$y' = \frac{\Phi_y}{\Phi_x}.$$

The solutions of this differential equation are called the *orthogonal trajectories* of the original family of curves $\Phi(x, y) = c$. The curves $\Phi = c$ (the level lines of the function $\Phi$) and their orthogonal trajectories intersect everywhere at right angles. Hence, if a family of curves is given by the differential equation $y' = f(x, y)$, we can find the differential equation of the orthogonal trajectories without integrating the given differential equation, for the equation of the orthogonal trajectories is

$$y' = -\frac{1}{f(x, y)}.$$

In the example (31a) discussed above, from the differential equation satisfied by the circles $x^2 + y^2 = c$ we find that the differential equation of the orthogonal trajectories is $y' = y/x$. The orthogonal trajectories are therefore straight lines through the origin [see (31b)].

If $p > 0$, the family of confocal parabolas (cf. Chapter 3, p. 234) $y^2 - 2p(x + p/2) = 0$ satisfies the differential equation

$$y' = \frac{1}{y}\, (-x + \sqrt{x^2 + y^2}).$$

Hence, the differential equation of the orthogonal trajectories of this family is

$$y' = \frac{-1}{(-x + \sqrt{x^2 + y^2})/y} = \frac{1}{y}\, (- x - \sqrt{x^2 + y^2}).$$

The solutions of this differential equation are the parabolas

$$y^2 - 2p(x + p/2) = 0,$$

where $p < 0$; these are parabolas confocal with one another and with the curves of the first family.

### c. Theorem of the Existence and Uniqueness of the Solution

We now prove the theorem of the existence and uniqueness of the solution of the differential equation $y' = f(x, y)$ that we stated on p. 698. Without loss of generality, we can assume that for the solution $y(x)$ in question the initial condition $f(x_0) = y_0$ reduces to $y(0) = 0$, for we could introduce $y - y_0 = \eta$ and $x - x_0 = \xi$ as new variables and should then obtain a new differential equation, $d\eta/d\xi = f(\xi + x_0, \eta + y_0)$, of the same type, satisfying the desired condition.

In the proof, we may confine ourselves to a sufficiently small neighborhood of the point $x = 0$. If we have proved the existence and uniqueness of the solution for such an interval about the point $x = 0$, we can then prove the existence and uniqueness for a neighborhood of one of its end points, and so on.

Let us then consider a rectangle $|x| \leq a$, $|y| \leq b$ contained in the domain of the function $f(x, y)$. There exist bounds $M$, $M_1$ such that

(32d)        $|f_y(x, y)| \leq M$, $|f(x, y)| \leq M_1$ for $|x| \leq a$, $|y| \leq b$.

Replacing, if necessary, $a$ by a smaller positive value, we can always bring about that

(32e)                        $M_1\, a < b$, $Ma < 1$.

The inequalities (32d) will still be valid in the smaller rectangle. For any solution $y(x)$ of $y' = f(x, y)$ with initial value $y(0) = 0$ we then have the estimate $|y(x)| \leqq b$ for $|x| \leqq a$. For otherwise there would exist values $\xi$ for which $|\xi| \leqq a$, $|y(\xi)| = b$. There would be such a $\xi$ of smallest absolute value. Then the relation

$$b = |y(\xi)| = \left| \int_0^\xi (x, y(x)) \, dx \right| \leqq M_1 |\xi| \leqq M_1 a < b$$

would lead to a contradiction.

We first convince ourselves that there cannot be *more* than one solution of the differential equation satisfying the initial conditions, for if there were two solutions $y_1(x)$ and $y_2(x)$, the difference $d(x) = y_1 - y_2$ would satisfy

$$d'(x) = f(x, y_1(x)) - f(x, y_2(x)).$$

By the mean value theorem, the right side of this equation can be put in the form $(y_1 - y_2) f_y(x, \bar{y}) = d(x) f_y(x, \bar{y})$, where $\bar{y}$ is a value intermediate between $y_1$ and $y_2$. In a neighborhood $|x| \leqq a$ of the origin, $y_1$ and $y_2$ are continuous functions of $x$ that vanish at $x = 0$. Here $b$ is an upper bound of the absolute values of the two functions in this neighborhood, so that $|\bar{y}| \leqq b$ whenever $|x| \leqq a$. Furthermore, $M$ is a bound of $|f_y|$ in the region $|x| \leqq a$, $|y| \leqq b$. Finally, let $D$ be the greatest value of $|d(x)|$ in the interval $|x| \leqq a$ and suppose that this value is assumed at $x = \xi$. Then, for $|x| \leqq a$,

$$|d'(x)| = |d(x) f_y(x, \bar{y})| \leqq DM,$$

and therefore,

$$D = |d(\xi)| = \left| \int_0^\xi d'(x) \, dx \right| \leqq |\xi| DM \leqq aDM.$$

But since $a M < 1$, it follows that $D = 0$. That is, in such an interval $|x| \leqq a$ we have[1] $y_1(x) = y_2(x)$.

By a similar integral estimate we arrive at a proof of the existence for the solution. We construct the solution by a method that has other important applications, in particular, to the numerical solution of differential equations and to the inversion of mappings (see p. 266). This is the process of *iteration or successive approximations*. Here we

---

[1] The root idea of this proof is the fact that for bounded integrands integration gives a quantity that vanishes to the same order as the interval of integration, as that interval tends to zero.

obtain the solution as the limit function of a sequence of approximate solutions $y_0(x)$, $y_1(x)$, $y_2(x)$, . . .. As a first approximation $y_0(x)$, we take $y_0(x) = 0$. Using the differential equation, we take

$$y_1(x) = \int_0^x f(\xi, 0) \, d\xi$$

as the second approximation: from this we obtain the next approximation $y_2(x)$,

$$y_2(x) = \int_0^x f(\xi, y_1(\xi)) \, d\xi,$$

and in general the $(n + 1)$-th approximation is obtained from the $n$-th by the equation

$$(33a) \qquad y_n(x) = \int_0^x f(\xi, y_{n-1}(\xi)) \, d\xi.$$

If in an interval $|x| \leq a$ these approximating functions converge uniformly to a limit function $y(x)$, we can at once perform the passage to the limit under the integral sign and obtain for the limit function the equation

$$(33b) \qquad y(x) = \int_0^x f(\xi, y(\xi)) \, d\xi.$$

From this it follows by differentiation that $y' = f(x, y)$, so that $y$ is actually the required solution.

We prove convergence for a sufficiently small interval $|x| \leq a$ by means of the following estimate. We put $y_{n+1}(x) - y_n(x) = d_n(x)$ and by $D_n$ denote the maximum of $|d_n(x)|$ in the interval $|x| \leq a$.

From the equation

$$d_n'(x) = y_{n+1}' - y_n' = f(x, y_n) - f(x, y_{n-1})$$

the mean value theorem gives

$$(33c) \qquad d_n'(x) = d_{n-1}(x) f_y(x, \bar{y}_{n-1}(x)),$$

where $\bar{y}_{n-1}$ is a value intermediate between $y_n$ and $y_{n-1}$. Let the inequalities $|f_y(x, y)| = M$, $|f(x, y)| \leq M_1$ hold in the rectangular region $|x| \leq a$, $|y| \leq b$. If we assume that for the function $y_n$ the relation $|y_n| \leq b$ holds in the interval $|x| \leq a$, then, by the definition of $y_{n+1}$, we have

$$|y_{n+1}(x)| = \left| \int_0^x f(\xi, \ y_n(\xi)) \ d\xi \right| \leq |x| \ M_1 \leq aM_1.$$

We shall therefore choose the bound $a$ for $x$ so small that $aM_1 \leq b$. Then, in the interval $|x| \leq a$, we shall certainly have $|y_{n+1}(x)| \leq b$. Since for $y_0(x) = 0$ it is obvious that $|y_0| \leq b$, it follows by induction that in the interval $|x| \leq a$ we have $|y_n(x)| \leq b$ for every $n$. Hence, in (33c) we may use the estimate $|f_y| \leq M$ and integrate to obtain

$$|d_n(x)| = \left| \int_0^x d_n'(\xi) \ d\xi \right| \leq \left| \int_0^x M \ |d_{n-1}(\xi)| \ d\xi \right|.$$

Thus, we may bound the maximum $D_n$ of $|d_n(x)|$ in the interval $|x| \leq a$ by

$$D_n \leq aMD_{n-1}.$$

We now take $a$ so small that $aM \leq q < 1$, where $q$ is a fixed proper fraction, say $q = \frac{1}{4}$. Then $D_{n+1} \leq qD_n \leq q^n \ D_0$.

Let us now consider the series

$$d_0(x) + d_1(x) + d_2(x) + \cdots + d_{n-1}(x) + \cdots.$$

The $n$th partial sum of this series is $y_n(x)$. The absolute value of the $n$th term is not greater than the number $D_0 q^{n-1}$ when $|x| \leq a$. Our series is therefore dominated by a convergent geometric series with constant terms. Hence (cf. Volume I, p. 535), it converges uniformly in the interval $|x| \leq a$ to a limit function $y(x)$, and thus, we see that an interval $|x| \leq a$ exists in which the differential equation has a unique solution.

All that now remains to be shown is that this solution can be extended step by step until it reaches the boundary of the (closed bounded) region $R$ in which we assume $f(x, y)$ to be defined. The proof so far shows that if the solution has been extended to a certain point, it can be continued onward over an $x$-interval of length $a$, where $a$, however, depends on the coordinates $(x, y)$ of the end point of the portion already constructed. It might be imagined that in this advance $a$ diminishes from step to step so rapidly that the solution cannot be extended by more than a small amount, no matter how many steps are made. This, as we shall show, is not the case.

Suppose that $R'$ is a closed bounded region interior to $R$. Then we can find a number $b$ so small that for very point $(x_0, y_0)$ in $R'$ the whole square $x_0 - b \leqq x \leqq x_0 + b$, $y_0 - b \leqq y \leqq y_0 + b$ lies in $R$. If by $M$ and $M_1$ we denote the upper bounds of $|f_y(x, y)|$ and $|f(x, y)|$ in the region $R$, then we find that in the preceding proof all the conditions imposed on $a$ are certainly satisfied if we take $a$ to be, say, the smallest of the numbers $b$, $M/2$, and $b/M_1$. This no longer depends on $(x_0, y_0)$; hence, at each step we can advance by an amount $a$ that is a constant. Thus, we can proceed step by step until we reach the boundary of $R'$. Since $R'$ can be chosen as any closed region in $R$, we see that the solution can be extended to the boundary of $R$.[1]

### Exercises    6.4

1. Let

$$f(x, y, c) = 0$$

be a family of plane curves. By eliminating the constant $c$ between this and the equation

$$\frac{\partial f}{\partial x} + \frac{\partial f}{\partial y} y' = 0,$$

we get the differential equation

$$F(x, y, y') = 0$$

of the family of curves (cf. p. 700). Now let $\phi(p)$ be a given function of $p$; a curve $C$ satisfying the differential equation

$$F(x, y, \phi(y')) = 0$$

is called a *trajectory* of the family of curves $f(x, y, c) = 0$. The second and third equations show that

$$y' = \phi(Y')$$

is the relation between the slope $Y'$ of $C$ at any given point, and the slope

---

[1]It is essential in this theorem that $R$ be a *closed and bounded* region and not, for example, the whole $x, y$-plane. This is shown by the differential equation

$$y' = 1 + y^2$$

for which $f(x, y)$ is defined and continuously differentiable for all $x, y$. The unique solution of this equation with initial condition $y = 0$ for $x = 0$ is the function $y = \tan x$ for $|x| < \pi/2$. The solution ceases to exist at $x = \pm \pi/2$, in spite of the fact that $f(x, y)$ is regular for all $x$ and $y$. In agreement with the general theorem proved, the graph of the solution leaves any prescribed bounded and closed subset of $R$, for example, any rectangle $|x| \leqq a$, $|y| \leqq b$, before ceasing to exist. The function $y = \tan x$ either exists in the whole interval $|x| \leqq a$ or exists and becomes larger than $b$ in absolute value in some subinterval.

$y'$ of the curve $f(x, y, c) = 0$ passing through this point. The most important case is $\phi(p) = -1/p$, leading to the equation

$$F\left(x, y, -\frac{1}{y'}\right) = 0,$$

which is the differential equation of the *orthogonal trajectories* of the family of curves (cf. p. 701).

Use this method to find the orthogonal trajectories of the following families of curves:

(a) $x^2 + y^2 + cy - 1 = 0$

(b) $y = cx^2$

(c) $\dfrac{x^2}{a^2 + c} + \dfrac{y^2}{b^2 + c} = 1, (a > b > 0, -b^2 < c < \infty)$

(d) $y = \cos x + c$

(e) $(x - c)^2 + y^2 = a^2$.

In each case draw the graphs of the two orthogonal families of curves.

2. For the family of lines $y = cx$, find the two families of trajectories in which (a) the slope of the trajectory is twice as large as the slope of the line; (b) the slope of the trajectory is equal and of opposite sign to the slope of the line.

3. Differential equations of the type

$$y = xp + \psi(p), \quad p = y'$$

were first investigated by Clairaut. Differentiating, we get

$$[x + \psi'(p)]\frac{dp}{dx} = 0,$$

which gives $p = c = \text{constant}$, so that

$$y = xc + \psi(c)$$

is the general integral of the differential equation; it represents a family of straight lines. Another solution is

$$x = -\psi'(p),$$

which together with

$$y = -p\psi'(p) + \psi(p)$$

gives a parametric representation of the so-called *singular integral*. Note that the curve given by the last two equations is the envelope of the family of lines.

Use this method to find the singular solution of the equations

(a) $y = xp - \dfrac{p^2}{4}$

(b) $y = xp + e^p$.

4. Find the differential equation of the tangents to the catenary

$$y = a \cosh \frac{x}{a}.$$

5. Lagrange investigated the most general differential equation linear in both $x$ and $y$, namely,

$$y = x\phi(p) + \psi(p).$$

Differentiating, we get

$$p = \phi(p) + [x\phi'(p) + \psi'(p)]\frac{dp}{dx}$$

which is equivalent to the linear differential equation

$$\frac{dx}{dp} + \frac{\phi'(p)}{\phi(p) - p}x + \frac{\psi'(p)}{\phi(p) - p} = 0,$$

provided $\phi(p) - p \neq 0$ and $p$ is not constant. Integrating and using the first equation, we get a parametric representation of the general integral. From the second equation we see that the equations $\phi(p) - p = 0$, $p =$ constant lead to a certain number of singular solutions representing straight lines.

The solutions can be interpreted geometrically as follows: Consider the Clairaut equation

$$y = xp + [\psi(\phi^{-1}(p)],$$

where $\phi^{-1}(p)$ is the inverse function of $\phi(p)$, that is, $\phi^{-1}(\phi(p)) \equiv p$. From this we see that the solutions of the differential equation are a family of trajectories of the family of straight lines

$$y = xc + \psi[\phi^{-1}(c)]$$

or

$$y = x\phi(c) + \psi(c) \qquad\qquad (c = \text{constant}).$$

Thus, for example,

$$y = -\frac{x}{p} + \psi(p)$$

is the differential equation of the involutes (orthogonal trajectories of the tangents) of the curve that represents the singular integral of the Clairaut equation

$$y = xp + \psi\left(-\frac{1}{p}\right).$$

Use this method to integrate the equation

$$y = x(p + a) - \frac{1}{4}(p + a)^2.$$

6. Express, when possible, the integrals of the following differential equations by elementary functions:

(a) $\left[\dfrac{dy}{dx}\right]^2 = 1 - y^2$

(c) $\left[\dfrac{dy}{dx}\right]^2 = \dfrac{2a - y}{y}$

(b) $\left[\dfrac{dy}{dx}\right]^2 = \dfrac{1}{1 - y^2}$

(d) $\left[\dfrac{dy}{dx}\right]^2 = \dfrac{1 - y^2}{1 + y^2}.$

In each case, draw a graph of the family of integral curves, and detect the singular solutions if any, from the figures.

7. Integrate the homogeneous equation

$$\left[xy' - y\right]^2 = \left[x^2 - y^2\right]\left[\arcsin \dfrac{y}{x}\right]^2$$

and find the singular solutions.

8. As mentioned in Exercise 3, a curve is the envelope of its tangents, hence, it is the singular integral of the Clairaut equation satisfied by its tangent lines. With this in mind, ascertain what kind of curve satisfies each of the following properties and give the corresponding Clairaut equation:
   (a) The sum of the $x$- and $y$-intercepts of a tangent line is constant.
   (b) The length of the segment intercepted on a tangent by the axes is constant.
   (c) The area bounded by the tangent line and the axes is constant.

## 6.5. Systems of Differential Equations and Differential Equations of Higher Order

The above arguments extend to systems of differential equations of the first order with as many unknown functions of $x$ as there are equations. As an example of sufficient generality, we shall consider here the system of two differential equations for two functions $y(x)$ and $z(x)$,

$$y' = f(x, y, z),$$
$$z' = g(x, y, z),$$

where the functions $f$ and $g$ are continuously differentiable. This system of differential equations can be interpreted by a field of directions in $x$, $y$, $z$-space. To the point $(x, y, z)$ of space a direction is assigned whose direction cosines are in the proportion $dx : dy : dz = 1 : f : g$. The problem of integrating the differential equation again amounts geometrically to finding curves in space that belong to this field of directions. As in the case of a single differential equation, we again have the fundamental theorem that through every point $(x_0, y_0, z_0)$ of a region $R$ in which the given functions $f$ and $g$ are continuously differentiable, there passes one, and only one, integral curve of the system of differential equations.[1] The region $R$ is covered by a two-parameter family of curves in space. These give the solutions of the system of differential equations as two functions $y(x)$ and $z(x)$ that both depend on the independent variable $x$ and also on two arbitrary parameters $c_1$ and $c_2$, the constants of integration.

Systems of differential equations of the first order are particularly important because differential equations of higher order, that is, differential equations in which derivatives higher than the first occur, can always be reduced to such systems.

For example, the differential equation of the second order

$$y'' = h(x, y, y')$$

can be written as a system of two differential equations of the first order. We have only to take the first derivative of $y$ with respect to $x$ as a new unknown function $z$ and then write down the system of differential equations

$$y' = z,$$

$$z' = h(x, y, z).$$

This is exactly equivalent to the given differential equation of the second order, in the sense that every solution of the one problem is at the same time a solution of the other.

The reader may use this as a starting point for the discussion of the linear differential equation of the second order and thus prove the fundamental existence theorem for linear differential equations used on p. 687.

## Exercises 6.5

1. Solve the following differential equations:

   (a) $y'y'' = x$

   (b) $2y''' y'' = 1$

---

[1]For $x_0 = y_0 = z_0 = 0$ the proof again can be given by a suitable iteration scheme with the recursion formulae

$$y_{n+1}(x) = \int_0^x f(\xi, y_n(\xi), z_n(\xi))\, d\xi,$$

$$z_{n+1}(x) = \int_0^x g(\xi, y_n(\xi), z_n(\xi))\, d\xi$$

taking the place of the single relation (33a).

(c) $xy'' - y' = 2$

(d) $2xy''' y'' = y''^2 - 2$

2. A differential equation of the form

$$f(y, y', y'') = 0$$

(note that $x$ does not occur explicitly) may be reduced to an equation of the first order as follows: Choose $y$ as the independent variable and $p = y'$ as the unknown function. Then

$$y' = p, \qquad y'' = \frac{dp}{dx} = \frac{dp}{dy}\frac{dy}{dx} = p'p,$$

and the differential equation becomes $f(y, p, pp') = 0$.

Use this method to solve the following equations.

(a) $2yy'' + y'^2 = 0$

(b) $yy'' + y'^2 - 1 = 0$

(c) $y^3 y'' = 1$

(d) $y'' - y'^2 + y^2 y' = 0$

(e) $y^{iv} = (y''')^{1/2}$

(f) $y^{iv} + y'' = 0.$

3. Use the method of Exercise 2 to solve the following problem: At a variable point $M$ of a plane curve $\Gamma$ draw the normal to $\Gamma$; mark on this normal the point $N$ where the normal meets the $x$-axis and $C$, the center of curvature of $\Gamma$ at $M$. Find the curves such that

$$MN \cdot MC = \text{constant} = k.$$

Discuss the various possible cases for $k > 0$ and $k < 0$, and draw the graphs.

4. Find the differential equation of the third order satisfied by all circles

$$x^2 + y^2 + 2ax + 2by + c = 0.$$

## 6.6 Integration by the Method of Undetermined Coefficients

In conclusion, we mention yet another general device that can frequently be applied to the integration of differential equations. This is the method of integration in terms of power series. We assume that in the differential equation

$$y' = f(x, y)$$

the function $f(x, y)$ can be expanded as a power series in the variables $x$ and $y$ and accordingly possesses derivatives of any order with respect to $x$ and $y$. We can then attempt to find the solutions of the differential equation in the form of a power series

$$y = c_0 + c_1 x + c_2 x^2 + \cdots$$

and to determine the coefficients of this power series by means of the differential equation.[1] To do this we proceed by forming the differentiated series

$$y' = c_1 + 2c_2 x + 3c_3 x^2 + \cdots,$$

replacing $y$ in the power series for $f(x, y)$ by its expression as a power series, and then equating the coefficients of like powers of $x$ on the right and on the left (*method of undetermined coefficients*). Then, if $c_0 = c$ is given any arbitrary value, we can attempt to determine the coefficients

$$c_1, c_2, c_3, c_4, \ldots$$

successively.

The following process, however, is often simpler and more elegant. We assume that we are seeking that solution of the differential equation for which $y(0) = 0$, that is, for which the integral curve passes through the origin. Then $c_0 = c = 0$. If we recall that by Taylor's theorem the coefficients of the power series are given by the expressions

$$c_\nu = \frac{1}{\nu!} y^{(\nu)}(0),$$

we can calculate them easily. In the first place, $c_1 = y'(0) = f(0, 0)$. To obtain the second coefficient $c_2$ we differentiate both sides of the differential equation with respect to $x$ and obtain

$$y''(x) = f_x + f_y y'.$$

If we here substitute $x = 0$ and the already known values $y(0) = 0$ and $y'(0) = f(0, 0)$, we obtain the value $y''(0) = 2c_2$. In the same way, we can continue the process and determine the other coefficients $c_3, c_4, \ldots$, one after the other.

It can be shown that this process always gives a solution if the power series for $f(x, y)$ converges absolutely in the interior of a circle about $x = 0$, $y = 0$. We shall not give the proof here.

---

[1] The first few terms of the series then form a polynomial of approximation to the solution.

## Exercises 6.6

1. Obtain the power series expansions to the indicated number of terms for the solution passing through the given point of each of the following differential equations.

   (a) $y' = x + y$, $k$ terms, $(0, a)$

   (b) $y' = \sin(x + y)$, four terms, $(0, \pi/2)$

   (c) $y' = e^{xy}$, four terms, $(0, 0)$

   (d) $y' = \sqrt{x^2 + y^2}$, four terms, $(0, 1)$.

2. Solve the differential equation

$$y'' + \frac{1}{x} y' + y = 0,$$

with $y(0) = 1$, $y'(0) = 0$, by means of a power series. Prove that this function is identical with the Bessel function $J_0(x)$ defined in Section 4.12, Exercise 7, p. 475.

## 6.7 The Potential of Attracting Charges and Laplace's Equation

Differential equations for functions of a single independent variable, such as we have discussed above, are usually called *ordinary* differential equations, to indicate that they involve only "ordinary" derivatives, those of functions of one independent variable. In many branches of analysis and its applications, however, an important part is played by *partial* differential equations for the function of several variables, that is, equations between the variables and the *partial* derivatives of the unknown function. Here we shall touch upon some typical applications that involve Laplace's differential equation.

We have already considered the field of force produced by masses according to Newton's law of attraction, and we have represented it as the gradient of a potential $\Phi$ (cf. Chapter 4, pp. 439 ff.). In this section we shall study the potential in somewhat greater detail.[1]

### a. Potentials of Mass Distributions

As an extension of the cases considered previously, we now take $m$ as a positive or negative mass or charge. Negative masses do not enter into the ordinary Newtonian law of attraction, but in the theory

---

[1] An extensive literature is devoted to this important branch of analysis (see, e.g., O.D. Kellogg *Foundations of Potential Theory* Frederick Ungar Publ. Co.).

of electricity, where mass is replaced by electric charge, we distinguish between positive and negative electricity; there, Coulomb's law of attracting charges has the same form as the law of gravitational attraction of masses. If a charge $m$ is concentrated at a single point of space with coordinates $(\xi, \eta, \zeta)$, we call the expression $m/r$, where

$$r = \sqrt{(x - \xi)^2 + (y - \eta)^2 + (z - \zeta)^2},$$

the *potential*[1] of this mass at the point $(x, y, z)$. By adding up a number of such potentials for different *sources* or *poles* $(\xi_i, \eta_i, \zeta_i)$, we obtain as before (cf. p. 439) the potential of a system of particles or point charges

$$\Phi = \sum_i \frac{m_i}{r_i}.$$

The corresponding fields of force are given by the expression $\mathbf{f} = \gamma \operatorname{grad} \Phi$, where $\gamma$ is a constant independent of the masses and of their positions.

For masses that are not concentrated at single points but are distributed continuously with density $\mu(\xi, \eta, \zeta)$ over a definite portion $R$ of $\xi$, $\eta$, $\zeta$-space, we defined the potential of this mass-distribution to be

(34a) $$\Phi = \iiint \frac{\mu}{r} \, d\xi \, d\eta \, d\zeta.$$

If the masses are distributed over a surface $S$ with surface density $\mu$, the potential of this surface is the surface integral

(34b) $$\iint \frac{\mu(u, v)}{r} \, d\sigma$$

taken over the surface $S$ with surface element $d\sigma$.

For the potential of a mass distributed along a curve, we likewise obtain an expression of the form

(34c) $$\int \frac{\mu(s)}{r} \, ds,$$

---

[1] We could call this a potential of the mass. Any function obtained by adding an arbitrary constant to this could equally well be called a potential of the mass, since it would give the same field of force.

where $s$ is the length of arc on this curve and $\mu\ (s)$ is the linear density of the mass.

For every such potential the level surfaces of $\Phi$ defined by $\Phi =$ constant represent the *equipotential surfaces*.[1]

One example of the potential of a line-distribution is that of a mass of constant linear density $\mu$ distributed along the segment $-l \leqq z \leqq +l$ of the $z$-axis. We consider a point $P$ with coordinates $(x, y)$ in the plane $z = 0$. For brevity we introduce $\rho = \sqrt{x^2 + y^2}$, the distance of the point $P$ from the origin. The potential at $P$ is then

$$\Phi(x, y) = \mu \int_{-l}^{+l} \frac{dz}{\sqrt{\rho^2 + z^2}} + C.$$

Here we have added a constant $C$ to the integral, which does not affect the field of force derived from the potential. The indefinite integral on the right can be evaluated as in Volume I [p. 270 (26)], and we obtain

$$\int \frac{dz}{\sqrt{\rho^2 + z^2}} = \text{ar sinh} \frac{z}{\rho} = \log \frac{z + \sqrt{z^2 + \rho^2}}{\rho},$$

so that the potential in the $x$, $y$-plane is given by

$$\Phi(x, y) = 2\mu \log \frac{l + \sqrt{l^2 + \rho^2}}{\rho} + c.$$

To obtain the potential of a line extending to infinity in both directions, we give the value $-2\mu \log 2l$ to the constant[2] $C$ and thus obtain

$$\Phi(x, y) = 2\mu \log \frac{l + \sqrt{l^2 + \rho^2}}{2l} - 2\mu \log \rho.$$

If we now let the length $l$ increase without limit, that is, if we let the length of the line tend to infinity, the expression $\{l + \sqrt{l^2 + \rho^2}\}/2l$

---

[1]Curves that at every point have the direction of the force vector are called *lines of force*. Since the force here has the direction of the gradient of $\Phi$, the lines of force are curves that everywhere intersect the level surfaces at right angles. We thus see that the families of lines of force corresponding to potentials generated by a single pole or by a finite number of poles run out from these poles as if from a source. In the case of a single pole, for example, the lines of force are simply the straight lines passing through the pole.

[2]We make this choice in order that in the passage to the limit $l \to \infty$ the potential $\Phi$ shall remain finite.

tends to unity, and for the limiting value of $\Phi(x, y)$ we obtain the expression

(35a)  $$\Phi(x, y) = -2\mu \log \rho.$$

We thus see that, apart from the factor $-2\mu$, *the expression*

(35b)  $$\log \rho = \log \sqrt{x^2 + y^2}$$

is the *potential of a straight line perpendicular to the x, y-plane over which a mass is distributed uniformly.* The equipotential surfaces here are the circular cylinders

$$\rho = \sqrt{x^2 + y^2} = \text{constant.}$$

On p. 441 we already calculated the potential of a spherical surface of constant density (i.e., mass per unit area) $\mu$. We found that for a sphere of radius $a$ and center at the origin the potential $\Phi$ at a point $P = (x, y, z)$ is given by

(36a)  $$\Phi = \frac{4\pi a^2}{r} \mu \qquad\qquad (r > a)$$

(36b)  $$\Phi = 4\pi a\mu \qquad\qquad (r < a)$$

where

(36c)  $$r = \sqrt{x^2 + y^2 + z^2}$$

is the distance of $P$ from the origin. The potential of a solid sphere of density $\mu$ can be obtained by decomposing the ball into spherical surfaces of radius $a$ and surface density $\mu$ $da$. Accordingly, the potential of a solid sphere of radius $A$ is obtained from formulae (36a, b) by integrating with respect to $a$ from 0 to $A$. One finds (cf. p. 442) that

(37a)  $$\Phi = \frac{4\pi A^3}{3r} \mu \qquad\qquad (r > A)$$

(37b)  $$\Phi = \left(2\pi A^2 - \frac{2}{3} \pi r^2\right) \mu \qquad\qquad (r < A).$$

The corresponding gravitational force

(37c)  $$\mathbf{f} = \gamma \,\text{grad}\, \phi$$

exerted by the solid sphere on a unit mass at $P$ is directed toward the origin and has magnitude

(37d) $\qquad \dfrac{4\pi A^3}{3r^2}\gamma\mu \qquad$ for $\qquad r > A, \qquad \dfrac{4\pi r}{3}\gamma\mu \qquad$ for $\qquad r < A.$

In addition to the distributions previously considered, potential theory also deals with so-called *double layers,* which we obtain in the following way: We suppose that point charges $M$ and $-M$ are located at the points $(\xi,\,\eta,\,\zeta)$ and $(\xi + h,\,\eta,\,\zeta)$, respectively. The potential of this pair of charges is given by

$$\Phi = \frac{M}{\sqrt{(x - \xi)^2 + (y - \eta)^2 + (z - \zeta)^2}}$$
$$- \frac{M}{\sqrt{(x - \xi - h)^2 + (y - \eta)^2 + (z - \zeta)^2}}.$$

If we let $h$, the distance between the two poles, tend to zero and at the same time let the charge $M$ increase indefinitely in such a way that $M$ is always equal to $-\mu/h$, where $\mu$ is a constant, $\Phi$ tends to the limit

$$\mu\,\frac{\partial}{\partial\xi}\!\left(\frac{1}{r}\right)\!.$$

We call this expression the *potential of a dipole or doublet* with its axis in the $\xi$-direction and with "moment" $\mu$. Physically it represents the potential of a pair of equal and opposite charges lying very close to one another. In the same way, we can express the potential of a dipole in the form

$$\mu\,\frac{\partial}{\partial\nu}\!\left(\frac{1}{r}\right)\!,$$

where $\partial/\partial\nu$ denotes differentiation in an arbitrary direction $\nu$, that of the axis of the dipole.

If we imagine dipoles distributed over a surface $S$ with moment-density $\mu$ and if we assume that at each point the axis of the dipole is normal to the surface, we obtain an expression of the form

$$\iint_S \mu(\xi,\,\eta,\,\zeta)\,\frac{\partial}{\partial\nu}\!\left(\frac{1}{r}\right) d\sigma,$$

where $\partial/\partial v$ denotes differentiation in the direction of the normal to the surface (we can, as before, choose either direction for the normal) and $r$ is the distance of the point $(\xi, \eta, \zeta)$ that ranges over the surface from the point $(x, y, z)$. This potential of a *double layer* can be thought of as arising in the following way: On each side of the surface and at a distance $h$ we construct surfaces, and we give one of these surfaces a surface-density $\mu/2h$ and the other a surface-density $-\mu/2h$. At an external point these two layers together create a potential that tends to the expression above as $h \to 0$.

### b.    *The Differential Equation of the Potential*

We shall assume that in all our expressions the point $(x, y, z)$ considered is at a point in space at which no charge is present, so that the integrands and their derivatives with respect to $x, y, z$ are continuous. By virtue of this hypothesis we can obtain a relation that all the foregoing potentials satisfy, namely, *Laplace's differential equation*

$$(38a) \qquad\qquad \Phi_{xx} + \Phi_{yy} + \Phi_{zz} = 0,$$

which is abbreviated

$$(38b) \qquad\qquad \Delta\Phi = 0.$$

As can easily be verified by simple calculation (p. 59), this equation is satisfied by the expression $1/r$. It therefore holds also for all the other expressions formed from $1/r$ by summation or integration, since we can perform the differentiations with respect to $x, y, z$ under the integral sign.[1] This differential equation is also satisfied by the potential of a double layer, for by virtue of the reversibility of the order of differentiation[2] we find that for the potential of a single dipole the equation

---

[1]Observe that the differentiation under the integral sign is only legitimate as long as $r \neq 0$, that is in regions where no charge is present. Laplace's equation does not have to hold otherwise. For example, within a solid sphere, its potential satisfies, by (37b), the equation

$$\Delta\Phi = \Delta(2\pi A^2 - \frac{2}{3}\,\pi r^2)\mu = -\,4\pi\mu \neq 0.$$

[2]Note that the differentiation $\partial/\partial v$ refers to the variables $(\xi, \eta, \zeta)$ and the expression $\Delta$ to the variables $(x, y, z)$. Incidentally, the function $1/r$, considered as a function of the six variables $(x, y, z; \xi, \eta, \zeta)$, is symmetrical in the two sets of variables and therefore satisfies the Laplace equation

$$\Phi_{\xi\xi} + \Phi_{\eta\eta} + \Phi_{\zeta\zeta} = 0$$

with respect to the variables $(\xi, \eta, \zeta)$ also.

(38c)
$$\Delta \frac{\partial}{\partial \nu}\left(\frac{1}{r}\right) = \frac{\partial}{\partial \nu} \Delta \frac{1}{r} = 0$$

holds.

Laplace's equation is also satisfied by the expression $\log \sqrt{x^2 + y^2}$ obtained for the potential of a vertical line, as we can readily verify (cf. also Chapter 5, p. 569). Since this no longer depends on the variable $z$, it also satisfies the simpler Laplace's equation in two dimensions,

(38d)
$$\Phi_{xx} + \Phi_{yy} = 0.$$

The study of these and related partial differential equations forms one of the most important branches of analysis. We point out that potential theory is not by any means chiefly directed to the search for general solutions of the equation $\Delta \Phi = 0$ but rather to the question of the existence and to the investigation of those solutions that satisfy preassigned conditions. Thus, a central problem of the theory is the boundary value problem, in which we seek a solution $\Phi$ of $\Delta \Phi = 0$ that, together with its derivatives up to the second order, is continuous in a region $R$ and that has preassigned continuous values on the boundary of $R$.

### c. *Uniform Double Layers*

We cannot enter here into a detailed study of *potential functions*,[1] that is, of functions that satisfy Laplace's equation $\Delta u = 0$. In this subject Gauss's theorem and Green's theorem (pp. 601, 608) are among the chief tools employed. It will be sufficient to show by some examples how such investigations are carried out.

We shall first consider the potential of a double layer with constant moment-density $\mu = 1$, that is, an integral of the form

(39)
$$V = \iint_S \frac{\partial}{\partial \nu}\left(\frac{1}{r}\right) d\sigma.$$

This integral has a simple geometrical meaning. Let us assume that each point of the surface carrying the double layer can be "seen" from the point $P$ with coordinates $(x, y, z)$, meaning that it can be joined to this point $P$ by a straight line that meets the surface nowhere else. The surface $S$, together with the rays joining its boundary to the point $P$, forms a conical region $R$ of space. We now state that *the*

---

[1] also called *harmonic functions*.

*potential of the uniform double layer, except perhaps for sign, is equal to the solid angle that the boundary of the surface S subtends at the point P.* By this solid angle we mean the area of that portion of the spherical surface of unit radius about the point $P$ as center that is cut out of the spherical surface by the rays going from $P$ to the boundary of $S$. We give this solid angle the positive sign when the rays pass through the surface $S$ in the same direction as the positive normal v, otherwise we give it the negative sign.

To prove this, we recall that the function $u = 1/r$, when considered not only as a function of $(x, y, z)$ but also as a function of $(\xi, \eta, \zeta)$ still satisfies the Laplace equation

$$\Delta u = u_{\xi\xi} + u_{\eta\eta} + u_{\zeta\zeta} = 0.$$

We fix the point $P$ with coordinates $(x, y, z)$ and denote the rectangular coordinates in the conical region $R$ by $(\xi, \eta, \zeta)$; we use a small sphere of radius $\rho$ about the point $P$ to cut off the vertex from $R$; the residual region we call $R_\rho$. To the function $u = 1/r$, considered as a function of $(\xi, \eta, \zeta)$ in the region $R_\rho$, we now apply Green's theorem (Chapter 5, p. 608) in the form

$$\iiint_{R_\rho} \Delta u \, d\xi \, d\eta \, d\zeta = \iint_{S'} \frac{\partial u}{\partial n} \, d\sigma.$$

Here $S'$ is the boundary surface of $R_\rho$ and $\partial/\partial n$ denotes differentiation in the direction of the outward normal. Since $\Delta u = 0$, the left side is zero.[1] If we have chosen the positive normal direction v on $S$ so as to coincide with the outward normal $n$, the surface integral on the right side consists of three parts: (1) the surface integral

$$\iint_S \frac{\partial}{\partial n}\left(\frac{1}{r}\right) d\sigma = \iint_S \frac{\partial}{\partial v}\left(\frac{1}{r}\right) d\sigma$$

over the surface $S$, which is the expression $V$ considered in (39); (2) an integral over the lateral surface formed by the linear rays; (3) an integral over a portion $\Gamma_\rho$ of the surface of the small sphere of radius $\rho$. The second part is zero, since there the normal direction $n$ is per-

---

[1] From this form of Green's theorem it follows in general that the surface integral

$$\iint \frac{\partial u}{\partial n} \, d\sigma$$

taken over a closed surface must always vanish when the function $u$ satisfies Laplace's equation $\Delta u = 0$ everywhere in the interior of the surface.

pendicular to the radius, and therefore is tangential to the sphere $r = $ constant. For the inner sphere with radius $\rho$ the symbol $\partial/\partial n$ is equivalent to $-\partial/\partial\rho$, since the outward direction of the normal points in the direction of diminishing values of $r$. We thus obtain the equation

$$V - \iint_{\Gamma_\rho} \frac{\partial}{\partial\rho}\left(\frac{1}{\rho}\right)d\sigma = 0$$

or

$$V = -\frac{1}{\rho^2} \iint_{\Gamma_\rho} d\sigma,$$

where on the right we have to integrate over the portion $\Gamma_\rho$ of the small spherical surface that belongs to the boundary of $R_\rho$. We now write the surface element on the sphere with radius $\rho$ in the form $d\sigma = \rho^2\, d\omega$, where $d\omega$ is the surface element on the unit sphere, to obtain

$$V = - \iint d\omega.$$

The integral on the right is to be taken over the portion of the spherical surface of unit radius lying in the cone of rays, and we see at once that the right side has the geometrical meaning stated above; it is the negative of the apparent *angular magnitude* if the normal direction on $S$ is chosen so that it points outward[1] from the conical region $R$. Otherwise, the positive sign is to be taken.

If the surface $S$ is not in the simple position relative to $P$ described above but instead is intersected several times by some of the rays through $P$, we have only to divide the surface into a number of portions of the simpler kind in order to see that the statement still holds good. *The potential of the uniform double layer (of moment 1) on a bounded surface is therefore, except perhaps for sign, equal to the "apparent" magnitude that the boundary has when looked at from the point $(x, y, z)$.*

For a *closed surface* we see by subdividing it into two bounded portions that our expression is equal to zero if the point $P$ is outside and equal to $-4\pi$ if it is inside.

---

[1]The negative sign is explained by the fact that with this choice of the normal direction the negative charge lies on the side of the surface facing the point $P$.

A similar argument shows in the case of two independent variables that the integral

$$\int_C \frac{\partial}{\partial v} (\log r)\, ds$$

along the curve $C$, except possibly for sign, is equal to the angle that
this curve subtends at the point $P$ with the coordinates $(x, y)$.

This result, like the corresponding result in space, can also be
explained geometrically as follows. Let the point $Q$ with the coordinates $(\xi, \eta)$ lie on the curve $C$. Then the derivative of $\log r$ at the point
$Q$ in the direction of the normal to the curve is given by the equation

$$\frac{\partial}{\partial v} (\log r) = \frac{\partial}{\partial r} (\log r) \cos (v, r) = \frac{1}{r} \cos (v, r),$$

where the symbol $(v, r)$ denotes the angle between this normal and the
direction of the radius vector $r$. On the other hand, when written in
polar coordinates $(r, \theta)$, the element of arc $ds$ of the curve has the
form

$$ds = \sqrt{\dot{x}^2 + \dot{y}^2}\, d\theta = \frac{r\sqrt{\dot{x}^2 + \dot{y}^2}}{-\dot{y}x + \dot{x}y} r\, d\theta = \frac{r\, d\theta}{\cos (v, r)}$$

(cf. Volume I, p. 351), so that the integral is transformed as follows:

$$\int \frac{\partial}{\partial v} (\log r)\, ds = \int \frac{1}{r} \cos (v, r) \frac{r\, d\theta}{\cos (v, r)} = \int d\theta.$$

The final integral on the right is the analytical expression for the
angle.

### d.  The Mean Value Theorem

As a second application of Green's transformation, we prove the
following mean value property of potential functions:

*Let $u$ satisfy the differential equation $\Delta u = 0$ in a certain region
R. Then the value of the potential function at the center $P$ of an arbitrary solid sphere of radius r lying completely in the region R is equal
to the mean value of the function $u$ on the surface $S_r$ of the sphere; that
is,*

(40a)
$$u(x, y, z) = \frac{1}{4\pi r^2} \iint_{S_r} \bar{u}\, d\sigma,$$

*where u(x, y, z) is the value at the center P and ū the value on the surface $S_r$ of the sphere of radius r.*

To prove this we proceed as follows: Let $S_\rho$ be a sphere concentric to, and inside of, $S_r$ with radius $0 < \rho \leq r$. Since $\Delta u = 0$ everywhere in the interior of $S_\rho$, by the footnote on p. 720 we have

$$\iint_{S_\rho} \frac{\partial u}{\partial n} \, d\sigma = 0,$$

where $\partial u/\partial n$ is the derivative of $u$ in the direction of the outward normal to $S_\rho$. If $(\xi, \eta, \zeta)$ are running coordinates and if with the point $(x, y, z)$ as pole we introduce spherical coordinates by the equations

$$\xi - x = \rho \cos \phi \sin \theta, \quad \eta - y = \rho \sin \phi \sin \theta, \quad \zeta - z = \rho \cos \theta,$$

the above equation becomes

$$\iint_{S_\rho} \frac{\partial u(\rho, \theta, \phi)}{\partial \rho} \, d\sigma = 0.$$

Since the surface element $d\sigma$ of the sphere $S_\rho$ is equal to $\rho^2 \, d\bar\sigma$, where $d\bar\sigma$ is the element of surface of the sphere $S$ of unit radius (cf. (30e) p. 429), we find that

$$\iint_S \frac{\partial u}{\partial \rho} \, d\bar\sigma = 0,$$

where the region of integration no longer depends on $\rho$. Consequently,

$$\int_0^r d\rho \iint_S \frac{\partial u}{\partial \rho} \, d\bar\sigma = 0,$$

and on interchanging the order of integration and performing the integration with respect to $\rho$, we have

$$\iint_S \{u(r, \theta, \phi) - u(0, \theta, \phi)\} \, d\bar\sigma = 0.$$

Since $u(0, \theta, \phi) = u(x, y, z)$ is independent of $\theta$ and $\phi$,

$$\iint_S u(r, \theta, \phi) \, d\sigma = u(x, y, z) \iint_S d\bar\sigma = 4\pi u(x, y, z).$$

Because

$$\iint_S u(r, \theta, \phi) \, d\bar{\sigma} = \frac{1}{r^2} \iint_{S_r} u(r, \theta, \phi) \, d\sigma,$$

where the integral on the right is to be taken over the surface of $S_r$, the mean value property of $u$ is proved.

In exactly the same way, a function $u$ of two variables that satisfies Laplace's equation $u_{xx} + u_{yy} = 0$ has the *mean value property* expressed by the formula

$$(40b) \qquad\qquad 2\pi r u(x, y) = \int_{S_r} \bar{u} \, ds,$$

where $\bar{u}$ denotes the value of the potential function on a circle $S_r$ with radius $r$ centered at the point $(x, y)$ and $ds$ is the element of arc of this circle.

### e.  *Boundary Value Problem for the Circle. Poisson's Integral*

A boundary value problem that we can treat rather completely is that of Laplace's equation in two independent variables $x$, $y$ for the case of a circular boundary. Within the circular region $x^2 + y^2 \leq R^2$ we introduce polar coordinates $(r, \theta)$. We wish to find a function $u(x, y)$ continuous within the circle and on the boundary, possessing continuous derivatives of the first and second order within the region, satisfying Laplace's equation $\Delta u = 0$, and having prescribed values $u(R, \theta) = f(\theta)$ on the boundary. Here we assume that $f(\theta)$ is a continuous periodic function of $\theta$ with sectionally continuous first derivatives.

The solution of this problem, in terms of polar coordinates, is given by the so-called *Poisson integral:*

$$(41) \qquad u = \frac{R^2 - r^2}{2\pi} \int_0^{2\pi} \frac{f(\alpha)}{R^2 - 2Rr \cos(\theta - \alpha) + r^2} \, d\alpha.$$

To prove this, we begin by constructing special solutions of Laplace's equations in the following way. We transform Laplace's equation to polar coordinates, obtaining

$$\Delta u = \frac{1}{r} (r u_r)_r + \frac{1}{r^2} u_{\theta\theta} = 0,$$

and seek solutions that can be expressed in the "separated" form $u = \phi(r) \, \psi(\theta)$, that is, as a product of a function of $r$ and a function of θ. If we substitute this expression for $u$ in Laplace's equation, the equation becomes

$$r \frac{[r\phi'(r)]_r}{\phi(r)} = -\frac{\psi''(\theta)}{\psi(\theta)}.$$

Since the left side does not involve θ and the right side does not involve $r$, the two sides must each be independent of both variables, that is, must be equal to the same constant $k$. Accordingly, $\psi(\theta)$ satisfies the differential equation $\psi'' + k\psi = 0$.

Since the function $u$ and, hence, $\psi(\theta)$ must be periodic with period $2\pi$, the constant $k$ is equal to $n^2$, where $n$ is an integer. Hence,

$$\psi(\theta) = a \cos n\theta + b \sin n\theta,$$

where $a$ and $b$ are arbitrary constants.

The differential equation for $\phi(r)$,

$$r^2\phi''(r) + r\phi'(r) - n^2\phi(r) = 0,$$

is a linear differential equation, and as we can immediately verify, the functions $r^n$ and $r^{-n}$ are independent solutions. Since the second solution becomes infinite at the origin, while $u$ is to be continuous there, we are left with the first solution $\phi = r^n$ and obtain the separated solutions of Laplace's equation

$$r^n(a \cos n\theta + b \sin n\theta).$$

We can now generate other solutions by linear combination of such solutions according to the principle of superposition (cf. p. 684)

$$\frac{1}{2} a_0 + \sum r^n(a_n \cos n\theta + b_n \sin n\theta).$$

Even an infinite series of this form will be a solution, provided that the series converges uniformly and can be differentiated term by term twice in the interior of the circle.

The Fourier expansion of the prescribed boundary function $f(\theta)$

$$f(\theta) = \frac{1}{2} a_0 + \sum_{n=1}^{\infty} (a_n \cos n\theta + b_n \sin n\theta),$$

regarded as a series in θ, certainly converges absolutely and uniformly (cf. Volume I, p. 604). Hence, the series

$$u(r, \theta) = \frac{1}{2} a_0 + \sum_{n=1}^{\infty} \frac{r^n}{R^n} (a_n \cos n\theta + b_n \sin n\theta)$$

a fortiori converges uniformly and absolutely in the interior of the circle. This series, however, can be differentiated term by term, provided $r < R$, because the resulting series again converge uniformly (cf. Volume I, p. 539). The function $u(r, \theta)$ is, therefore, a potential function. Since it has the prescribed value on the boundary, it is a solution of our boundary value problem.

We can reduce this solution to the integral form (41) by introducing the integrals for the Fourier coefficients,

$$a_n = \frac{1}{\pi} \int_0^{2\pi} f(\alpha) \cos n\alpha \, d\alpha, \ b_n = \frac{1}{\pi} \int_0^{2\pi} f(\alpha) \sin n\alpha \, d\alpha.$$

Since the convergence is uniform, we can interchange integration and summation and obtain

$$u(r, \theta) = \frac{1}{\pi} \int_0^{2\pi} f(\alpha) \left\{ \frac{1}{2} + \sum_{n=1}^{\infty} \frac{r^n}{R^n} \cos n(\theta - \alpha) \right\} d\alpha.$$

Poisson's integral formula will be proved if we can establish the relation

$$\frac{1}{2} + \sum_{n=1}^{\infty} \frac{r^n}{R^n} \cos n\tau = \frac{1}{2} \frac{R^2 - r^2}{R^2 - 2Rr \cos \tau + r^2}.$$

But this can be proved by the method used in Volume I (p. 586), that is, by reduction to a geometric series, using the complex representation

$$\cos n\tau = \frac{1}{2} (e^{in\tau} + e^{-in\tau}).$$

We leave the details of the proof to the reader.

### Exercises 6.7

1. By applying inversion to Poisson's formula, find a potential function $u(x, y)$ that is bounded in the region *outside* the unit circle and assumes given values $f(\theta)$ on its boundary (the so-called *outer* boundary value problem).
2. Find (a) the equipotential surfaces and (b) the lines of force for the potential of the segment $x = y = 0, -l \le z \le +l$, of constant linear density $\mu$.

3. Prove that if the values of a harmonic $u(x, y, z)$ and of its normal derivative $\partial u/\partial n$ are given on a closed surface $S$, then the value of $u$ at any interior point is given by the expression

$$u(x,y,z) = \frac{1}{4\pi} \iint_S \left( \frac{1}{r} \frac{\partial u}{\partial n} - u \frac{\partial(1/r)}{\partial n} \right) d\sigma ,$$

where $r$ is the distance from the point $(x, y, z)$ to the variable point of integration (apply Green's theorem to the functions $u$ and $1/r$).

## 6.8 Further Examples of Partial Differential Equations from Mathematical Physics

### a. The Wave Equation in One Dimension

The phenomena of wave propagation (e.g., of light or sound) are governed by the so-called *wave equation.* We begin by considering the simple idealized case of a so-called *one-dimensional wave.* Such a wave involves the magnitude $u$ of some property—for example, pressure, position of a particle, or intensity of an electric field—which depends not only on the coordinate of position $x$ (we take the direction of propagation as the $x$-axis) but also on the time $t$.

A wave function $u(x, t)$ then satisfies a partial differential equation of the form

(42a)
$$u_{xx} = \frac{1}{a^2} u_{tt},$$

where $a$ is a constant depending on the physical nature of the medium.[1]

We can find solutions of equation (42a) of the form

$$u = f(x - at),$$

where $f(\xi)$ is an arbitrary function of $\xi$, which we only assume to have continuous derivatives of the first and second order. If we put $\xi = x - at$, we see at once that our differential equation is actually satisfied, for

$$u_{xx} = f''(\xi), \qquad u_{tt} = a^2 f''(\xi).$$

In the same way, using an arbitrary function $g(\xi)$, we obtain a solution of the form

[1] For example, for transverse vibrations of a string, $u$ represents the lateral displacement of a particle, and $a^2 = T/\rho$, where $T$ is the tension and $\rho$ the mass per unit length.

$$u = g(x + at).$$

Both solutions represent wave motions propagated with the velocity $a$ along the $x$-axis; the first represents a wave traveling in the positive $x$-direction, the second a wave traveling in the negative $x$-direction. Let $u = f(x - at)$ have the value $u(x_1, t_1)$ at any point $x_1$ at time $t_1$; then $u$ has the same value at time $t$ at the point $x = x_1 - a(t - t_1)$, for then $x - at = x_1 - at_1$, so that $f(x - at) = f(x_1 - at_1)$. In the same way we can see that the function $g(x + at)$ represents a wave traveling in the negative $x$-direction with velocity $a$.

We shall now solve the following *initial value problem* for this wave equation. From all possible solutions of the differential equation we wish to select those for which the *initial state* (at $t = 0$) is given by two prescribed functions $u(x, 0) = \phi(x)$ and $u_t(x, 0) = \psi(x)$. To solve this problem, we merely write

(42b) $$u = f(x - at) + g(x + at)$$

and determine the functions $f$ and $g$ from the two equations

$$\phi(x) = f(x) + g(x),$$

$$\frac{1}{a} \psi(x) = -f'(x) + g'(x).$$

The second equation gives

$$c + \frac{1}{a} \int_0^x \psi(\tau)\, d\tau = -f(x) + g(x),$$

where $c$ is an arbitrary constant of integration. From this we readily obtain the required solution in the form

(42c) $$u(x, t) = \frac{\phi(x + at) + \phi(x - at)}{2} + \frac{1}{2a} \int_{x-at}^{x+at} \psi(\tau)\, d\tau.$$

The reader should prove for himself, by introducing new independent variables $\xi = x - at$, $\eta = x + at$ instead of $x$ and $t$, that no solutions of the differential equation exist other than those given.

### b. The Wave Equation in Three-Dimensional Space

In space of three dimensions the wave function $u$ depends on four independent variables, namely, the three space coordinates $x$, $y$, $z$ and the time $t$. The wave equation is then

(43a)
$$u_{xx} + u_{yy} + u_{zz} = \frac{1}{a^2} u_{tt},$$

or, more briefly,

(43b)
$$\Delta u = \frac{1}{a^2} u_{tt}.$$

Here again we can easily find solutions that represent the propagation of a plane wave in the physical sense. Namely, any function $f(\xi)$ that is twice continuously differentiable yields a solution of the differential equation if we make $\xi$ a linear expression of the form

$$\xi = \alpha x + \beta y + \gamma z \pm at,$$

whose coefficients satisfy the relation

$$\alpha^2 + \beta^2 + \gamma^2 = 1.$$

For, since

$$\Delta u = (\alpha^2 + \beta^2 + \gamma^2) f''(\xi) = f''(\xi)$$

and

$$u_{tt} = a^2 f''(\xi),$$

we see that $u = f(\alpha x + \beta y + \gamma z \pm at)$ really is a solution of the equation (43b).

If $q$ is the distance of the point $(x, y, z)$ from the plane $\alpha x + \beta y + \gamma z = 0$, we know by analytical geometry (cf. p. 135) that

$$q = \alpha x + \beta y + \gamma z.$$

Hence, in the first place, we see from the expression

$$u = f(q + at)$$

that at all points of a plane at a distance $q$ from the plane $\alpha x + \beta y + \gamma z = 0$ and parallel to it the property that is being propagated (represented by $u$) has the same value at a given moment. The property is propagated in space in such a way that planes parallel to $\alpha x + \beta y + \gamma z = 0$ are always surfaces on which the property is constant; the velocity of propagation is $a$ in the direction perpendicular to the planes. In theoretical physics a propagated phenomenon of this kind is referred to as a *plane wave*.

A case of particular importance is that in which the property varies periodically with time. If the frequency of the vibration is $\omega$, a phenomenon of this kind may be represented by

$$u = \exp[ik(\alpha x + \beta y + \gamma z + at)] = \exp[ik(\alpha x + \beta y + \gamma z)] \exp(i\omega t),$$

where $k/2\pi$ is the reciprocal of the wavelength $\lambda$: $k = 2\pi/\lambda = \omega/a$.

The wave equation with four independent variables has other solutions, which represent *spherical waves* spreading out from a given point, say the origin. A spherical wave is defined by the statement that the property is the same at a given instant at every point of a sphere with its center at the origin, that is, that $u$ has the same value at all points of the sphere. To find solutions satisfying this condition, we transform $\Delta u$ to polar coordinates $(r, \theta, \phi)$, and then assume that $u$ depends only on $r$ and $t$ but not on $\theta$ and $\phi$. If we accordingly equate the derivatives of $u$ with respect to $\theta$ and $\phi$ to zero (cf. p. 610), the differential equation (43b) becomes

$$u_{rr} + \frac{2}{r} u_r = \frac{1}{a^2} u_{tt}$$

or

$$(ru)_{rr} = \frac{1}{a^2} (ru)_{tt}.$$

For the moment we replace $ru$ by $w$ and observe that $w$ is a solution of the equation

$$w_{rr} = \frac{1}{a^2} w_{tt},$$

which we have already discussed; hence, $w$ must be expressible in the form

$$w = f(r - at) + g(r + at).$$

Consequently,

(43c)     $$u = \frac{1}{r} [f(r - at) + g(r + at)].$$

The reader should now verify for himself directly that a function of this type is actually a solution of the differential equation (43b).

Physically the function $u = f(r - at)/r$ represents a wave propagated with velocity $a$ from a center outward into space.

### c. *Maxwell's Equations in Free Space*

As a concluding example we shall discuss the system of equations known as *Maxwell's equations*, which form the foundations of electrodynamics. However, we shall not attempt to approach the equations from the physical point of view but shall merely use them to illustrate the various mathematical concepts developed above.

The electromagnetic state in free space is determined by two vectors given as functions of position and time, an electric vector **E** with components $E_1$, $E_2$, $E_3$ and a magnetic vector **H** with components $H_1$, $H_2$, $H_3$. These vectors satisfy Maxwell's equations:

(44a)
$$\operatorname{curl} \mathbf{E} + \frac{1}{c} \frac{\partial \mathbf{H}}{\partial t} = 0,$$

(44b)
$$\operatorname{curl} \mathbf{H} - \frac{1}{c} \frac{\partial \mathbf{E}}{\partial t} = 0,$$

where $c$ is the velocity of light in free space. Expressed in terms of the components of the vectors, the equations are:

$$\frac{\partial E_3}{\partial y} - \frac{\partial E_2}{\partial z} + \frac{1}{c} \frac{\partial H_1}{\partial t} = 0,$$

$$\frac{\partial E_1}{\partial z} - \frac{\partial E_3}{\partial x} + \frac{1}{c} \frac{\partial H_2}{\partial t} = 0,$$

$$\frac{\partial E_2}{\partial x} - \frac{\partial E_1}{\partial y} + \frac{1}{c} \frac{\partial H_3}{\partial t} = 0,$$

and

$$\frac{\partial H_3}{\partial y} - \frac{\partial H_2}{\partial z} - \frac{1}{c} \frac{\partial E_1}{\partial t} = 0,$$

$$\frac{\partial H_1}{\partial z} - \frac{\partial H_3}{\partial x} - \frac{1}{c} \frac{\partial E_2}{\partial t} = 0,$$

$$\frac{\partial H_2}{\partial x} - \frac{\partial H_1}{\partial y} - \frac{1}{c} \frac{\partial E_3}{\partial t} = 0,$$

We thus have a system of six partial differential equations of the first order, that is, of equations involving the first partial derivatives of the components with respect to the space coordinates and to the time.

We shall now deduce some distinctive consequences of Maxwell's equations. If we form the *divergence* of both equations, and remember that div curl $\mathbf{A} = 0$ (see p. 211) and that the order of differentiation with respect to the time and formation of the divergence is interchangeable, we obtain from (44a, b)

(45a)                          $\operatorname{div} \mathbf{E} = \text{constant},$

(45b)                          $\operatorname{div} \mathbf{H} = \text{constant};$

this is, the two divergences are independent of the time. In particular, if initially div $\mathbf{E}$ and div $\mathbf{H}$ are zero, they remain zero for all time.

We now consider any closed surface $S$ lying in the field and take the volume integrals

$$\iiint \operatorname{div} \mathbf{E} \, d\tau$$

and

$$\iiint \operatorname{div} \mathbf{H} \, d\tau$$

throughout the volume enclosed by it. If we apply Gauss's theorem (p. 601) to these integrals, they become integrals of the normal components $E_n$, $H_n$ over the surface $S$. That is, the equations

$$\operatorname{div} \mathbf{E} = 0, \qquad \operatorname{div} \mathbf{H} = 0$$

give

$$\iint_S E_n \, d\sigma = 0, \qquad \iint_S H_n \, d\sigma = 0.$$

In electrical theory, surface integrals

$$\iint_S E_n \, d\sigma \qquad \text{or} \qquad \iint_S H_n \, d\sigma$$

are called the *electric* or *magnetic flux* across the surface $S$, and our result may accordingly be stated as follows:

*The electric flux and the magnetic flux across a closed surface, subject to the zero initial conditions on div* **E** *and div* **H**, *are zero.*

We obtain a further deduction from Maxwell's equations if we consider a portion of surface $S$ bounded by the curve $\Gamma$, as follows:

If we denote the components of a vector normal to the surface $S$ by the suffix $n$, it immediately follows from Maxwell's equations (44a, b) that

$$(\text{curl } \mathbf{E})_n = -\frac{1}{c}\frac{\partial H_n}{\partial t}.$$

$$(\text{curl } \mathbf{H})_n = +\frac{1}{c}\frac{\partial E_n}{\partial t}.$$

If we integrate these equations over the surface with surface element $d\sigma$, we can transform the left sides into line integrals taken round the boundary $\Gamma$ by Stokes's theorem (cf. p. 611). Doing this, and taking the differentiation with respect to $t$ outside the integral sign, we obtain the equations

$$\int_\Gamma E_s \, ds = -\frac{1}{c}\frac{d}{dt}\iint_S H_n \, d\sigma,$$

$$\int_\Gamma H_s \, ds = +\frac{1}{c}\frac{d}{dt}\iint_S E_n \, d\sigma,$$

where the symbols $E_s$ and $H_s$ under the integral signs on the left are the *tangential components* of the electric and magnetic vectors in the direction of increasing arc and the sense of description of the curve $\Gamma$ in conjunction with the direction of the normal **n** forms a right-handed screw.

The facts expressed by these equations may be expressed in words as follows:

*The line integral of the electric or the magnetic force round an element of surface is proportional to the rate of change of the electric or magnetic flux across the element of surface, the constant of proportionality being* $-1/c$ *or* $+1/c$.

Finally, we shall establish the connection betweene Maxwell's equations and the wave equation. We find, in fact, that each of the vectors **E** and **H**, that is, each component of the vectors, satisfies the wave equation

$$\Delta u = \frac{1}{c^2}\, u_{tt}.$$

To show this, we eliminate the vector **H**, say, from the two equations, by differentiating the second equation with respect to the time and substituting for $\partial \mathbf{H}/\partial t$ from the first equation.

It then follows that

$$c \text{ curl (curl } \mathbf{E}) + \frac{1}{c}\,\frac{\partial^2 \mathbf{E}}{\partial t^2} = 0.$$

If we now use the vector relation[1]

(46)                    $\text{curl (curl } \mathbf{A}) = -\Delta \mathbf{A} + \text{grad(div } \mathbf{A}),$

and recall that

$$\text{div } \mathbf{E} = 0,$$

we at once obtain

(47a)                    $$\Delta \mathbf{E} = \frac{1}{c^2}\,\frac{\partial^2 \mathbf{E}}{\partial t^2}.$$

In the same way we can show that the vector **H** satisfies the same equation:

(47b)                    $$\Delta \mathbf{H} = \frac{1}{c^2}\,\frac{\partial^2 \mathbf{H}}{\partial t^2}.$$

## Exercises 6.8

1. Integrate the following partial differential equations:
   (a) $u_{xy} = 0$
   (b) $u_{xyz} = 0$
   (c) $u_{xy} = a(x,y).$
2. Find a solution of the equation
   $$u_{xy} = u,$$
   for which $u(x, 0) = u(0, y) = 1$, in the form of a power series.
3. Find the partial differential equation satisfied by the two-parameter family of spheres
   $$z^2 = 1 - (x - a)^2 - (y - b)^2.$$
4. Prove that if

---

[1]This vector relation follows immediately from its expression in terms of coordinates.

$$z = u(x, y, a, b)$$

is a solution depending on two parameters $a$, $b$, of the partial differential equation of the first order

$$F(x, y, z, z_x, z_y) = 0,$$

then the envelope of every one-parameter family of solutions chosen from $z = u(x, y, a, b)$ is again a solution.

5. (a) Find particular solutions of the equation

$$u_x^2 + u_y^2 = 1$$

of the form $u = f(x) + g(y)$.

   (b) Find particular solutions of the equation

$$u_x u_y = 1$$

of the forms $u = f(x) + g(y)$ and $u = f(x) g(y)$.

   (c) Use the result of Exercise 4 to obtain other solutions of the equation in part (b) by putting $b = ka$ in

$$u = ax + \frac{1}{a}y + b,$$

where $k$ is a constant.

6. Solve the equation

$$u_{xx} + 5u_{xy} + 6u_{yy} = e^{x+y}$$

by reducing it to one of the form of Exercise 1(c).

7. Prove that if $K$ is a homogeneous function of $x$, $y$, $z$ the equation

$$\frac{\partial}{\partial x}\left(K\frac{\partial u}{\partial x}\right) + \frac{\partial}{\partial y}\left(K\frac{\partial u}{\partial y}\right) + \frac{\partial}{\partial z}\left(K\frac{\partial u}{\partial z}\right) = 0$$

has a solution that is a power of $(x^2 + y^2 + z^2)$.

8. Determine the solutions of the equation

$$\frac{\partial^2 z}{\partial t^2} = a^2 \frac{\partial^2 z}{\partial x^2}$$

that are also solutions of

$$\left(\frac{\partial z}{\partial t}\right)^2 = a^2 \left(\frac{\partial z}{\partial x}\right)^2.$$

9. (a) Obtain particular solutions of the wave equation

$$u_{xx} = \frac{1}{c^2} u_{tt}$$

in the form $u(x, t) = \phi(x)\psi(t)$ satisfying the boundary conditions

$$u(0, t) = u(\pi, t) = 0.$$

   (b) Express the solution of part (a) in the form $f(x + ct) + g(x - ct)$.

   (c) *Plucked string problem:* By expanding $f(x)$ over the interval $[0, \pi]$ in a Fourier sine series (which *defines* $f(-x) = -f(x)$ for $0 \leq x \leq \pi$), find a solution of the foregoing type that satisfies the initial conditions, for $0 \leq x \leq \pi$,

$$u(x,0) = f(x)$$
$$u_t(x,0) = 0,$$

where

(i) $$f(x) = \begin{cases} x, & 0 \leq x \leq \pi/2 \\ \pi - x, & \pi/2 \leq x \leq \pi \end{cases}$$

(ii) $$f(x) = \sum_{n=1}^{\infty} a_n \sin nx.$$

10 . Let $u(x, t)$ denote a solution of the wave equation

$$u_{xx} = \frac{1}{a^2} u_{tt} \qquad\qquad (a > 0)$$

that is twice continuously differentiable. Let $\phi(t)$ be a given function that is twice continuously differentiable and such that

$$\phi(0) = \phi'(0) = \phi''(0) = 0.$$

Find the solution $u$ for $x \geq 0$ and $t \geq 0$ that is determined by the boundary conditions

$$u(x,0) = u_t(x,0) = 0 \qquad\qquad (x \geq 0),$$
$$u(0,t) = \phi(t) \qquad\qquad (t \geq 0).$$

# CHAPTER
# 7

# *Calculus of Variations*

## 7.1 Functions and Their Extrema

In the theory of ordinary maxima and minima of a differentiable function $f(x_1, \ldots, x_n)$ of $n$ independent variables, the necessary condition (pp. 326–7) for the occurrence of an extreme value at a point of the domain of $f$ is

(1)   $df = 0$   or   $\operatorname{grad} f = 0$   or   $f_{x_i} = 0$   $(i = 1, \ldots, n)$.

These equations express the *stationary character* of the function $f$ at the point in question. Whether these stationary points are actually maximum or minimum points can only be decided upon further investigation. In contrast to the equations (1), sufficient conditions for extrema take the form of *inequalities* (see p. 349).

The calculus of variations is likewise concerned with the problem of extreme values (*respectively stationary values*) but in a completely new situation. Now the functions whose extrema we seek no longer depend on one independent variable or a finite number of independent variables within a certa n region but are so-called *functionals,* or functions of functions. Specifically, in order to determine them we must know one or more functions or curves (or surfaces, as the case may be), the so-called *argument functions.*

General attention was first drawn to problems of this type in 1696 by John Bernoulli's statement of the *brachistochrone problem.*

In a vertical $x, y$-plane a point $A = (x_0, y_0)$ is to be joined to a point $B = (x_1, y_1)$, such that $x_1 > x_0$, $y_1 > y_0$, by a smooth curve $y = u(x)$ in such a way that the time taken by a particle sliding without friction from $A$ to $B$ along the curve under gravity (which is taken as acting in the direction of the positive $y$-axis) is as short as possible.

The mathematical expression of the problem is based on the physical assumption that along such a curve $y = \phi(x)$ the velocity $ds/dt$ ($s$ being the length of arc of the curve) is proportional to $\sqrt{2g(y - y_0)}$, the square root of the height of fall. The time taken in the fall of the particle is therefore given by

$$T = \int_{x_0}^{x_1} \frac{dt}{ds} \frac{ds}{dx} dx = \frac{1}{\sqrt{2g}} \int_{x_0}^{x_1} \frac{\sqrt{1 + y'^2}}{\sqrt{y - y_0}} dx$$

(cf. Volume I, p. 408). If we drop the unimportant factor $\sqrt{2g}$ and take $y_0 = 0$ (which we can do without loss of generality), we obtain the following problem: Among all continuously differentiable functions $y = \phi(x_0)$, $y \geq 0$ for which $\phi(x_0) = 0$, $\phi(x_1) = y_1$, find the one for which the integral

(2a)
$$I\{\phi\} = \int_{x_0}^{x_1} \sqrt{\frac{1 + y'^2}{y}} dx$$

has the least possible value.

On p. 751 we shall obtain the result—very surprising to Bernoulli's contemporaries—that the curve $y = \phi(x)$ must be a *cycloid*. Here we wish to emphasize that Bernoulli's problem and the elementary problems of maxima and minima are quite different. The expression $I\{\phi\}$ depends on the whole course of the function $\phi$. Since $\phi$ cannot be described by the values of a finite number of independent variables, $I$ is a function of a new kind. We indicate its character of "function of a function $\phi(x)$" by means of braces.

The following is another problem of a similar nature: Two points $A = (x_0, y_0)$ and $B = (x_1, y_1)$, where $x_1 > x_0$, $y_0 > 0$, $y_1 > 0$, are to be joined by a curve $y = u(x)$ lying above the $x$-axis, in such a way that the area of the surface of revolution formed when the curve is rotated about the $x$-axis is *as small as possible*.

Using the expression given on p. 429 for the area of a surface of revolution and dropping the unimportant factor $2\pi$, we have the following mathematical statement of the problem: Among all continuously differentiable functions $y = \phi(x)$ for which $\phi(x_0) = y_0$, $\phi(x_1) = y_1$, $\phi(x) > 0$, find the one for which the integral

(2b)
$$I\{\phi\} = \int_{x_0}^{x_1} y \sqrt{1 + y'^2} dx \qquad\qquad [y = \phi(x)]$$

has the least possible value. It will be found that the solution is a *catenary*.

The elementary geometrical problem of finding the shortest curve joining two points $A$ and $B$ in the plane belongs to the same category. Analytically, the problem is that of finding two functions $x(t)$, $y(t)$ of a parameter $t$ in an interval $t_0 \leq t \leq t_1$, for which the values $x(t_0) = x_0$, $x(t_1) = x_1$ and $y(t_0) = y_0$, $y(t_1) = y_1$ are prescribed and for which the integral

(2c)
$$\int_{t_0}^{t_2} \sqrt{\dot{x}^2 + \dot{y}^2}\, dt \qquad \left( \dot{x} = \frac{dx}{dt}, \; \dot{y} = \frac{dy}{dt} \right)$$

has the least possible value. The solution is, of course, a straight line.

Less trivial is the solution of the corresponding problem of finding the *geodesics on a given surface* $G(x, y, z) = 0$, that is, of joining two points on the surface with coordinates $(x_0, y_0, z_0)$ and $(x_1, y_1, z_1)$ by the shortest possible curve lying in the surface. In analytical language, we have the following problem: Among all triads of functions $x(t)$, $y(t)$, $z(t)$ of the parameter $t$ that make the equation

(3a)
$$G(x, y, z) = 0$$

an identity in $t$ and for which $x(t_0) = x_0$, $y(t_0) = y_0$, $z(t_0) = z_0$ and $x(t_1) = x_1$, $y(t_1) = y_1$, $z(t_1) = z_1$, find that for which the integral

(3b)
$$\int_{t_0}^{t_1} \sqrt{\dot{x}^2 + \dot{y}^2 + \dot{z}^2}\, dt$$

has the least possible value.

The *isoperimetric problem* of finding a closed curve of given length enclosing the largest possible area, already discussed on p. 366, also belongs to the same category. We have proved above that the solution is a circle.[1]

The general formulation of the type of problem encountered here is as follows: We are given a function $F(x, \phi, \phi')$ of three arguments

---

[1]The proof given there applied only to convex curves; the following remark, however, enables us to extend the result immediately to any curve: We consider the *convex hull* of the curve $C$ (i.e., the smallest convex set enclosing $C$). Its boundary $K$ consists of convex arcs of $C$ and rectilinear portions of tangents to $C$ that touch $C$ at two points and bridge over concave parts of $C$ by straight lines. It is evident that the area of $K$ exceeds that of $C$, provided $C$ is not convex, and, on the other hand, that the perimeter of $K$ is less than that of $C$. If we now make $K$ expand uniformly so that it always retains the same shape, until the resulting curve $K'$ has the prescribed perimeter, $K'$ will be a curve of the same perimeter as $C$ but enclosing a greater area. Hence, in the isoperimetric problem we may from the outset confine ourselves to *convex* curves, in order to obtain the maximum area.

that in the region of the arguments considered is continuous and has continuous derivatives of the first and second orders. If in this function $F$ we replace $\phi$ by a function $y = \phi(x)$ and $\phi'$ by the derivative $y' = \phi'(x)$, $F$ becomes a function of $x$, and an integral of the form

$$(4) \qquad I\{\phi\} = \int_{x_0}^{x_1} F(x, y, y') \, dx$$

becomes a definite number depending on the function $y = \phi(x)$; that is, it is a "functional evaluated for the function $\phi(x)$."

The fundamental problem of the calculus of variations is the following:

*Among all the functions that are defined and continuous and possess continuous first and second derivatives in the interval $x_0 \leq x \leq x_1$ and for which the boundary values $y_0 = \phi(x_0)$ and $y_1 = \phi(x_1)$ are prescribed find the one for which the functional $I\{\phi\}$ has the least possible value (or the greatest possible value).*

In discussing this problem, an essential point is the nature of the *admissibility conditions* imposed on the functions $\phi(x)$. Forming the value $I\{\phi\}$ merely requires that when $\phi(x)$ is substituted, $F$ shall be a sectionally continuous function of $x$, and this is assured if the derivative $\phi'(x)$ is sectionally continuous. But we have made the conditions for admission more stringent by requiring that the first derivatives, and even the second derivatives, of the functions $\phi(x)$ shall be continuous. The field in which the maximum or minimum is to be sought is of course thereby restricted. It will, however, be found that this restriction does not, in fact, affect the solution, that is, that the function that is most favorable when the wider field is available will always be found in the more restricted field of functions with continuous first and second derivatives.

Problems of this type occur very frequently in geometry and physics. Here we mention only one example: the fundamental principle of geometrical optics. We consider a ray of light in the $x, y$-plane and assume that the velocity of light is a given function $v(x, y, y')$ of the point $(x, y)$ and of the direction $y'$ [$y = \phi(x)$ being the equation of the light-path and $y' = \phi'(x)$ the corresponding derivative]. Then *Fermat's principle of least time* states:

*The actual path of a ray of light between two given points A, B is such that the time taken by the light in traversing it is less than the time that light would take to traverse any other path from A to B.*

In other words, if $t$ is the time and $s$ the length of arc of *any* curve $y = \phi(x)$ joining the points $A$ and $B$, the time that light would take to traverse the portion of curve between $A$ and $B$ is given by the integral

$$(5) \qquad I\{\phi\} = \int_{x_0}^{x_1} \frac{dt}{ds}\frac{ds}{dx}\,dx = \int_{x_0}^{x_1} \frac{\sqrt{1 + y'^2}}{v(x, y, y')}\,dx.$$

The actual path of the light is determined by the function $y = \phi(x)$ for which this integral has the least possible value.

We see that the optical problem of finding the light ray is a special case of the general problem stated above, corresponding to

$$F = \frac{\sqrt{1 + y'^2}}{v}$$

In most optical cases the velocity of light $v$ is independent of the direction and is merely a function of position $v(x, y)$.

## 7.2 Necessary Conditions for Extreme Values of a Functional

### a. *Vanishing of the First Variation*

Our object is to find necessary conditions that a function $y = \phi(x)$ may yield a maximum or minimum or, to use a general term, an extreme value, of the integral $I\{\phi\}$ defined by (4). We proceed by a method quite analogous to that used in the elementary problem of finding the extreme values of a function of one or more variables. We assume that $y = \phi = u(x)$ is the solution. Then we have to express the fact that (for a minimum) $I$ must *increase* when $u$ is replaced by another admissible function $\phi$. Moreover, because we are merely concerned with obtaining necessary conditions, we may confine ourselves to the consideration of any special class of functions $\phi$ that are close to $u$, that is, functions for which the absolute value of the difference $\phi - u$ remains between prescribed bounds.

We think of the function $u$ as a member of a one-parameter family with parameter $\varepsilon$, constructed as follows: We take any function $\eta(x)$ that vanishes on the boundary of the interval—that is, for which $\eta(x_0) = 0$, $\eta(x_1) = 0$—and that has continuous first and second derivatives everywhere in the closed interval. We then form the family of functions

$$\phi(x, \varepsilon) = u(x) + \varepsilon\eta(x).$$

The expression $\varepsilon\eta(x) = \delta u$ is called a *variation of the function u.* [since $\eta(x) = \partial\phi/\partial\varepsilon$, the symbol $\delta$ denotes the differential obtained when $\varepsilon$ is regarded as the independent variable and $x$ as a parameter.] Then, if we regard the function $u$ as well as the function $\eta$ as fixed, the value of the functional

$$I\{u + \varepsilon\eta\} = G(\varepsilon) = \int_{x_0}^{x_1} F(x,\, u + \varepsilon\eta,\, u' + \varepsilon\eta')\, dx$$

becomes a function of $\varepsilon$; and the postulate that $u$ shall give a minimum of $I\{\phi\}$ implies that the function above shall possess a minimum for $\varepsilon = 0$, so that as necessary conditions we have the equation

(6a) $$G'(0) = 0$$

and also the inequality

(6b) $$G''(0) \geqq 0.$$

The corresponding necessary conditions for a maximum are the same equation $G'(0) = 0$ and the reversed inequality $G''(0) \leqq 0$. The condition $G'(0) = 0$ must be satisfied for every function $\eta$ that satisfies the above conditions but is otherwise arbitrary.

Putting aside the question of discriminating between maxima and minima, we say that if a function $u$ satisfies the equation $G'(0) = 0$, for all functions $\eta$, the integral $I$ is *stationary* for $\phi = u$. If, as before, we use the symbol $\delta$ to denote differentiation with respect to $\varepsilon$, we also say that the equation

$$\delta I = \varepsilon G'(0) = 0,$$

when satisfied by a function $\phi = u$ and arbitrary $\eta$, expresses the stationary character of $I$. The expression

(6c) $$\varepsilon G'(0) = \varepsilon \left\{ \frac{d}{d\varepsilon} \int_{x_0}^{x_1} F(x,\, u + \varepsilon\eta,\, u' + \varepsilon\eta')\, dx \right\}_{\varepsilon=0}$$

is called the *variation* or, more accurately, the *first variation,*[1] of the integral. *Stationary character of an integral* and *vanishing of the first variation,* therefore, mean exactly the same thing.

---

[1] From this comes the use of the term *calculus of variations,* which is meant to indicate that in this subject we are concerned with the behavior of functions of a function when this independent function, or *argument function,* is made to vary by altering a parameter $\varepsilon$.

Stationary character is *necessary* for the occurrence of maxima or minima, but as in the case of ordinary maxima or minima, it is not a *sufficient* condition for the occurrence of either of these possibilities. We shall not treat the problem of sufficiency here; in what follows, we confine ourselves to the problem of stationary character.

Our main object is to transform the condition $G'(0) = 0$ for the stationary character of the integral in such a way that it becomes a condition for $u$ only and no longer contains the arbitrary function $\eta$.

## Exercises 7.2a

1. In connection with the brachistochrone problem (see pp. 737–738), calculate the time of fall when the points $A$ and $B$ are joined by a straight line.
2. Let the velocity of a particle with spherical coordinates $(r, \theta, \phi)$ moving in three-dimensional space be $v = 1/f(r)$. What time does the particle take to describe the portion of a curve given by a parameter $\sigma$ [the coordinates of a point on the curve being $r(\sigma)$, $\theta(\sigma)$, $\phi(\sigma)$] between the points $A$ and $B$?

### b. Derivation of Euler's Differential Equation

The fundamental criterion of the calculus of variations is constituted by the following theorem:

*Necessary and sufficient for the integral*

(7a)
$$I\{\phi\} = \int_{x_0}^{x_1} F(x, \phi, \phi') \, dx$$

*to be stationary when $\phi = u$ is that $u$ shall be an admissible function satisfying Euler's differential equation*

(7b)
$$L[u] = F_u - \frac{d}{dx} F_{u'} = 0,$$

*or, in full,*

(7c)
$$F_{u'u'} u'' + F_{uu'} u' + F_{xu'} - F_u = 0.$$

To prove this we note that we can differentiate the expression

$$G(\varepsilon) = \int_{x_0}^{x_1} F(x, u + \varepsilon\eta, u' + \varepsilon\eta') \, dx$$

with respect to $\varepsilon$ under the integral sign (cf. p. 74), provided that the differentiation yields a function of $x$ that is continuous or at least sectionally continuous. In this case, on putting $u + \varepsilon\eta = y$ and differentiating, we obtain under the integral sign the expression $\eta F_y + \eta' F_{y'}$, which, owing to the assumptions made about $f$, $u$, and $\eta$, satisfies the conditions just stated. Hence, we immediately obtain

(7d)       $$G'(0) = \int_{x_0}^{x_1} [\eta F_u(x, u, u') + \eta' F_{u'}(x, u, u')] \, dx.$$

For subsequent purposes, we note that in deriving this equation we have used nothing beyond the continuity of the functions $u$ and $\eta$ and the sectional continuity of their first derivatives. In this equation the arbitrary function appears under the integral sign in a twofold form, namely, as $\eta$ and $\eta'$. We can, however, immediately get rid of $\eta'$ by integration by parts; we have

$$\int_{x_0}^{x_1} \eta' F_{u'} \, dx = \eta \, F_{u'} \Big|_{x_0}^{x_1} - \int_{x_0}^{x_1} \eta \left( \frac{d}{dx} F_{u'} \right) dx = - \int_{x_0}^{x_1} \eta \left( \frac{d}{dx} F_{u'} \right) dx,$$

for by hypotheses $\eta(x_0)$ and $\eta(x_1)$ vanish. In this integration by parts we have to assume that the expression $(d/dx)F_{u'}$ is defined and integrable, but this is certainly the case since we assumed continuity of the second derivatives of $F$. Hence, if we write

(7e)       $$L[u] = F_u - \frac{d}{dx} F_{u'}$$

for brevity, we have the equation

(7f)       $$\int_{x_0}^{x_1} \eta L[u] \, dx = 0.$$

This equation must be satisfied for every function $\eta$ that satisfies our conditions but is otherwise arbitrary. From this, we conclude that

(7g)       $$L[u] = 0,$$

by virtue of the following:

LEMMA I.   *If a function $C(x)$ that is continuous in the interval under consideration satisfies the relation*

$$\int_{x_0}^{x_1} \eta(x) \, C(x) \, dx = 0$$

*for an arbitrary function $\eta(x)$ such that $\eta(x_0) = \eta(x_1) = 0$ and $\eta''(x)$*

*is continuous, then C(x) = 0 for every value of x in the interval.* (The proof of this lemma will be postponed to p. 747.)

We could, however, obtain condition (7g) in a different way,[1] by getting rid of the term in $\eta$ in the quation

$$\int_{x_0}^{x_1} (\eta\, F_u + \eta'\, F_{u'})\, dx = 0$$

by integration by parts, for if we write $F_{u'} = A$, $F_u = b = B'$ for brevity and remember the boundary condition for $\eta$, on integrating by parts we obtain

$$\int_{x_0}^{x_1} \eta\, F_u\, dx = \int_{x_0}^{x_1} \eta\, B'\, dx = -\int_{x_0}^{x_1} \eta'B\, dx.$$

If we put $\zeta = \eta'$, we have, in analogy to (7f), the condition

(7h) $$\int_{x_0}^{x_1} \zeta(A - B)\, dx = 0.$$

In deriving this formula we need not make any assumptions about the second derivatives of $\eta$ and $u$. On the contrary, it is sufficient to assume that $\phi$ (or $u$ and $\eta$) are continuous and have sectionally continuous first derivatives. Now equation (7h) must hold, not, it is true, for any arbitrary (sectionally continuous) function $\zeta$ but only for those functions $\zeta$ that are derivatives of a function $\eta(x)$ satisfying our conditions at the end points. However, if $\zeta(x)$ is any given sectionally continuous function satisfying the relation

(7i) $$\int_{x_0}^{x_1} \zeta(x)\, dx = 0,$$

we can put

$$\eta = \int_{x_0}^{x} \zeta(t)\, dt;$$

we have then constructed an admissible $\eta$, for $\eta' = \zeta$ and $\eta(x_0) = \eta(x_1) = 0$. We thus obtain the following result:

*A necessary condition that the integral should be stationary is*

(7j) $$\int_{x_0}^{x_1} \zeta(A - B)\, dx = 0,$$

---

[1]The first method is Lagrange's, and the second, P. Du Bois Reymond's.

*where $\zeta$ is an arbitrary sectionally continuous function merely satisfying the condition (7i).*

We now require the help of the following:

LEMMA II.  *If a sectionally continuous function $S(x)$ satisfies the condition*

(8a)
$$\int_{x_0}^{x_1} \zeta S \, dx = 0,$$

*for all functions $\zeta(x)$ that are sectionally continuous in the interval and for which*

(8b)
$$\int_{x_0}^{x_1} \zeta \, dx = 0,$$

*then $S(x)$ is a constant c.*

This lemma will also be proved below on p. 747. If meanwhile we assume its truth, it follows from (7h)—if we substitute the above expressions for $A$ and $B$—that

$$\int_{x_0}^{x} F_u \, dx + c = F_{u'}.$$

Since $F_u$ is sectionally continuous, the left side regarded as an indefinite integral may be differentiated with respect to $x$ and has $F_u$ as its derivative; the same is therefore true of the right side. Hence, the expression $(d/dx) \, F_{u'}$ for the supposed solution $u$ exists, and the equation

(9a)
$$F_u = \frac{d}{dx} F_{u'}$$

holds at all points of continuity of $u'$.

Thus, Euler's equation remains the necessary condition for an extreme value, or the condition that the integral should be stationary, when the class of admissible functions $\phi(x)$ is extended from the outset by requiring only sectional continuity of the first derivative of $\phi(x)$.

Euler's equation is an *ordinary differential equation of the* second order. Its solutions are called the *extremals* of the minimum problem. To solve the minimum problem, we must find among all the extremals that one that satisfies the prescribed boundary conditions.

If *Legendre's condition*

(9b) $$F_{u'u'} \neq 0$$

is satisfied for $\phi = u(x)$, the differential equation can be brought into the "regular" form $u'' = f(x, u, u')$, where the right side is a known expression involving $x, u, u'$.

### c. Proofs of the Fundamental Lemmas

We now prove the two lemmas used above. To prove Lemma I, we assume that at some point, say $x = \xi$, $C(x)$ is not zero and is positive. Then, since $C(x)$ is continuous, we can certainly mark off a subinterval of $(x_0, x_1)$,

(9c) $$\xi - a \leqq x \leqq \xi + a,$$

within which $C(x)$ remains positive. We now choose a twice continuously differentiable $\eta$, positive in the interior of this subinterval and zero elsewhere, say, by setting for $x$ in (9c)

$$\eta(x) = (x - \xi + a)^4 (x - \xi - a)^4 = \{(x - \xi)^2 - a^2\}^4.$$

This function $\eta$ certainly fulfills all the prescribed conditions; $\eta(x)C(x)$ is positive inside the subinterval and zero outside it. The integral

$$\int_{x_0}^{x_1} \eta C \, dx$$

therefore cannot be zero.[1] Since this contradicts our hypothesis, $C(\xi)$ cannot be positive. For the same reasons, $C(\xi)$ cannot be negative. Hence, $C(\xi)$ must vanish for all values of $\xi$ within the interval, as was stated in the lemma.

To prove Lemma II, we note that our assumption (8b) about $\zeta(x)$ immediately leads to the relation

(10) $$\int_{x_0}^{x_1} \zeta(x) \{S(x) - c\} \, dx = 0,$$

where $c$ is an arbitrary constant. We now choose $c$ in such a way that $S(x) - c$ is an admissible function $\zeta(x)$; that is, we determine $c$ by the equation

---

[1] The integral of a continuous nonnegative function is positive except when the integrand vanishes everywhere; this follows immediately from the definition of integral.

$$0 = \int_{x_0}^{x_1} \zeta \, dx = \int_{x_0}^{x_1} \{S(x) - c\} \, dx = \int_{x_0}^{x_1} S(x) \, dx - c(x_1 - x_0).$$

Substituting this value of $c$ in equation (10) and taking $\zeta = S(x) - c$, we at once have

$$\int_{x_0}^{x_1} \{S(x) - c\}^2 \, dx = 0.$$

Since by hypothesis the integrand is continuous, or at least sectionally continuous, it follows that

$$S(x) - c = 0$$

is an identity in $x$, as was stated in the lemma.

### d.  Solution of Euler's Differential Equation in Special Cases. Examples.

To find the solutions $u$ of the minimum problem, we must find a particular solution of Euler's differential equation for the interval $x_0 \leq x \leq x_1$ that assumes the prescribed boundary values $y_0$ and $y_1$ at the end points. Since the complete integral of Euler's differential equation of the second order contains two constants of integration, we expect to determine a unique solution by making these two constants fit the boundary conditions, the latter giving two equations that the constants of integration must satisfy.

In general, it is not possible to solve Euler's differential equation explicitly in terms of elementary functions or quadratures, and we have to be content to show that the variational problem does reduce to a problem in differential equations. On the other hand, for important special cases and, in fact, for most of the classical examples, the equation can be solved by means of quadratures.

The first case is that in which $F$ does not contain the derivative $y' = \phi'$ explicitly: $F = F(\phi, x)$. Here Euler's differential equation is simply $F_u(u, x) = 0$; that is, it is no longer a differential equation at all but forms an implicit definition of the solution $y = u(x)$. Here, of course, there is no question of integration constants or the possibility of satisfying boundary conditions.

The second important special case is that in which $F$ does not contain the function $y = \phi(x)$ explicitly: $F = F(y', x)$. Here Euler's differential equation is $(d/dx) (F_{u'}) = 0$, which at once gives

$$F_{u'} = c,$$

where $c$ is an arbitrary constant of integration. We may use this equation to express $u'$ as a function $f(x, c)$ of $x$ and $c$, and we then have the equation

$$u' = f(x, c),$$

from which by a simple integration (quadrature) we obtain

$$u = \int_0^x f(\xi, c)\, d\xi + a;$$

that is, $u$ is expressed as a function of $x$ and $c$, together with an additional arbitrary constant of integration $a$. In this case, therefore, Euler's differential equation can be completely solved by quadrature.

The third case, which is the most important in examples and applications, is that in which $F$ does not contain the independent variable $x$ explicitly: $F = F(y, y')$. In this case, we have the following important theorem:

*If the independent variable $x$ does not occur explicitly in the variational problem, then*

(11) $$E = F(u, u') - u'\, F_{u'}(u, u') = c$$

*is an integral of Euler's differential equation. That is, if we substitute in this expression a solution $u(x)$ of Euler's differential equation for $F$, the expression becomes a constant independent of $x$.*

The truth of this statement follows at once if we form the derivative $dE/dx$. We have

$$\frac{dE}{dx} = F_u u' + F_{u'} u'' - u'' F_{u'} - u'^2 F_{uu'} - u' u'' F_{u'u'},$$

or by (7c)

$$\frac{dE}{dx} = u'\, L[u] = 0;$$

hence, for every solution $u$ of Euler's differential equation, we have $E = c$, where $c$ is a constant.

If we think of $u'$ as calculated from the equation $E = c$, say $u' = f(u, c)$, a simple quadrature applied to the equation

$$\frac{dx}{du} = \frac{1}{f(u, c)}$$

gives $x = g(u, c) + \alpha$ (where $\alpha$ is another constant of integration); that is, $x$ is expressed as a function of $u$, $c$, and $\alpha$. By solving for $u$, we then obtain the function $u(x, c, \alpha)$. Hence, the general solution of Euler's differential equation, depending on two arbitrary constants of integration, is obtained by a quadrature.

We shall now use these methods to discuss a number of examples.

*General Note*

There is a general class of examples in which $F$ is of the form

$$F = g(y) \sqrt{1 + y'^2},$$

where $g(y)$ is a function depending explicitly on $y$ only. For the extremals $y = u$, our last rule gives at once

$$g(u) \sqrt{1 + u'^2} - \frac{g(u) \, u'^2}{\sqrt{1 + u'^2}} = c$$

or

$$\frac{g(u)}{\sqrt{1 + u'^2}} = c;$$

whence,

$$\frac{dx}{du} = \frac{1}{\sqrt{(\{g(u)\}^2/c^2) - 1}},$$

and on integrating we have the equation

(12)
$$x - b = \int \frac{du}{\sqrt{(\{g(u)\}^2/c^2) - 1}},$$

where $b$ is another constant of integration. By evaluating the integral on the right and solving the equation for $u$, we obtain $u$ as a function of $x$ and of the two constants of integration $c$ and $b$.[1]

*The Surface of Revolution of Least Area*

In this case, by (2b), p. 738, $g = y$. The integral (11) becomes

$$x - b = \int \frac{du}{\sqrt{u^2/c^2 - 1}} = c \operatorname{ar cosh} \frac{u}{c};$$

---

[1] Of course, we may not be able to solve for $u$ in terms of elementary functions, but for all practical purposes, these procedures define $u$ well enough.

hence, the result is

$$y = u = c \cosh \frac{x - b}{c}.$$

That is, the solution of the problem of finding a curve that on rotation gives a surface of revolution with stationary area is a *catenary* (see Volume I, p. 378).

A necessary condition for the occurrence of such a stationary curve is that the two given points $A$ and $B$ can be joined by a catenary for which $y > 0$. The question whether the catenary really represents a minimum will not be discussed here.

### The Brachistochrone

Another example is obtained by taking $g = 1/\sqrt{y}$. This, according to (2a), p. 738, is the problem of the *brachistochrone*. By means of the substitutions $1/c^2 = k$, $u = k\tau$, $\tau = \sin^2\theta/2$, the integral (12)

$$\int \frac{du}{\sqrt{1/(uc^2) - 1}}$$

is immediately transformed into

$$x - b = k \int \sqrt{\frac{\tau}{1 - \tau}} \, d\tau = \frac{1}{2} k \int (1 - \cos\theta) \, d\theta,$$

whence

$$x - b = \frac{1}{2} k(\theta - \sin\theta),$$

$$y = u = \frac{1}{2} k(1 - \cos\theta).$$

The brachistochrone is accordingly (cf. Volume I, p. 329) a common cycloid with its cusps on the $x$-axis.

### Exercises 7.2d

1. Find the extremals for the following integrands:
   (a) $F = \sqrt{y(1 + y'^2)}$
   (b) $F = \sqrt{1 + y'^2}/y$
   (c) $F = y\sqrt{1 - y'^2}$

2. Find the extremals for the integrand $F = x^n y'^2$, and prove that if $n \geq 1$, two points lying on opposite sides of the $y$-axis cannot be joined by an extremal.

3. Find the extremals for the integrand $y^n y'^m$, where $n$ and $m$ are even integers.

4. Find the extremals for the integrand $F = ay'^2 + 2byy' + cy^2$, where $a$, $b$, $c$ are given continuously differentiable functions of $x$. Prove that Euler's differential equation is a linear differential equation of the second order. Why is it that when $b$ is constant, this constant does not enter into the differential equation at all?

5. Show that the extremals for the integrand $F = e^x \sqrt{1 + y'^2}$ are given by the equations $\sin(y - b) = e^{-(x-a)}$ and $y = b$, where $a$, $b$ are constants. Discuss the form of these curves, and investigate how the two points $A$ and $B$ must be situated if they can be joined by an extremal arc of the form $y = f(x)$.

6. For the case where $F$ does not contain the derivative $y'$, deduce Euler's condition $F_y = 0$ by an elementary method.

7. Find a function giving the absolute minimum of

$$I\{y\} = \int_0^1 y'^2 \, dx$$

with the boundary conditions

(a) $y(0) = y(1) = 0$

(b) $y(0) = 0$, $y(1) = 1$.

8. Find the extremals for $\int \sqrt{r^2 + r'^2} \, d\theta$, that is, the paths of shortest distance in polar coordinates.

### e. Identical Vanishing of Euler's Expression

Euler's differential equation (7c), p. 743 for $F(x,y,y')$ may degenerate into an identity that tells us nothing, that is, into a relation that is satisfied by every admissible function $y = \phi(x)$. In other words, the corresponding integral may be stationary for any admissible function $y = \phi(x)$. If this degenerate case is to occur, Euler's expression

$$F_y - F_{xy'} - F_{yy'}y' - F_{y'y'}y''$$

must vanish at every point $x$ of the interval, no matter what function $y = \phi(x)$ is substituted in it. We can, however, always find a curve for which $y = \phi$, $y' = \phi'$, and $y'' = \phi''$ have arbitrary prescribed values for a prescribed value of $x$. Euler's expression must therefore vanish for every quadruple of numbers $x$, $y$, $y'$, $y''$. We conclude that the coefficient of $y''$, (i.e., $F_{y'y'}$) must vanish identically. $F$ must therefore be a linear function of $y'$, say $F = ay' + b$, where $a$ and $b$ are functions of $x$ and $y$ only. If we substitute this in the remaining part of the differential equation,

$$F_{yy'}y' + F_{xy'} - F_y = 0,$$

it follows at once that

$$0 = a_y y' + a_x - a_y y' - b_y$$

or that

$$a_x - b_y$$

must vanish identically in $x$ and $y$. In other words, Euler's expression vanishes identically if, and only if, the integral is of the form

$$I = \int \{a(x, y) y' + b(x, y)\} \, dx = \int a \, dy + b \, dx,$$

where $a$ and $b$ satisfy the condition of integrability that we have already met with on p. 104, that is, where $a \, dy + b \, dx$ is a exact differential.

## 7.3 Generalizations

### a. Integrals with More Than One Argument Function

The problem of finding the extreme values (stationary values) of an integral can be extended to the case where this integral depends not on a single argument function but on a number of such functions $\phi_1(x), \phi_2(x), \ldots, \phi_n(x)$.

The typical problem of this type may be formulated as follows: Let $F(x, \phi_1, \ldots, \phi_n, \phi_1', \ldots, \phi_n')$ be a function of the $(2n + 1)$ arguments $x, \phi_1, \ldots, \phi_n'$, which is continuous and has continuous derivatives up to, and including, the second order in the region under consideration. If we replace $y_i = \phi_i$ by a function of $x$ with continuous first and second derivatives, and $\phi_i'$ by its derivative, $F$ becomes a function of the single variable $x$, and the integral

$$(13) \qquad I\{\phi_1, \ldots, \phi_n\} = \int_{x_0}^{x_1} F(x, \phi_1, \ldots, \phi_n, \phi_1', \ldots, \phi_n') \, dx$$

over a given interval $x_0 \le x \le x_1$ has a definite value determined by the choice of these functions.

In the comparison with the extreme value, we regard as admissible all functions $\phi_i(x)$ that satisfy the above continuity conditions and for which the boundary values $\phi_i(x_0)$ and $\phi_i(x_1)$ have prescribed fixed values. In other words, we consider the curves $y_i = \phi_i(x)$ joining two given points $A$ and $B$ in $(n + 1)$-dimensional space with coordinates $y_1, y_2, \ldots, y_n, x$. The variational problem now requires us to find, among all these systems of functions $\phi_i(x)$, one $[y_i = \phi_i(x) = u_i(x)]$ for which the integral (13) has an extreme value (a maximum or a minimum).

Again, we shall not discuss the actual nature of the extreme value but shall confine ourselves to inquiring for what systems of argument functions $\phi_i(x) = u_i(x)$ the integral is stationary.

We define the concept of stationary value in exactly the same way as we did on p. 742. We embed the system of functions $u_i(x)$ in a one-parameter family of functions depending on the parameter $\varepsilon$, in the following way: Let $\eta_1(x), \ldots, \eta_n(x)$ be $n$ arbitrarily chosen functions that vanish for $x = x_0$ and $x = x_1$, are continuous in the interval, and possess continuous first and second derivatives there. We embed the $u_i(x)$ in the family of functions $y_i = \phi_i(x) = u_i(x) + \varepsilon\eta_i(x)$.

The term $\varepsilon\eta_i(x) = \delta u_i$ is called the *variation* of the function $u_i$. If we substitute the expressions for $\phi_i$ in $I\{\phi_1, \ldots, \phi_n\}$, this integral is transformed into

$$G(\varepsilon) = \int_{x_0}^{x_1} F(x, u_1 + \varepsilon\eta_1, \ldots, u_n + \varepsilon\eta_n, u_1' + \varepsilon\eta_1', \ldots, u_n' + \varepsilon\eta_n')\, dx,$$

which is a function of the parameter $\varepsilon$. A necessary condition that there may be an extreme value when $\phi_i = u_i$ (i.e., when $\varepsilon = 0$) is

$$G'(0) = 0.$$

Exactly as for the case of one independent function, we say that the integral $I$ has a stationary value for $\phi_i = u_i$ if the equation $G'(0) = 0$ holds or

$$\delta I = \varepsilon G'(0) = 0$$

holds, no matter how the functions $\eta_i$ are chosen subject to the conditions stated above. In other words, *stationary character of the integral for a fixed system of functions $u_i(x)$ and vanishing of the first variation $\delta I$ mean the same thing.*

We have still the problem of setting up conditions for the stationary character of the integral that do not involve the arbitrary variations $\eta_i$. This requires no new ideas. We proceed as follows: First we take $\eta_2, \eta_3, \ldots, \eta_n$ as identically zero (i.e., we do not let the functions $u_2, \ldots, u_n$ vary). We thus consider only the first function $\phi_1(x)$ as variable and then the condition $G'(0) = 0$, by p. 744, is equivalent to Euler's differential equation

$$F_{u_1} - \frac{d}{dx} F_{u_1'} = 0.$$

Since we can pick out any one of the functions $u_i(x)$ in the same way, we obtain the following result:

*A necessary and sufficient condition that the integral (13) may be stationary is that the $n$ functions $u_i(x)$ shall satisfy the system of Euler's equations*

$$(13a) \qquad\qquad F_{u_i} - \frac{d}{dx} F_{u_i'} = 0 \qquad\qquad (i = 1, 2, \ldots, n).$$

This is a system of $n$ differential equations of the second order for the $n$ functions $u_i(x)$. All solutions of this system of differential equations are said to be *extremals* of the variational problem. Thus, the problem of finding stationary values of the integral reduces to the problem of solving these differential equations and adapting the general solution to the given boundary conditions.[1]

### b. Examples

The possibility of giving a general solution of the system of Euler's differential equations is even more remote than in the case in Section 7.2. Only in very special cases can we find all the extremals explicitly. Here the following theorem, analogous to the particular case of formula (11) on p. 749, is often useful:

---

[1]Using Lemma II (Section 7.2, p. 746), we can prove that these differential equations must hold under the general assumption that the admissible functions merely have sectionally continuous first derivatives. However, if we wish to concentrate on the formalism of the subject, it is more convenient to include continuity of the second derivatives in the conditions of admissibility of the functions $\phi_i(x)$. We can then write out the expressions $d/dx\ F_{u_i'}$ in the form

$$(13b) \qquad\qquad \sum_{k=1}^{n} F_{u_k' u_i'} u_k'' + \sum_{k=1}^{n} F_{u_k u_i'} u_k' + F_{x u_i'}.$$

*If the function F does not contain the independent variable x explicit-
ly, i.e. $F = F(\phi_1, \ldots, \phi_n, \phi_1', \ldots, \phi_n')$, then the expression*

$$E = F(u_1, \ldots, u_n, u_1', \ldots, u_n') - \sum_{i=1}^{n} u_i' \, F_{u_i'}$$

*is an integral of Euler's system of differential equations.* That is, if we
consider any system of solutions $u_i(x)$ of Euler's equations (13a), we
have

(13c)                    $$E = F - \sum u_i' \, F_{u_i'} = \text{constant} = c,$$

where, of course, the value of this constant depends upon the system
of solutions substituted.

The proof follows the same lines as on p. 749; we differentiate the
left side of our expression with respect to $x$ and, using (13b), verify
that the result is zero.

A trivial example is the problem of finding the shortest distance
between two points in three-dimensional space. Here we have to
determine two functions $y = y(x)$, $z = z(x)$ such that the integral

$$\int_{x_0}^{x_1} \sqrt{1 + y'^2 + z'^2} \, dx$$

has the least possible value, the values of $y(x)$ and $z(x)$ at the end
points of the interval being prescribed. Euler's differential equations
(13a) give

$$\frac{d}{dx} \frac{y'}{\sqrt{1 + y'^2 + z'^2}} = \frac{d}{dx} \frac{z'}{\sqrt{1 + y'^2 + z'^2}} = 0,$$

whence it follows at once that the derivatives $y'(x)$ and $z'(x)$ are
constant; hence, the extremals must be straight lines.

Somewhat less trivial is the problem of the *brachistochrone in three
dimensions*. (Gravity is again taken as acting along the positive
$y$-axis.) Here we have to determine $y = y(x)$, $z = z(x)$ in such a way that
the integral

$$\int_{x_0}^{x_1} \sqrt{\frac{1 + y'^2 + z'^2}{y}} \, dx = \int_{x_0}^{x_1} F(y, y', z') \, dx$$

is stationary. One of Euler's differential equations gives

$$\frac{z'}{\sqrt{y}} \frac{1}{\sqrt{1 + y'^2 + z'^2}} = a.$$

In addition, we have from (13c) that

$$F - y' F_{y'} - z' F_{z'} = \frac{1}{\sqrt{y}} \frac{1}{\sqrt{1 + y'^2 + z'^2}} = b,$$

where $a$ and $b$ are constants. By division it follows that $z' = a/b = k$ is likewise constant. The curve for which the integral is stationary must therefore lie in a plane $z = kx + h$. From the further equation

$$\frac{1}{\sqrt{y}} \frac{1}{\sqrt{1 + k^2 + y'^2}} = b,$$

there follows, as is obvious from p. 751, that this curve must again be a cycloid.

### Exercises 7.3b

1. Write down the differential equations for the path of a ray of light in three dimensions in the case where (spherical coordinates $r$, $\theta$, $\phi$ being used) the velocity of light is a function of $r$ (cf. Exercise 2, p. 743). Show that the rays are plane curves.
2. Show that the geodesics (curves of shortest length joining two points) on a sphere are great circles.
3. Find the geodesics on a right circular cone.
4. Show that the path minimizing the distance between two nonintersecting smooth closed curves is their common normal line.
5. Show that the path for the least time of fall from a given point to a given curve is the cycloid that meets the curve perpendicularly.
6. Prove that the extremals of $\int F(x, y) \sqrt{1 + y'^2}\, dx$, with end points freely movable on two curves, meet those curves orthogonally.

### c. Hamilton's Principle. Lagrange's Equations

Euler's system of differential equations has a very important bearing on many branches of applied mathematics, especially dynamics. In particular, the motion of a mechanical system consisting of a finite number of particles can be expressed by the condition that a certain expression, the so-called Hamilton's integral, is stationary. Here we shall briefly explain this connection.

A mechanical system has $n$ degrees of freedom if its position is determined by $n$ independent coordinates $q_1, q_2, \ldots, q_n$. If, for example, the system consists of a single particle, we have $n = 3$, since for $q_1, q_2, q_3$ we can take the three rectangular coordinates or the three spherical coordinates. Again, if the system consists of two particles held at unit distance apart by a rigid connection—assumed to have no mass—then $n = 5$, since for the coordinates $q_i$ we can take the three rectangular coordinates of one particle and two other coordinates determining the direction of the line joining the two particles.

A dynamical system can be described with sufficient generality by means of two functions, the *kinetic energy* and the *potential energy*. If the system is in motion, the coordinates $q_i$ will be functions $q_i(t)$ of the time $t$, the *components of velocity* being $\dot{q}_i = dq_i/dt$. The kinetic energy associated with the dynamical system is a function of the form

$$(14a) \qquad T(q_1, \ldots, q_n, \dot{q}_1, \ldots, \dot{q}_n) = \sum_{i, k=1} \alpha_{ik} \dot{q}_i \dot{q}_k \qquad (\alpha_{ik} = \alpha_{ki}).$$

The kinetic energy, therefore, is a homogeneous quadratic expression in the components of velocity, the coefficients $\alpha_{ik}$ being taken as known functions, not depending explicitly on the time, of the co-ordinates $q_1, \ldots, q_n$ themselves.[1]

In addition to the kinetic energy, the dynamical system is supposed to be characterized by another function, the potential energy $U(q_1, \ldots, q_n)$, which depends on the coordinates of position $q_i$ only and not on the velocities or the time.[2]

Hamilton's principle states that *the motion of a dynamical system in the interval of time $t_0 \leqq t \leqq t_1$ from a given initial position to a given final position is such that for this motion the integral*

$$(14b) \qquad H\{q_1, \ldots, q_n\} = \int_{t_0}^{t_1} (T - U) \, dt$$

*is stationary, in the class of all continuous functions $q_i(t)$ that have continuous derivatives up to, and including, the second order and that have the prescribed boundary values for $t = t_0$ and $t = t_1$*

---

[1] We obtain this expression for the kinetic energy $T$ by thinking of the individual rectangular coordinates of the particles of the system as expressed as functions of the coordinates $q_1 \ldots, q_n$. Then the rectangular velocity components of the individual particles can be expressed as linear homogeneous functions of the $\dot{q}_i$'s; from these we form the elementary expression for the kinetic energy, namely, half the sum of the products of the individual masses and the squares of the corresponding velocities.

[2] We restrict ourselves here to mechanical systems in which the forces acting are conservative and independent of time. As is shown in dynamical textbooks, the potential energy determines the external forces acting on the system (see p. 0000 for the case of a single particle). In bringing the system from one position into another, mechanical work is done; this is equal to the difference between the corresponding values $U$ and does not depend on the particular motion from one position to another.

This principle of Hamilton's is a fundamental principle of dynamics. It contains in condensed form the laws of dynamics. When applied to Hamilton's principle, the Euler equations (13a), give *Lagrange's equations,*

$$(14c) \qquad \frac{d}{dt}\frac{\partial T}{\partial \dot{q}_i} - \frac{\partial T}{\partial q_i} = -\frac{\partial U}{\partial q_i} \qquad (i = 1, 2, \ldots, n),$$

which are the fundamental equations of theoretical dynamics.

Here we shall only make one noteworthy deduction, namely, the law of *conservation of energy.*

Since the integrand in Hamilton's integral does not depend explicitly on the independent variable $t$, for the solution $q_i(t)$ of the differential equations of dynamics the expression

$$T - U - \sum \dot{q}_i \frac{\partial(T - U)}{\partial \dot{q}_i}$$

must be constant [see (13c)]. Since $U$ does not depend on the $\dot{q}_i$ and $T$ is a homogeneous quadratic function in them (cf. p. 119),

$$\sum \dot{q}_i \frac{\partial(T - U)}{\partial \dot{q}_i} = \sum \dot{q}_i \frac{\partial T}{\partial \dot{q}_i} = 2T.$$

Hence

$$T + U = \text{constant};$$

that is, *during the motion the sum of the kinetic energy and the potential energy does not vary with time.*

### d. Integrals Involving Higher Derivatives

Analogous methods can be used to attack the problem of the extreme values of integrals in which the integrand $F$ not only contains the required function $y = \phi$ and its derivative $\phi'$ but also involves higher derivatives. For example, suppose we wish to find the extreme values of an integral of the form

$$(15a) \qquad I\{\phi\} = \int_{x_0}^{x_1} F(x, \phi, \phi', \phi'')\, dx,$$

where in the comparison those functions $y = \phi(x)$ are admissible that, together with their first derivatives, have prescribed values at the end points of the interval and that have continuous derivatives up to, and including, the fourth order.

To find necessary conditions for an extreme value, we again assume that $y = u(x)$ is the desired function. We embed $u(x)$ in a family of functions $y = \phi(x) = u(x) + \varepsilon\eta(x)$, where $\varepsilon$ is an arbitrary parameter and $\eta(x)$ an arbitrarily chosen function with continuous derivatives up to, and including, the fourth order that vanishes together with its first derivatives at the end points. The integral then takes the form $G(\varepsilon)$, and the necessary condition

$$(15b) \qquad\qquad G'(0) = 0$$

must be satisfied for all choices of the function $\eta(x)$. Proceeding in a way analogous to that on p. 744, we differentiate under the integral sign and thus obtain the above condition in the form

$$(15c) \qquad\qquad \int_{x_0}^{x_1} (\eta F_u + \eta' \, F_{u'} + \eta'' \, F_{u''}) \, dx = 0,$$

which must be satisfied if $u$ is substituted for $\phi(x)$. Integrating once by parts, we reduce the term in $\eta'(x)$ to one in $\eta$, and integrating twice by parts, we reduce the term in $\eta''(x)$ to one in $\eta$; taking the boundary conditions into account, we easily obtain

$$(15d) \qquad\qquad \int_{x_0}^{x_1} \eta\left(F_u - \frac{d}{dx} F_{u'} + \frac{d^2}{dx^2} F_{u''}\right) dx = 0.$$

Hence, the necessary condition for an extreme value (i.e., that the integral may be stationary) is Euler's differential equation

$$(15e) \qquad\qquad L[u] = F_u - \frac{d}{dx} F_{u'} + \frac{d^2}{dx^2} F_{u''} = 0.$$

The reader can verify for himself that this is a differential equation of the fourth order.[1]

### e. Several Independent Variables

The general method for finding necessary conditions for an extreme value can equally well be applied when the integral is no longer a simple integral but a multiple integral. Let $D$ be a given region

---

[1] In deriving (15e) from (15d) we have to restrict $\eta$ in Lemma I (p. 744) to functions of class $C^4$ for which $\eta$ and $\eta'$ vanish at the end points. It is clear from the proof of the lemma on p. 747 that the conclusion is valid under these more restrictive conditions.

bounded by a curve $\Gamma$ in the $x$, $y$-plane. We assume that $D$ and $\Gamma$ are sufficiently regular to permit application of the rule for integration by parts (p. 557). Let $F(x, y, \phi, \phi_x, \phi_y)$ be a function that is continuous and twice continuously differentiable with respect to all five of its arguments. If in $F$ we substitute for $\phi$ a function $\phi(x, y)$ that has continuous derivatives up to, and including, the second order in the region $D$ and has prescribed boundary values on $\Gamma$ and if we replace $\phi_x$ and $\phi_y$ by the partial derivatives of $\phi$, $F$ becomes a function of $x$ and $y$, and the integral

$$(16a) \qquad I\{\phi\} = \iint_D F(x, y, \phi, \phi_x, \phi_y) \, dx \, dy$$

has a value depending on the choice of $\phi$. The problem is that of finding a function $\phi = u(x, y)$ for which this value is an extreme value.

To find necessary conditions we again use the old method. We choose a function $\eta(x, y)$ that vanishes on the boundary $\Gamma$; has continuous derivatives up to, and including, the second order; and is otherwise arbitrary. We assume that $u$ is the required function and then substitute $\phi = u + \varepsilon\eta$ in the integral, where $\varepsilon$ is an arbitrary parameter. The integral again becomes a function $G(\varepsilon)$, and a necessary condition for an extreme value is

$$G'(0) = 0.$$

As before, this condition takes the form

$$(16b) \qquad \iint_D (\eta F_u + \eta_x F_{u_x} + \eta_y F_{u_y}) \, dx \, dy = 0.$$

To get rid of the terms in $\eta_x$ and $\eta_y$ under the integral sign we integrate one term by parts with respect to $x$ and the other with respect to $y$. Since $\eta$ vanishes on $\Gamma$, the boundary values on $\Gamma$ fall out, and we have

$$(16c) \qquad \iint \eta \left\{ F_u - \frac{\partial}{\partial x} F_{u_x} - \frac{\partial}{\partial y} F_{u_y} \right\} dx \, dy = 0.$$

Lemma I (p. 744) can be extended at once to more dimensions than one, and we immediately obtain *Euler's partial differential equation of the second order*,

$$(16d) \qquad F_u - \frac{\partial}{\partial x} F_{u_x} - \frac{\partial}{\partial y} F_{u_y} = 0.$$

*Examples*

1. $F = \phi_x{}^2 + \phi_y{}^2$. If we omit the factor 2, Euler's differential equation becomes

$$\Delta u = u_{xx} + u_{yy} = 0.$$

That is, Laplace's equation has been obtained from a variation problem.

2. Minimal surfaces. *Plateau's problem* is this: To find, over a region $D$, a surface $z = f(x, y)$ that passes through a prescribed curve in space whose projection is $\Gamma$ and whose area

$$\iint_D \sqrt{1 + \phi_x{}^2 + \phi_y{}^2} \; dx \; dy$$

is a minimum.

Here Euler's differential equation is

$$\frac{\partial}{\partial x} \frac{u_x}{\sqrt{1 + u_x{}^2 + u_y{}^2}} + \frac{\partial}{\partial y} \frac{u_y}{\sqrt{1 + u_x{}^2 + u_y{}^2}} = 0$$

or, in expanded form,

$$u_{xx}(1 + u_y{}^2) - 2u_{xy}u_xu_y + u_{yy}(1 + u_x{}^2) = 0.$$

This is the celebrated differential equation of minimal surfaces, which we have treated extensively elsewhere.[1]

## 7.4  Problems Involving Subsidiary Conditions. Lagrange Multipliers

In discussing ordinary extreme values for functions of several variables in Chapter 3 (p. 332) we considered the case where these variables are subject to certain subsidiary conditions. In this case the method of undetermined multipliers led to a particularly clear expression for the conditions that the function may have a stationary value. An analogous method is even more important in the calculus of variations. Here we shall briefly discuss only the simplest cases.

### a.  *Ordinary Subsidiary Conditions*

A typical case is that of finding a curve $x = x(t)$, $y = y(t)$, $z = z(t)$, where $t_0 \leq t \leq t_1$, in three-dimensional space, expressed in terms of

---

[1]R. Courant, *Dirichlet's Principle, Conformal Mapping and Minimal Surfaces,* Interscience: New York, 1950.

the parameter $t$, subject to the subsidiary condition that the curve shall lie on a given surface $G(x, y, z) = 0$ and shall pass through two given points $A$ and $B$ on that surface. The problem is then to make an integral of the form

(17) $$\int_{t_0}^{t_1} F(x, y, z, \dot{x}, \dot{y}, \dot{z})\, dt$$

stationary by suitable choice of the functions $x(t)$, $y(t)$, $z(t)$, subject to the subsidiary condition $G(x, y, z) = 0$ and the usual boundary and continuity conditions.

This problem can be immediately reduced to the cases discussed on p. 753. We assume that $x(t)$, $y(t)$, $z(t)$ are the required functions. We assume further that on the portion of surface on which the required curve is to lie $z$ can be expressed in the form $z = g(x, y)$; this is certainly possible if $G_z$ differs from zero on this portion of the surface. If we assume that on the surface in question the three equations $G_x = 0$, $G_y = 0$, $G_z = 0$ are not simultaneously true and if we confine ourselves to a sufficiently small portion of surface, we can suppose without loss of generality that $G_z \neq 0$. Substituting $z = g(x, y)$ and $\dot{z} = g_x \dot{x} + g_y \dot{y}$ under the integral sign, we obtain a problem in which $x(t)$ and $y(t)$ are functions independent of one another. Thus, we can immediately apply the results of p. 755 and write down the conditions that the integral I may be stationary, by applying equations (13a) to the integrand

$$F(x, y, g(x, y), \dot{x}, \dot{y}, \dot{x}g_x + \dot{y}g_y) = H(x, y, \dot{x}, \dot{y}).$$

We then have the two equations

$$\frac{d}{dt}H_{\dot{x}} - H_x = \frac{d}{dt}F_{\dot{x}} - F_x + \frac{d}{dt}(F_{\dot{z}}g_x) - F_z g_x - F_{\dot{z}}\frac{\partial \dot{z}}{\partial x} = 0,$$

$$\frac{d}{dt}H_{\dot{y}} - H_y = \frac{d}{dt}F_{\dot{y}} - F_y + \frac{d}{dt}(F_{\dot{z}}g_y) - F_z g_y - F_{\dot{z}}\frac{\partial \dot{z}}{\partial y} = 0.$$

But

$$\frac{d}{dt}g_x = \frac{\partial \dot{z}}{\partial x}, \qquad \frac{d}{dt}g_y = \frac{\partial \dot{z}}{\partial y},$$

as we see at once on differentiation. Hence,

$$\frac{d}{dt}F_{\dot{x}} - F_x + g_x\left(\frac{d}{dt}F_{\dot{z}} - F_z\right) = 0,$$

$$\frac{d}{dt} F_{\dot{y}} - F_y + g_y \left( \frac{d}{dt} F_{\dot{z}} - F_z \right) = 0.$$

If, for brevity, we write

(18a)
$$\frac{d}{dt} F_{\dot{z}} - F_z = \lambda G_z,$$

with a suitable multiplier $\lambda(t)$ and use the relations (p. 229) $g_x = -G_x/G_z$, $g_y = -G_y/G_z$, we obtain the two further equations

(18b)
$$\frac{d}{dt} F_{\dot{x}} - F_x = \lambda G_x,$$

(18c)
$$\frac{d}{dt} F_{\dot{y}} - F_y = \lambda G_y.$$

We thus have the following condition that the integral may be stationary: If we assume that $G_x$, $G_y$, $G_z$ do not all vanish simultaneously on the surface $G = 0$, the necessary condition for an extreme value is the existence of a multiplier $\lambda(t)$ such that the three equations (18a, b, c) are simultaneously satisfied in addition to the subsidiary condition $G(x, y, z) = 0$. That is, we have four symmetrical equations determining the functions $x(t)$, $y(t)$, $z(t)$ and the multiplier $\lambda$.

The most important special case is the problem of finding the shortest line joining two points $A$ and $B$ on a given surface $G = 0$, on which it is assumed that the gradient of $G$ does not vanish. Here

$$F = \sqrt{\dot{x}^2 + \dot{y}^2 + \dot{z}^2},$$

and Euler's differential equations are

$$\frac{d}{dt} \frac{\dot{x}}{\sqrt{\dot{x}^2 + \dot{y}^2 + \dot{z}^2}} = \lambda G_x,$$

$$\frac{d}{dt} \frac{\dot{y}}{\sqrt{\dot{x}^2 + \dot{y}^2 + \dot{z}^2}} = \lambda G_y,$$

$$\frac{d}{dt} \frac{\dot{z}}{\sqrt{\dot{x}^2 + \dot{y}^2 + \dot{z}^2}} = \lambda G_z.$$

These equations are invariant with respect to the introduction of a new parameter $t$. That is, as the reader may easily verify for himself, they retain the same form if $t$ is replaced by any other parameter $\tau = \tau(t)$, provided that the transformation is 1–1, reversible, and continuously differentiable. If we take the arc length $s$ as the new parameter, so that $\dot{x}^2 + \dot{y}^2 + \dot{z}^2 = 1$, our differential equations take the form

$$(19) \qquad \frac{d^2x}{ds^2} = \lambda G_x, \qquad \frac{d^2y}{ds^2} = \lambda G_y, \qquad \frac{d^2z}{ds^2} = \lambda G_z.$$

The geometrical meaning of these differential equations is that the principal normal vectors[1] of the extremals of our problem are orthogonal to the surface $G = 0$. We call these curves *geodesics* of the surface. The shortest distance between two points on a surface, then, is necessarily given by an arc of a geodesic.

### Exercises 7.4a

1. Show that the same geodesics are also obtained as the paths of a particle constrained to move on the given surface $G = 0$, subject to no external forces. In this case the potential energy $U$ vanishes and the reader may apply Hamilton's principle (p. 758).
2. Let $C$ be a curve on a given surface $G(x, y, z) = 0$. At each point of $C$ take a perpendicular geodesic segment of fixed length and fixed orientation relative to $C$. The free end of the geodesic segment generates a curve $C'$. Show that $C'$, too, is perpendicular to the geodesic segment.

#### b. *Other Types of Subsidiary Conditions*

In the problem discussed above we were able to eliminate the subsidiary condition by solving the equation determining the subsidiary condition and thus reducing the problem directly to the type discussed previously. With the other kinds of subsidiary conditions that frequently occur, however, it is not possible to do this. The most important case of this type is the case of *isoperimetric* subsidiary conditions. The following is a typical example: With the previous boundary conditions and continuity conditions, the integral

$$(20a) \qquad I\{\phi\} = \int_{x_0}^{x_1} F(x, \phi, \phi')\, dx$$

is to be made stationary, the argument function $\phi(x)$ being subject to the further subsidiary condition

$$(20b) \qquad H\{\phi\} = \int_{x_0}^{x_1} G(x, \phi, \phi')\, dx = \text{a given constant } c.$$

---

[1] That is, the vectors $(\ddot{x}, \ddot{y}, \ddot{z})$; see p. 213.

The particular case $F = \phi$, $G = \sqrt{1 + \phi'^2}$ is the classical isoperimetric problem.

This type of problem cannot be attacked by our previous method of forming the "varied" function $\phi = u + \varepsilon\eta$ by means of an arbitrary function $\eta(x)$ vanishing on the boundary only, for in general, these functions do not satisfy the subsidiary condition in a neighborhood of $\varepsilon = 0$, except at $\varepsilon = 0$. We can attain the desired result, however, by a method similar to that used in the original problem, by introducing, instead of one function $\eta$ and one parameter $\varepsilon$, two functions $\eta_1(x)$ and $\eta_2(x)$ that vanish on the boundary and two parameters $\varepsilon_1$ and $\varepsilon_2$. Assuming that $\phi = u$ is the required function, we then form the varied function

$$\phi = u + \varepsilon_1\eta_1 + \varepsilon_2\eta_2.$$

If we introduce this function into the two integrals, we reduce the problem to the derivation of a necessary condition for the stationary character of the integral

$$I = \int_{x_0}^{x_1} F(x, u + \varepsilon_1\eta_1 + \varepsilon_2\eta_2, u' + \varepsilon_1\eta_1' + \varepsilon_2\eta_2')\, dx = K(\varepsilon_1, \varepsilon_2),$$

subject to the subsidiary condition

$$H = \int_{x_0}^{x_1} G(x, u + \varepsilon_1\eta_1 + \varepsilon_2\eta_2, u' + \varepsilon_1\eta'_1 + \varepsilon_2\eta_2')\, dx = M(\varepsilon_1, \varepsilon_2) = c;$$

the function $K(\varepsilon_1, \varepsilon_2)$ is to be stationary for $\varepsilon_1 = 0$, $\varepsilon_2 = 0$, where $\varepsilon_1, \varepsilon_2$ satisfy the subsidiary condition

$$M(\varepsilon_1, \varepsilon_2) = c.$$

A simple discussion, based on the previous results for ordinary extreme values with subsidiary conditions, and in other respects following the same lines as the account given on p. 743, then leads to this result:

*Stationary character of the integral is equivalent to the existence of a constant multiplier $\lambda$ such that the equation $H = c$ and Euler's differential equation*

$$\frac{d}{dx}(F_{u'} + \lambda G_{u'}) - (F_u + \lambda G_u) = 0$$

*are satisfied. An exception to this can only occur if the function u satisfies the equation*

$$\frac{d}{dx} G_{u'} - G_u = 0.$$

The details of the proof may be left to the reader, who may consult the literature on this subject.[1]

## Exercises 7.4b

1. Show that the geodesics on a cylinder are helices.
2. Find Euler's equations in the following cases:

   (a) $F = \sqrt{1 + y'^2} + yg(x)$

   (b) $F = \dfrac{y''^2}{(1 + y'^2)^3} + yg(x)$

   (c) $F = y''^2 - y'^2 + y^2$

   (d) $F = \sqrt[4]{1 + y'^2}$

3. If there are two independent variables, find Euler's equations in the following cases:

   (a) $F = a\phi_x^2 + 2b\phi_x\phi_y + c\phi_y^2 + \phi^2 d$

   (b) $F = (\phi_{xx} + \phi_{yy})^2 = (\Delta\phi)^2$

   (c) $F = (\Delta\phi)^2 + (\phi_{xx}\phi_{yy} - \phi_{xy}^2)$.

4. Find Euler's equations for the isoperimetric problem in which

$$\int_{x_0}^{x_1} (au'^2 + 2buu' + cu^2)\, dx$$

   is to be stationary subject to the condition

$$\int_{x_0}^{x_1} u^2\, dx = 1.$$

5. Let $f(x)$ be a given function. The integral

$$I(\phi) = \int_0^1 f(x)\phi(x)\, dx$$

   is to be made a maximum subject to the integral condition

$$H(\phi) = \int_0^1 \phi^2\, dx = K^2$$

   where $K$ is a given constant.

   (a) Find the solution $u(x)$ from Euler's equation.

   (b) Prove by applying Cauchy's inequality that the solution found in (a) gives the absolute maximum for $I$.

---

[1] See, for example, M. R. Hestenes, *Calculus of Variations and Optimal Control Theory*. John Wiley and Sons, New York, 1966. R. Courant and D. Hilbert: *Methods of Mathematical Physics*, Interscience Publishers, New York, 1953, Vol. I, Chapter IV.

6. Use the method of Lagrange's multiplier to prove that the solution of the classical isoperimetric problem is a cricle.

7. A thread of uniform density and given length is stretched between two points $A$ and $B$. If gravity acts in the direction of the negative $y$-axis. the equilibrium position of the thread is that in which the center of gravity has the lowest possible position. It is accordingly a question of making an integral of the form $\int_{x_0}^{x_1} y \sqrt{1 + y'^2}\, dx$ a minimum, subject to the subsidiary condition that $\int_{x_0}^{x_1} \sqrt{1 + y'^2}\, dx$ has a given constant value. Show that the thread will hang in a catenary.

8. Let $y = u(x)$ yield the smallest value for the integral $\int_{x_0}^{x_1} F(x, y, y')\, dx$ among all continuously differentiable functions $y(x)$ with prescribed boundary values $y(x_0) = y_0$, $y(x_1) = y_1$. Prove that $u(x)$ satisfies the inequality $F_{y'y'}(x, u(x), u'(x)) \geq 0$ (Legendre's condition) for all $x$ in the interval $x_0 \leq x \leq x_1$.

9. Let $(x_0, y_0)$ and $(x_1, y_1)$ be points lying above the $x$-axis. Find the extremals for the area under the graph of a function passing through the two points subject to the condition that the path between the two points has a fixed length.

# CHAPTER
# 8

# *Functions of a Complex Variable*

In Section 7.7 of Volume I we touched on the theory of functions of a complex variable and saw that this theory throws new light on the structure of functions of a real variable. Here we shall give a brief, but more systematic, account of the elements of that theory.

## 8.1 Complex Functions Represented by Power Series

### a. Limits and Infinite Series with Complex Terms

We start from the elementary concept of a complex number $z = x + iy$ (cf. Volume I, p. 104) formed from the imaginary unit $i$ and any two real numbers $x, y$. We operate with these complex numbers just as we do with real numbers, with the additional rule that $i^2$ may always be replaced by -1. We represent $x$, the *real part,* and $y$, the *imaginary part* of $z$, by rectangular coordinates in an $x, y$-plane or a complex $z$-plane. The number $\bar{z} = x - iy$ is called the complex number *conjugate* to $z$. We introduce polar coordinates $(r, \theta)$ by means of the relations $x = r \cos \theta$, $y = r \sin \theta$ and call $\theta$ the *argument* (or *amplitude*) of the complex number and

$$r = \sqrt{x^2 + y^2} = \sqrt{z\bar{z}} = |z|$$

its *absolute value (or modulus)*. We recall that

$$|z_1 z_2| = |z_1| |z_2|.$$

We can immediately establish the so-called *triangle inequality* satisfied by the complex numbers $z_1$, $z_2$, and $z_1 + z_2$,

$$|z_1 + z_2| \leq |z_1| + |z_2|,$$

and the further inequality

$$|u_1| - |u_2| \leq |u_1 - u_2|,$$

which follows immediately from it if we put $z_1 = u_1 - u_2$, $z_2 = u_2$.

The triangle inequality may be interpreted geometrically if we represent the complex numbers $z_1$, $z_2$ by vectors in the $x$, $y$-plane with components $x_1$, $y_1$ and $x_2$, $y_2$, respectively. The vector that represents the sum $z_1 + z_2$ is then simply obtained by vector addition of the first two vectors. The lengths of the sides of the triangle formed by this addition (see Fig. 8.1) are

$$|z_1|, |z_2|, |z_1 + z_2|.$$

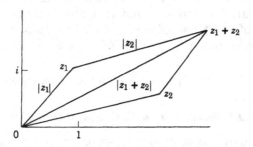

**Figure 8.1**    The triangle inequality for complex numbers.

Thus, the triangle inequality expresses the fact that any one side of a triangle is less than the sum of the other two.

The essentially new concept that we now consider is that of the *limit of a sequence of complex numbers*. We state the following definition: a sequence of complex numbers $z_n$ tends to a limit $z$ provided $|z_n - z|$ tends to zero. This, of course, means that the real part and the imaginary part of $z_n - z$ both tend to zero. It follows that Cauchy's test applies: the necessary and sufficient condition for the existence of a limit $z$ of a sequence $z_n$ is

$$\lim_{\substack{n \to \infty \\ m \to \infty}} |z_n - z_m| = 0.$$

A particularly important class of limits arises from *infinite series with complex terms*. We say that the infinite series with complex terms,

$$\sum_{v=0}^{\infty} c_v,$$

converges and has the sum $S$ if the sequence of partial sums,

$$S_n = \sum_{v=0}^{n} c_v,$$

tends to the limit $S$. If the real series with nonnegative terms

$$\sum_{v=0}^{\infty} |c_v|$$

converges, it follows, just as in Chapter 7 of Volume I (p. 514), that the original series with complex terms also converges. The latter series is then said to be *absolutely convergent*.

If the terms $c_v$ of the series, instead of being constants, depend on $(x, y)$, the coordinates of a point varying in a region $R$, the concept of *uniform convergence* acquires a meaning. The series is said to be uniformly convergent in $R$ if for an arbitrarily small prescribed positive $\varepsilon$ a fixed bound $N$ can be found, depending on $\varepsilon$ only, such that for every $n \geq N$ the relation $|S_n - S| < \varepsilon$ holds, no matter where the point $z = x + iy$ lies in the region $R$. *Uniform convergence of a sequence* of complex functions $S_n(z)$ depending on the point $z$ of $R$ is, of course, defined in exactly the same way. All these relations and definitions and the associated proofs correspond exactly to those with which we are already familiar from the theory of real variables.

The simplest example of a convergent series is the geometric series

$$1 + z + z^2 + z^3 + \cdots.$$

As for a real variable, the $n$th partial sum of this series is

$$S_n = \frac{1 - z^{n+1}}{1 - z},$$

and

(8.1) $\qquad 1 + z + z^2 + \cdots = \dfrac{1}{1 - z} \qquad$ for $|z| < 1.$

We see that the geometric series converges absolutely provided $|z| < 1$ and that the convergence is uniform provided $|z| \leq q$, where $q$ is any fixed positive number between 0 and 1. In other words, *the geometric series converges absolutely for all values of $z$ within the unit*

*circle and converges uniformly in every closed circle concentric with the*
*unit circle and with a radius less than unity.*

For the investigation of convergence the *comparison test* is again
available: If $|c_v| \leq p_v$, where $p_v$ is real and nonnegative and if the
infinite series

$$\sum_{v=0}^{\infty} p_v$$

converges, then the complex series $\sum c_v$ converges absolutely.

If the $p_v$'s are constants, while the $c_v$'s depend on a point $z$ varying
in $R$, the series $\sum c_v$ converges *uniformly* in the region in question.
The proofs are the same, word for word, as the corresponding proofs
for a real variable (Volume I, Chapter 7, p. 535) and therefore need
not be repeated here.

If $M$ is an arbitrary positive constant and $q$ a positive number
between 0 and 1, the infinite series with the positive terms $p_v = Mq^v$
or $Mq^{v-1}$ or

$$\frac{M}{v+1} q^{v+1}$$

also converge, as we know from Volume I, p. 543. We shall immediately make use of these series for purposes of comparison.

### b.  Power Series

The most important infinite series with complex terms are power
series, in which $c_v$ is of the form $c_v = a_v z^v$; that is, a power series
may be expressed in the form

$$P(z) = \sum_{v=0}^{\infty} a_v z^v$$

or, somewhat more generally, in the form

$$\sum_{v=0}^{\infty} a_v (z - z_0)^v,$$

where $z_0$ is a fixed point. As this form can, however, always be reduced to the preceding one by the substitution $z' = z - z_0$, we need
only consider the case where $z_0 = 0$.

The main theorem on power series is word for word the same as
the corresponding theorem for real power series in Chapter 7 of
Volume I (p. 541):

*If the power series converges for $z = \xi$, it converges absolutely for every value of z such that $|z| < |\xi|$. Further, if q is a positive number less than 1, the series converges uniformly within the circle $|z| \leq q|\xi|$.*

We can at once proceed to the following further theorem:

The two series

$$D(z) = \sum_{v=1}^{\infty} v a_v z^{v+1}$$

$$I(z) = \sum_{v=0}^{\infty} \frac{a_v}{v+1} z^{v+1}$$

*also converge absolutely and uniformly if $|z| \leq q|\xi|$.*

The proof follows exactly as before. Since the series $P(z)$ converges for $z = \xi$, it follows that the $n$th term, $a_n \xi^n$, tends to zero as $n$ increases. Hence, a positive constant $M$ certainly exists such that the inequality $|a_n \xi^n| < M$ holds for all values of $n$. If now $|z| = q|\xi|$, where $0 < q < 1$, we have

$$|a_n z^n| < M q^n, \quad |n a_n z^{n-1}| < \frac{M}{|\xi|} n q^{n-1}, \quad \left| \frac{a_n}{n+1} z^{n+1} \right| < \frac{M|\xi|}{n+1} q^{n+1}.$$

We thus obtain comparison series that, as we have seen already (p. 771), converge absolutely. Our theorem is thus proved.

In the case of a power series there are two possibilities: either it converges for all values of $z$ or there are values $z = \eta$ for which it diverges. Then, by the preceding theorem, the series must diverge for all values of $z$ for which $|z| > |\eta|$ (cf. Volume I, p. 541), and just as in the case of real power series, there is a *radius of convergence* $\rho$ such that the series converges when $|z| < \rho$ and diverges when $|z| > \rho$. The same applies to the two series $D(z)$ and $I(z)$, the value of $\rho$ being the same as for the original series. The circle $|z| = \rho$ is called the *circle of convergence* of the power series. No general statement can be made about the convergence or divergence of the series on the circumference of the circle itself, that is, for $|z| = \rho$.

#### c. Differentiation and Integration of Power Series

A convergent power series

$$P(z) = \sum_{v=0}^{\infty} a_v z^v$$

defines a function of the complex variable $z$ in the interior of its circle of convergence. In that region it is the limit to which the polynomials

$$P_n(z) = \sum_{\nu=0}^{n} a_\nu z^\nu$$

tend as $n$ tends to infinity.

A polynomial $f(z)$ may be differentiated with respect to the independent variable $z$ in exactly the same way as for a real variable. In the first place, we notice that the algebraic identity

$$\frac{z_1{}^n - z^n}{z_1 - z} = z_1{}^{n-1} + z_1{}^{n-2} z + \cdots + z^{n-1}$$

holds. If we now let $z_1$ tend to $z$, [1] we immediately have

$$\frac{d}{dz} z^n = \lim_{z_1 \to z} \frac{z_1{}^n - z^n}{z_1 - z} = n z^{n-1}.$$

In the same way, we immediately have

$$P_n'(z) = \frac{d}{dz} P_n(z) = \lim_{z_1 \to z} \frac{P_n(z_1) - P_n(z)}{z_1 - z} = \sum_{\nu=1}^{n} \nu a_\nu z^{\nu-1} = D_n(z).$$

We naturally call the expression $P_n'(z)$ the *derivative* of the complex polynomial $P_n(z)$.

We now have the following theorem, which is fundamental in the theory of power series:

*A convergent power series*

(8.2a)
$$P(z) = \sum_{\nu=0}^{\infty} a_\nu z^\nu$$

*may be differentiated term by term in the interior of its circle of convergence. That is, the limit*

(8.2b)
$$P'(z) = \lim_{z_1 \to z} \frac{P(z_1) - P(z)}{z_1 - z}$$

*exists, and*

---

[1] The concept of a limit for a *continuous* complex variable ($z_1 \to z$) can be introduced in exactly the same way as for a real variable.

(8.2c)     $P'(z) = \sum\limits_{v=1}^{\infty} v a_v z^{v-1} = \lim\limits_{n \to \infty} P_n'(z) = \lim\limits_{n \to \infty} D_n(z) = D(z).$

From this theorem it is at once clear that the power series

$$I(z) = \sum\limits_{v=0}^{\infty} \frac{a_v}{v+1} z^{v+1}$$

may be regarded as the *indefinite integral* of the first power series, that is, that $I'(z) = P(z)$.

The term-by-term differentiability of the power series is proved in the following way:

From p. 773 we know that the relation

$$D(z) = \lim\limits_{n \to \infty} D_n(z)$$

holds within the circle of convergence. We have to prove that the difference quotient

$$\frac{P(z_1) - P(z)}{z_1 - z}$$

differs in absolute value from $D(z)$ by less than a prescribed positive number $\varepsilon$ if only we take $z_1$ sufficiently close to $z$ within the circle of convergence. For this purpose, we form the difference quotient

$$D(z_1, z) = \frac{P(z_1) - P(z)}{z_1 - z} = \frac{P_n(z_1) - P_n(z)}{z_1 - z} + \sum\limits_{v=n+1}^{\infty} a_v \lambda_v,$$

where for brevity we write

$$\lambda_v = \frac{z_1{}^v - z^v}{z_1 - z} = z_1{}^{v-1} + z_1{}^{v-2} z + \cdots + z^{v-1}$$

If we keep to the notation used on p. 773 and if $|z| < q|\xi|$ and $|z_1| < q|\xi|$, then

$$|\lambda_v| \leqq v q^{v-1} |\xi|^{v-1}.$$

Hence,

$$|R_n| = \left| \sum\limits_{v=n+1}^{\infty} a_v \lambda_v \right| \leqq \sum\limits_{v=n+1}^{\infty} |a_v| v q^{v-1} |\xi|^{v-1} \leqq \frac{M}{|\xi|} \sum\limits_{v=n+1}^{\infty} v q^{v-1}.$$

Owing to the convergence of the series of positive terms $\sum \nu q^{\nu-1}$, the expression $|R_n|$ can therefore be made as small as we please, provided we make $n$ sufficiently large. We choose $n$ so large that this expression is less than $\varepsilon/3$ and so large—increasing $n$ further if necessary—that

$$|D(z) - D_n(z)| < \varepsilon/3.$$

We now choose $z_1$ so close to $z$ that the absolute value of

$$\frac{P_n(z_1) - P_n(z)}{z_1 - z}$$

also differs from $D_n(z)$ by less than $\varepsilon/3$. Then,

$$|D(z_1, z) - D(z)| \leqq \left| \frac{P_n(z_1) - P_n(z)}{z_1 - z} - D_n(z) \right|$$

$$+ |D_n(z) - D(z)| + |R_n|$$

$$< \frac{\varepsilon}{3} + \frac{\varepsilon}{3} + \frac{\varepsilon}{3} = \varepsilon,$$

and this inequality expresses the fact asserted.

Since the derivative of the function is again a power series with the same radius of convergence, we can differentiate again and repeat the process as often as we like. That is, *a power series can be differentiated as often as we please in the interior of its circle of convergence.*

*Power series are the Taylor series of the functions P(z) that they represent;* that is, *the coefficients $a_\nu$ may be expressed by the formula*

$$(8.3) \qquad\qquad a_\nu = \frac{1}{\nu!} P^{(\nu)}(0).$$

The proof is word for word the same as for a real variable (cf. Volume I, p. 545).

### d. *Examples of Power Series*

As we mentioned in Chapter 7 (p. 553) of Volume I, the power series for the elementary functions can immediately be extended to the complex variable; in other words, we can regard the power series for the elementary functions as complex power series and extend the definitions of these functions to the complex realm in this way. For example, the series

$$\sum_{v=0}^{\infty} \frac{z^v}{v!}, \quad \sum_{v=0}^{\infty} (-1)^v \frac{z^{2v}}{(2v)!}, \quad \sum_{v=0}^{\infty} \frac{(-1)^v z^{2v+1}}{(2v+1)!}, \quad \sum_{v=0}^{\infty} \frac{z^{2v}}{(2v)!}, \quad \sum_{v=0}^{\infty} \frac{z^{2v+1}}{(2v+1)!}$$

converge for all values of $z$. (This follows at once from comparison tests.) The functions represented by these power series are again denoted, respectively, by the symbols $e^z$, $\cos z$, $\sin z$, $\cosh z$, $\sinh z$, just as in the real case. The relations

(8.4a) $$\cos z + i \sin z = e^{iz},$$

(8.4b) $$\cosh z = \cos iz, \quad i \sinh z = \sin iz$$

now follow immediately from the power series. Again, by differentiating term by term, we obtain the relation

(8.4c) $$\frac{d}{dz} e^z = e^z.$$

As examples of power series with a finite radius of convergence, other than the geometric series, we consider the series

(8.4d) $$\log(1+z) = \sum_{v=1}^{\infty} (-1)^{v+1} \frac{z^v}{v}$$

$$\arctan z = \sum_{v=0}^{\infty} (-1)^v \frac{z^{2v+1}}{2v+1} = \frac{1}{2i} [\log(1+iz) - \log(1-iz)],$$

whose sums we again denote by *log* and *arc tan*. Here the radius of convergence is again 1. Differentiating term by term, we obtain geometric series and find

$$\frac{d \log(1+z)}{dz} = \frac{1}{1+z}, \quad \frac{d}{dz} (\arctan z) = \frac{1}{1+z^2}.$$

### Exercises 8.1

1. (a) Show that the operation of taking the conjugate of a complex number distributes over rational algebraic operations, for example,

$$\overline{\alpha \beta} = \overline{\alpha} \overline{\beta}.$$

   (b) Prove that if $f(z)$ is defined by a power series with real coefficients, then $\overline{f(z)} = f(\bar{z})$.

2. (a) Prove for a polynomial $P(z)$ with real coefficients that $\alpha$ is a root if and only if its complex conjugate is a root.
   (b) Prove under the assumption above that if $P(\alpha) = 0$ and $\alpha$ is not real, $\alpha = a + ib$ and $b \neq 0$, then $P(z)$ has the real quadratic factor.

$$(z - \alpha)(z - \bar{\alpha}) = z^2 - 2az + a^2 + b^2.$$

3. (a) Show that $|z - \alpha| = \lambda \, | \, z - \beta \, |$, $\lambda \neq 1$, $\lambda$ real is the equation of a circle. Determine the center $z_0$ and the radius $r$ of the circle. If $\lambda = 1$ what is the locus of this equation?

   (b) Show that the *general linear transformation*

$$z' = \frac{\alpha z + \beta}{\gamma z + \delta},$$

   where $\alpha\delta - \beta\gamma \neq 0$, transforms circles and straight lines into circles and straight lines.

4. For which points $z = x + iy$ is

$$\left| \frac{z - 1}{z + 1} \right| \leq 1?$$

5. Prove that if $\Sigma \, a_n \, z^n$ is *absolutely* convergent for $z = \zeta$, then it is uniformly convergent for every $z$ such that $|z| \leq |\zeta|$.

6. Using the power series for cos $z$ and sin $z$, show that

$$\cos^2 z + \sin^2 z = 1.$$

7. For what values of $z$ is

$$\sum_{\nu=1}^{\infty} \frac{z^\nu}{1 - z^\nu}$$

   convergent?

## 8.2 Foundations of the General Theory of Functions of a Complex Variable

### a. *The Postulate of Differentiability*

As we have seen above, all functions that are represented by power series possess a derivative and an indefinite integral. This fact may be made the starting point for the general theory of functions of a complex variable. The object of such a theory is to extend the differential and integral calculus to functions of a complex variable. In particular, it is important that the concept of function should be generalized for complex independent variables in such a way that it comprises any function that is differentiable in a complex region.

We could, of course, confine ourselves from the very beginning to the consideration of functions that are represented by power series and thus satisfy the postulate of differentiability. There are, however, two objections to this procedure. In the first place, we cannot tell a priori whether the postulate of the differentiability of a complex function necessarily implies that the function can be expanded in a power series. (In the case of the real variable we saw that functions even exist that possess derivatives of any order and yet cannot be expanded in a power series; cf. Volume I, p. 462.) In the second place, we learn even from the the simple function $1/(1 - z)$, whose power series, the geometric series, converges in the unit circle only, that even for simple functional expressions the power series does not everywhere represent the function, which in this particular case we already know in other ways.

These difficulties can be avoided by a method of Weierstrass, and the theory of functions of a complex variable can actually be developed on the basis of the theory of power series. It is desirable, however, to emphasize another point of view, that of Cauchy and Riemann. In their method, functions are characterized not by *explicit expressions* but by simple *properties*. More precisely, the property that a function shall be differentiable, and not that it shall be capable of being represented by a power series, is to be used to mark out the domain in which a function is defined.

We start from the general concept of a complex function $\zeta = f(z)$ of the complex variable $z$. If $R$ is a region of the $z$-plane and if with every point $z = x + iy$ in $R$ we associate a complex number $\zeta = u + iv$ by means of any relation, $\zeta$ is said to be a complex function of $z$ in $R$. This definition, therefore, merely expresses the fact that every pair of real numbers $x$, $y$, such that the point $(x, y)$ lies in $R$, has a corresponding pair of real numbers $u$, $v$, that is, that $u$ and $v$ are any two real functions $u(x, y)$ and $v(x, y)$, defined in $R$, of the two real variables $x$ and $y$.

This concept of function embraces too much for complex calculus. We limit it in the first place by the condition that $u(x, y)$ and $v(x, y)$ must be continuous functions in $R$ with continuous first derivatives $u_x$, $u_y$, $v_x$, $v_y$. Further, we insist that our expression $u + iv = \zeta = f(z)$ $= f(x + iy)$ shall be *differentiable in R with respect to the complex independent variable z;* that is, the limit

$$\lim_{z_1 \to z} \frac{f(z_1) - f(z)}{z_1 - z} = \lim_{h \to 0} \frac{f(z + h) - f(z)}{h} = f'(z)$$

shall exist for all values of $z$ in $R$. This limit is then called the *derivative* of $f(z)$.

In order that the function may be differentiable, it is by no means sufficient that $u$ and $v$ should possess continuous derivatives with respect to $x$ and $y$. Our postulate of differentiability implies far more than differentiability does for functions of real variables, since $h = r + is$ can tend to zero through both real values ($s = 0$) and purely imaginary values ($r = 0$) or in any other way, and the *same* limit $f'(z)$ must result in all cases if the function is to be differentiable.

If, for example, we put $u = x$, $v = 0$, that is, $f(z) = f(x + iy) = x$, we have a correspondence in which $u(x, y)$ and $v(x, y)$ are continuously differentiable. For the derivative of $f$ with respect to $z$, however, by putting $h = r$, we obtain

$$\lim_{r \to 0} \frac{f(z + r) - f(z)}{r} = \lim_{r \to 0} \frac{x + r - x}{r} = 1,$$

whereas if we put $h = is$, we have

$$\lim_{s \to 0} \frac{f(z + is) - f(z)}{is} = \lim_{s \to 0} \frac{0}{is} = 0;$$

that is, we obtain two entirely different limits. For $\zeta = u + iv = x + 2iy$ we similarly obtain different limits for the difference quotient as $h$ tends to zero in different ways.

Thus, in order to ensure the differentiability of $f(z)$ with respect to $z$ we have to impose yet another restriction. This fundamental fact in the theory of functions of a complex variable is expressed by the following theorem:

*If $\zeta = u(x, y) + iv(x, y) = f(z) = f(x + iy)$, where $u(x, y)$ and $v(x, y)$ are continuously differentiable, the necessary and sufficient conditions that the function $f(z)$ be differentiable in the complex region are the so-called Cauchy-Riemann differential equations.*

(8.5a) $$u_x = v_y, \quad u_y = - v_x.$$

*In every open set $R$ where $u$ and $v$ are continuously differentiable and satisfy these conditions, $f(z)$ is said to be an analytic[1] function of the complex variable $z$, and the derivative of $f(z)$ is given by*

(8.5b) $$f'(z) = u_x + iv_x = v_y - iu_y = \frac{1}{i}(u_y + iv_y).$$

We shall first show that the Cauchy-Riemann differential equations constitute a *necessary* condition. We assume that $f'(z)$ exists. Ac-

---

[1]The term *holomorphic* is also used. A deeper theorem, not proved here, asserts that for $f$ differentiable in a region, the derivatives of $u$ and $v$ not only exist but automatically are continuous. Hence, actually, differentiability of $f$ implies continuous differentiability. In what follows, however, we shall not make use of that theorem and always *assume* that the differentiable $f$ considered have continuously differentiable real and imaginary parts or, equivalently, that $f'(z)$ is a continuous function of $z$.

cordingly, we must obtain the limit $f'(z)$ by taking $h$ equal to a *real* quantity $r$. That is,

$$f'(z) = \lim_{r \to 0} \left( \frac{u(x + r, y) - u(x, y)}{r} + i\frac{v(x + r, y) - v(x, y)}{r} \right)$$

$$= u_x + iv_x.$$

In the same way, we must obtain $f'(z)$ if we take $h$ to be a pure imaginary $is$; that is, we must have

$$f'(z) = \lim_{s \to 0} \left( \frac{u(x, y + s) - u(x, y)}{is} + i\frac{v(x, y + s) - v(x, y)}{is} \right)$$

$$= \frac{1}{i}(u_y + iv_y).$$

Hence,

$$u_x + iv_x = \frac{1}{i}(u_y + iv_y).$$

By equating real and imaginary parts, we at once obtain the Cauchy-Riemann equations.

These equations, however, also form a *sufficient* condition for the differentiability of the function $f(z)$. To prove this, we form the difference quotient [see formula (13) p. 41]

$$\frac{f(z + h) - f(z)}{h} = \frac{u(x + r, y + s) - u(x, y) + i\{v(x + r, y + s) - v(x, y)\}}{r + is}$$

$$= \frac{ru_x + su_y + irv_x + isv_y + \varepsilon_1|h| + i\varepsilon_2|h|}{r + is},$$

where $\varepsilon_1$ and $\varepsilon_2$ are two real quantities that tend to zero with $|h| = \sqrt{r^2 + s^2}$. If now the Cauchy-Riemann equations hold, the above expression immediately becomes

$$u_x + iv_x + \varepsilon_1 \frac{|h|}{r + is} + i\varepsilon_2 \frac{|h|}{r + is}.$$

We see at once that as $h \to 0$, this expression tends to the limit $u_x + iv_x$ independently of the way in which the passage to the limit $h \to 0$ is carried out.

We now use the Cauchy-Riemann equations, or the property of differentiability that is equivalent to them, as the definition of an analytic function, on which we shall base our deduction of all the properties of such functions.

### b.  The Simplest Operations of the Differential Calculus

All polynomials and all power series in the interior of their circle of convergence are analytic functions (see p. 776). We see at once that the operations that lead to the elementary rules of the differential calculus can be carried out in exactly the same way as for the real variable (see Volume I, pp. 201–206, 218–220). In particular, the following rules hold: The sum, the difference, the product, and (provided the denominator does not vanish) the quotient of analytic functions can be differentiated according to the elementary rules of the calculus and, hence, are again analytic functions. Further, an analytic function of an analytic function can be differentiated according to the chain rule and therefore is itself an analytic function.

We also note the following theorem:

*If the derivative of an analytic function $\zeta = f(z)$ vanishes everywhere in a region R, the function is a constant.*

PROOF.  We have by (8.5a, b) $v_y - iu_y = 0$ everywhere in $R$. Hence, $v_y = 0$, $u_y = 0$, and by virtue of the Cauchy-Riemann equations, $v_x = 0$, $u_x = 0$; that is, $u$ and $v$ are constants; hence, $\zeta$ is a constant.

### Application to the Exponential Function

We use this theorem to derive some of the basic properties of the exponential function, defined for all complex $z$ by the power series

$$e^z = \sum_{k=0}^{\infty} \frac{z^k}{k!} = 1 + \frac{z}{1!} + \frac{z^2}{2!} + \cdots.$$

Since we may differentiate this series (see p. 776), we find that

(8.6) $$\frac{d}{dz} e^z = 1 + z + \frac{z^2}{2!} + \cdots = e^z.$$

Thus, the exponential function $f(z) = e^z$ is a solution of the differential equation

$$f'(z) = f(z)$$

for all $z$. By the chain rule of differentiation, it follows then for any fixed complex $\zeta$ that

$$\frac{d}{dz} e^{z+\zeta} e^{-z} = \frac{d}{dz} f(z + \zeta) f(-z)$$

$$= f'(z + \zeta) f(-z) - f(z + \zeta) f'(-z)$$

$$= f(z + \zeta) f(-z) - f(z + \zeta) f(-z) = 0.$$

Using the theorem above, we see that

$$e^{z+\zeta} e^{-z}$$

is a constant independent of $z$. We find the value of this constant by putting $z = 0$, and since $e^0 = 1$, obtain

(8.6a) $$e^{z+\zeta} e^{-z} = e^\zeta$$

for all $z$ and $\zeta$. For $\zeta = 0$ it follows that

(8.6b) $$e^z e^{-z} = 1.$$

Consequently, *the exponential function is different from zero for all complex $z$ and the reciprocal of $e^z$ is $e^{-z}$*. Multiplying both sides of the identity (8.6a) by $e^z$ we arrive at the *functional equation of the exponential function*

(8.6c) $$e^{z+\zeta} = e^z e^\zeta,$$

which could not be derived as easily directly from the power series representation.

If $f(z)$ is any solution of the differential equation

(8.7a) $$f'(z) = f(z)$$

we have

$$\frac{d}{dz} f(z)e^{-z} = f'(z)e^{-z} - f(z)e^{-z} = 0.$$

Hence,

$$f(z)e^{-z} = \text{constant} = c.$$

Thus, the most general solution of the differential equation (8.7a) has the form

(8.7b) $$f(z) = ce^z$$

where $c$ is a constant.

We found on p. 777 that

(8.8a) $$e^{iz} = \cos z + i \sin z,$$

where $\cos z$ and $\sin z$ are defined by their power series. Replacing $z$ by $-z$, we find, since $\sin(-z) = -\sin z$

$$e^{-iz} = \cos z - i \sin z.$$

Multiplying the two relations, we see that

$$e^{iz} e^{-iz} = \cos^2 z + \sin^2 z.$$

Since $e^{iz} e^{-iz} = e^{iz-iz} = 1$, we have proved the identity

(8.8b) $$\cos^2 z + \sin^2 z = 1$$

for all complex $z$.

By (8.6c) and (8.8a),

(8.8c) $$e^{x+iy} = e^x e^{iy} = e^x(\cos y + i \sin y).$$

If here $x$ and $y$ are real, we find that the absolute value of $e^z = e^{x+iy}$ is given by

$$
\begin{aligned}
(8.8d) \quad |e^z| = |e^{x+iy}| &= |e^x \cos y + i e^x \sin y| \\
&= \sqrt{(e^x \cos y)^2 + (e^x \sin y)^2} = \sqrt{e^{2x}(\cos^2 y + \sin^2 y)} \\
&= e^x.
\end{aligned}
$$

Another important consequence of the relation (8.8a) connecting the exponential and trigonometric functions is obtained if we put $z = 2\pi$:

(8.9a) $$e^{2\pi i} = \cos(2\pi) + i \sin(2\pi) = 1.$$

More generally, from (8.6c) for $\zeta = 2\pi i$, we have

(8.9b) $$e^{z+2\pi i} = e^z.$$

Thus, *for complex arguments the exponential function is periodic and has the period $2\pi i$.*

Formula (8.8a) shows that for any integer $n$

(8.9c) $$e^{2n\pi i} = \cos(2n\pi) + i \sin(2n\pi) = 1.$$

One easily sees that the values $z$ of the form

$$z = 2n\pi i \qquad (n = \text{integer})$$

are the only ones for which

$$e^z = 1,$$

for if $z = x + iy$, with real $x, y$, we find from $e^z = 1$ and (8.8d) that $e^x = 1$, and hence, $x = 0$. Then

$$1 = e^{iy} = \cos y + i \sin y,$$

which yields

$$\cos y = 1, \sin y = 0.$$

Hence, $y$ must be a multiple of $2\pi$.

We conclude that an equation

(8.9d) $$e^z = e^\zeta$$

can hold if and only if

(8.9e) $$z = \zeta + 2n\pi i,$$

where $n$ is an integer, for multiplying (8.9d) by $e^{-\zeta}$, we get

$$e^{z-\zeta} = e^z e^{-\zeta} = 1.$$

### c. Conformal Transformation. Inverse Functions

By means of the functions $u(x, y)$ and $v(x, y)$ the points of the $z$-plane or $x, y$-plane are made to correspond to points of the $\zeta$-plane or $u, v$-plane. Thus, we have a transformation or mapping of regions of the $x, y$-plane onto regions of the $u, v$-plane determined by $\zeta = f(z) = u + iv$. By (8.5a, b), p. 780, the Jacobian of the transformation is

$$D = \frac{d(u,v)}{d(x,y)} = u_x v_y - u_y v_x = u_x^2 + v_x^2 = |f'(z)|^2.$$

The Jacobian is therefore different from zero and is, in fact, positive wherever $f'(z) \neq 0$. If we assume that $f'(z) \neq 0$, our previous results (p. 261) show that a neighborhood of the point $z_0$ in the $z$-plane, if sufficiently small, is mapped 1–1 and continuously on a region of the

$\zeta$-plane in the neighborhood of the point $\zeta_0 = f(z_0)$. This mapping is *conformal* (i.e., angles are unchanged by it), for as we have seen in Chapter 3 (p. 288), the Cauchy-Riemann equations are the necessary and sufficient conditions for the transformation to be conformal and to preserve not only the magnitude but also the sign of angles. We thus have the following result:

*Conformality of the transformation given by u(x, y) and v(x, y) and analytic character of the function f(z) = u + iv mean exactly the same thing, provided we avoid points z₀ for which f'(z₀) = 0.*

The reader should study the examples of conformal representation discussed in Chapter 3 (pp. 243–244) and prove that all these transformations can be expressed by analytic functions of simple form.

For a 1–1 conformal representation of a neighborhood of $z_0$ on a neighborhood of $\zeta_0$, the reverse transformation is also conformal. It follows that $z = x + iy$ may also be regarded as an analytic function $\phi(\zeta)$ of $\zeta = u + iv$. This function is called the *inverse* of $\zeta = f(z)$.

Instead of using this geometrical argument, we can establish the analytic character of this inverse directly by calculating the derivatives of $x(u, v)$, $y(u, v)$ as in (24d) on p. 0000. We have

$$(8.10) \qquad x_u = \frac{v_y}{D}, \quad x_v = -\frac{u_y}{D}, \quad y_u = -\frac{v_x}{D}, \quad y_v = \frac{u_x}{D},$$

and we see that the Cauchy-Riemann equations $x_u = y_v$, $x_v = -y_u$ are satisfied by the inverse function. As we can at once verify, the derivative of the inverse $z = \phi(\zeta)$ of the function $\zeta = f(z)$ is given by the formula

$$(8.10b) \qquad \frac{dz}{d\zeta}\frac{d\zeta}{dz} = 1.$$

## Exercises 8.2

1. Prove that the product and the quotient of analytic functions and the function of an analytic function are again analytic, using not the property of differentiability but the Cauchy-Riemann differential equations.
2. Show that if $|f(z)|$ is constant in a region $R$, then $f(z)$ is constant.
3. Where are the following functions continuous? Which ones are differentiable?

   (a) $\bar{z}$;   (b) $|z|$;   (c) $\dfrac{z + \bar{z}}{1 + |z|}$;   (d) $\dfrac{z^2 + \bar{z}^2}{|z|^2}$.

4. Prove that in the transformation $\zeta = \frac{1}{2}(z + 1/z)$ the circles with centers at the origin and the straight lines through the origin of the z-plane

are respectively transformed into confocal ellipses and hyperbolas in the $\zeta$-plane.

5. For the general linear transformation

$$\zeta = \frac{az+b}{cz+d} \qquad (ad-bc \neq 0),$$

there may be as many as two *fixed points,* values of $z$ for which $\zeta = z$. Show that if the transformation does have two fixed points, the family of circles through the two fixed points and the family of circles orthogonal to them transform into themselves. (For this purpose the straight line through the points and the perpendicular bisector of the segment joining them are considered to be "circles" of the respective families.

6. Relate the inversion in the unit circle to the analytic function $f(z) = 1/z$ and thus derive the basic properties of inversion stated in Section 3.3d, Exercise 4, p. 256.

7. Prove that a substitution of the form

$$\zeta = \frac{\alpha z + \bar{\beta}}{\beta z + \bar{\alpha}},$$

where $\alpha$ and $\beta$ are any complex numbers satisfying the relation

$$\alpha\bar{\alpha} - \beta\bar{\beta} = 1,$$

transforms the circumference of the unit circle into itself and the interior of the circle into itself. Prove also that if

$$\beta\bar{\beta} - \alpha\bar{\alpha} = 1,$$

the interior is transformed into the exterior.

8. Prove that any circle may be transformed by a substitution of the form $\zeta = (\alpha z + \beta)/(\gamma z + \delta)$ into the upper half-plane bounded by the real axis. (Use Exercise 4, p. 778.)

9. Prove that a substitution $\zeta = (\alpha z + \beta)/(\gamma z + \delta)$, where $\alpha\delta - \beta\gamma \neq 0$, leaves the cross ratio

$$\frac{(z_1 - z_3)/(z_2 - z_3)}{(z_1 - z_4)/(z_2 - z_4)}$$

of four points $z_1, z_2, z_3, z_4$ unaltered.

## 8.3 The Integration of Analytic Functions

### a. *Definition of the Integral*

The central theorem of the differential and integral calculus of functions of a real variable is that the *indefinite integral* of a function (the upper limit being undetermined) may be regarded as the primitive function or *antiderivative* of the original function (Volume I, p. 188).

A corresponding relation forms the nucleus of the theory of analytic functions of a complex variable.

We begin by extending the definition of the definite integral of a given function $f(z)$. Here it is convenient to use $t = r + is$, instead of the independent variable $z$, to denote the variable of integration. Let the function $f(t)$ be analytic in a region $R$, and let $t = t_0$ and $t = z$ be two points in this region, joined by an oriented curve $C$ that is sectionally smooth (see p. 88) and lies wholly within $R$ (Fig. 8.2). We then subdivide the curve $C$ into $n$ portions by means of the successive points $t_0, t_1, \ldots, t_n = z$ and form the sum

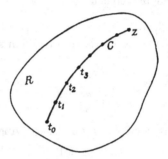

**Figure 8.2**

(8.11a)
$$S_n = \sum_{v=1}^{n} f(t_v')\,(t_v - t_{v-1}),$$

where $t_v'$ denotes any point lying on $C$ between $t_{v-1}$ and $t_v$. If we now make the subdivision finer and finer by letting the number of points increase without limit in such a way that the greatest of the lengths $|t_v - t_{v-1}|$ tends to zero, $S_n$ tends to a limit that is independent of the choice of the particular intermediate point $t_v'$ and of the points $t_v$.

This can be proved directly by a method analogous to that used to prove the corresponding theorem of the existence of the definite integral for real variables. For our purpose, however, it is more convenient to reduce the theorem to what we already know about real curvilinear integrals (cf. Chapter 1, p. 89) as follows: We put $f(t) = u(r, s) + iv(r, s)$, $t_v = r_v + is_v$, $t_v' = r_v' + is_v'$, $\Delta t_v = t_v - t_{v-1} = \Delta r_v + i\,\Delta s_v$. Then, we have

$$S_n = \sum_{v=1}^{n} u(r_v', s_v')\,\Delta r_v - v(r_v', s_v')\,\Delta s_v$$

$$+ i\left\{\sum_{r=1}^{n} v(r_v', s_v')\,\Delta r_v + u(r_v', s_v')\,\Delta s_v\right\}.$$

As $n$ increases the sums on the right side tend to the real curvilinear integrals

$$\int_C (u\ dx - v\ dy) \quad \text{and} \quad i\int_C (v\ dx + u\ dy),$$

respectively, and hence, as we asserted, $S_n$ tends to a limit. We call this limit the definite integral of the function $f(t)$ along the curve $C$ from $t_0$ to $z$ and write it

$$\int_{t_0}^z f(t)\ dt \quad \text{or} \quad \int_C f(t)\ dt.$$

Thus,

(8.11b) $$\int_C f(t)\ dt = \int_C (u\ dx - v\ dy) + i \int_C (v\ dx + u\ dy).$$

The definition of this definite integral at once gives an important estimate: *If $|f(t)| \leq M$ on the path of integration, where M is a constant and L is the length of the path of integration, then*

(8.11c) $$\left| \int_C f(t)\ dt \right| \leq ML,$$

*for by (8. 11a) and Volume I (p. 350),*

$$|S_n| \leq M \sum_v |t_r - t_{r-1}| \leq ML.$$

In addition, we point out that operations with complex integrals (in particular, combinations of different paths of integration) satisfy all the rules stated in this connection for curvilinear integrals in Chapter 1 (pp. 93–95).

### b.  Cauchy's Theorem

The most important property of functions of a complex variable is that the integral between $t_0$ and $z$ is largely independent of the choice of the path of integration $C$. In fact, we have Cauchy's theorem:

*If the function $f(t)$ is analytic in a simply connected region R, the integral*

$$\int_{t_0}^z f(t)\ dt = \int_C f(t)\ dt$$

*is independent of the particular choice of the path of integration C joining* $t_0$ *and z in R; the integral is an analytic function F(z) such that*

$$\frac{d}{dz} F(z) = \frac{d}{dz}\left[ \int_{t_0}^{z} f(t)\, dt\right] = f(z).$$

*F(z) is accordingly a primitive function or indefinite integral of f(z).*
    Cauchy's theorem may also be expressed as follows:

    *The integral of f(t) around a closed curve lying in a simply connected region in which f is analytic, has the value zero.*
    The proof that the integral is independent of the path follows immediately from (8.11b) and the main theorem on curvilinear integrals (cf. Chapter 1, p. 104); for both $u\, dx - v\, dy$, the integrand in the real part, and $v\, dx + u\, dy$, the integrand in the imaginary part, satisfy the condition of integrability, by virtue of the Cauchy-Riemann equations (8.5a). Thus the integral is a function of $x$, $y$ or of $x + iy = z$, $F(z) = U(x, y) + iV(x, y)$, and from our previous results for curvilinear integrals, we have the relations

$$U_x = u, \quad U_y = -v, \quad V_x = v, \quad V_y = u,$$

that is [see (8.5b), p. 780],

$$U_x = V_y, \quad U_y = -V_x, \quad U_x + iV_x = u + iv,$$

which shows that $F(z)$ is actually an analytic function in $R$ with the derivative $F'(z) = f(z)$.
    The assumption that the region is *simply-connected* is essential for the validity of Cauchy's theorem. For example, consider the function $1/t$, which is analytic everywhere in the $t$-plane except at the origin. We are not entitled to conclude from Cauchy's theorem that the integral of $1/t$, taken around a closed curve enclosing the origin, vanishes, for such a curve cannot be enclosed in a simply connected region in which the function is analytic. The simple connectivity of the region is destroyed by the exceptional point $t = 0$. If, for example, we take the integral around a circle $K$ given by $|t| = r$ or $t = re^{i\theta}$ in the positive sense and make $\theta$ the variable of integration ($dt = rie^{i\theta}\, d\theta$), we have

(8.12a) $$\int_K \frac{dt}{t} = \int_0^{2\pi} \frac{rie^{i\theta}}{re^{i\theta}}\, d\theta = 2\pi i;$$

that is, the value of the integral is not zero but $2\pi i$.

We can, however, extend Cauchy's theorem to multiply connected regions as follows:

*If a multiply connected region R is bounded by a finite number of sectionally smooth closed curves $C_1, C_2, \ldots$ and if $f(z)$ is analytic in the interior of this region and on its boundary,[1] then the sum of the integrals of the function along all the boundary curves is zero, provided that all the boundaries are described in the same sense relative to the interior of the region R, that is. that the region R is always on the same side, say the left-hand side, of the curve as it is described.*

The proof follows at once, on the model of the corresponding proofs for curvilinear integrals: We cut up the region $R$ into a finite number of simply-connected regions (Figs. 8.3 and 8.4), apply Cauchy's theorem

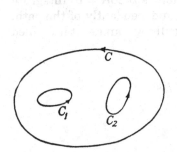

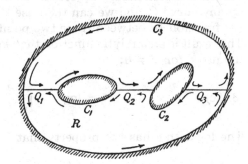

**Figure 8.3** $\int_C = \int_{C_1} + \int_{C_2}$.

**Figure 8.4** A multiply connected region $R$ subdivided by segments $Q_1, Q_2, \ldots$ into simply connected regions.

to these regions separately, and add the results. We can express this theorem in a somewhat different way:

*If the region R is formed from the interior of a closed curve C by cutting out of this interior the interiors of further curves $C_1, C_2, \ldots$, then*

(8.12b)
$$\int_C f(t)\, dt = \sum_\nu \int_{C_\nu} f(t)\, dt,$$

*where the integrals around the external boundary C and the internal boundaries are to be taken in the same sense.*

---

[1] A function is said to be analytic on a curve if it is analytic throughout a neighborhood, no matter how small, of this curve.

### c. Applications. The Logarithm, the Exponential Function, and the General Power Function

We can now use Cauchy's theorem as the basis for a satisfactory theory of the logarithm, the exponential function, and hence the other elementary functions, following a procedure similar to that adopted for a real variable (Volume I, Chapter 2, p. 145).

We begin by defining the logarithm as the integral of the function $1/t$. At first, we limit the path of integration by making it lie in a simply connected region of analyticity by making a cut along the negative real $x$-axis, that is, by permitting no path of integration to cross the negative real axis. More precisely, if we put $t = |t|(\cos \theta + i \sin \theta)$, we limit $\theta$ by the inequality $-\pi < \theta \leq \pi$. In the $t$-plane, after the cut has been made, we join the point $t = 1$ to an arbitrary point $z$ by any curve $C$, and we can then use Cauchy's theorem to integrate the function $1/t$ between these two points, independently of the path. The result is an analytic function that we call $\log z$ and that is defined uniquely for $z \neq 0$:

$$(8.12c) \qquad \zeta = \log z = \int_1^z \frac{dt}{t} = f(z).$$

The logarithm has the property that

$$(8.12d) \qquad \frac{d}{dz} (\log z) = \frac{1}{z}.$$

*The inverse of the logarithm can be identified with the exponential function.* We consider the function $e^{\log z}$ defined for $z \neq 0$ in the plane slit along the negative real axis, in accordance with the definition of the logarithm. Using the chain rule of differentiation, we find from (8.12d) and (8.6) for $z \neq 0$:

$$\frac{d}{dz} \frac{1}{z} e^{\log z} = - \frac{1}{z^2} e^{\log z} + \frac{1}{z^2} e^{\log z}.$$

Hence,

$$\frac{1}{z} e^{\log z} = \text{constant} = c.$$

If we take here $z = 1$, we find that

$$c = e^{\log 1} = e^0 = 1.$$

Thus,

(8.13a) $$e^{\log z} = z \qquad \text{for all } z \neq 0.[1]$$

Equation (8.13a) shows that the equation

(8.13b) $$e^w = z$$

has at least one solution $w$ for every $z \neq 0$, namely,

(8.13c) $$w = \log z.$$

Hence, *the exponential function assumes all complex values but zero.*

The solution, however, is not unique. We know from p. 785 that if $w$ is any particular solution of (8.13b), then the general solution has the form

$$w + 2n\pi i,$$

where $n$ is an integer. Hence:

*For any $z \neq 0$ the equation*

(8.13d) $$e^w = z$$

*is equivalent to*

(8.13e) $$w = \log z + 2n\pi i,$$

*where n is an integer.*

As an application we derive the *addition theorem* for logarithms. We have for any complex $z$, $\zeta$ that do not vanish, from (8.13a)

$$z\zeta = e^{\log z}\, e^{\log \zeta} = e^{\log z + \log \zeta}$$

and, on the other hand,

$$z\zeta = e^{\log(z\zeta)}.$$

---

[1]One is tempted to conclude similarly from

$$\frac{d}{dz} \log(e^z) = \frac{1}{e^z} e^z = 1$$

that

$$g(z) = \log(e^z) - z = \text{constant}.$$

But this is wrong, since $g(0) = 0$ and $g(2\pi i) = -2\pi i$. It is left to the reader to discover the fallacy of the argument.

Hence,

(8.14)                    $\log(z\zeta) = \log z + \log \zeta + 2n\pi i,$

where $n$ is an integer. Here, for positive real $z$, $\zeta$ we can always take $n = 0$ but not for others, as the following example shows.

The integral

$$\log z = \int_1^z \frac{dt}{t}$$

is easily evaluated explicitly by taking the straight line joining the points $t = 1$ and $t = |z|$ together with the circular arc $|t| = |z|$ as the path of integration. Setting $t = |z|e^{i\zeta}$ on the circle, we have

(8.15)        $\log z = \int_1^{|z|} \frac{dt}{t} + \int_0^\theta i\,d\zeta = \log |z| + i\theta,$

where $\theta$ is the argument of the complex number $z$ (Fig. 8.5) For example,

$$\log 1 = 0, \quad \log i = \frac{\pi i}{2}, \quad \log(-1) = \pi i.$$

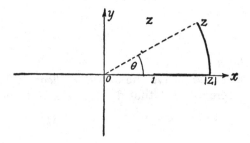

**Figure 8.5**   $\text{Log } z = \log |z| + i\theta.$

We notice that

$$\log [(-1)(-1)] = \log 1 = 0 = \log(-1) + \log (-1) - 2\pi i.$$

Thus, in formula (8.14), we cannot take $n = 0$ when $z = \zeta = -1$.

The value obtained in this way for the logarithm of any complex number $z$, whose argument lies in the interval $-\pi < \theta \leq \pi$, is often called the *principal value* of the logarithm. This terminology is justified by the fact that other values of the logarithm can be obtained by removing the condition that the negative real axis must not be crossed. We can then join the point 1 to the point $z$ by a path that encloses the origin $t = 0$. On this curve, the argument of $t$ will increase up to a value that is greater or less than the argument previously assigned to $z$ by $2\pi$. We then have the value

$$\log z = \log |z| + i\theta \pm 2\pi i$$

for the integral (Fig. 8.6). In the same way, by making the curve travel around the origin in one direction or the other any integral number of times $n$, we obtain the value

(8.16) $$\log z = \log |z| + i\theta + 2n\pi i.$$

This expresses the *many-valuedness of the logarithm*.[1] Formula (8.16) represents the general solution of the equation $e^{\log z} = z$.

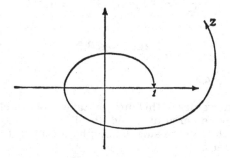

**Figure 8.6**  $\text{Log } z = \log |z| + i\theta + 2\pi i$.

Now that we have introduced the logarithm and the exponential function it is easy to define the general power functions $a^z$ and $z^a$, where $\alpha$ and $a$ are complex constants (cf. the corresponding discussion for the real variable in Volume I, p. 152). We define $a^z$ by the relation

(8.16a) $$a^z = e^{z \log a} \qquad (a \neq 0),$$

where the principal value of $\log a$ is to be taken. In the same way we define $z^\alpha$ by the relation

---

[1]Of course, the many-valued logarithm is not a function in the sense of a univalent assignment of a complex logarithm to each number $z$; the principal value is a function in that sense.

(8.16b) $$z^a = e^{a \log z} \qquad (z \neq 0).$$

While the function $a^z$ is defined uniquely if we use the principal value of log $a$ in the definition, the many-valuedness of the function $z^a$ goes deeper. Taking the many-valuedness of log $z$ into account, we see that along with any one value of $z^a$ we also have all the other values obtained by multiplying one value by $e^{2n\pi i a}$, where $n$ is any positive or negative integer. If $a$ is rational, say $a = p/q$, where $p$ and $q$ are integers prime to one another, among these multipliers there are only a finite number of different values (whose $q$th power must be unity). If, however, $a$ is irrational, we obtain an infinite number of different multipliers. The many-valuedness of the function $z^a$ will be discussed in greater detail on p. 815.

As we see from the chain rule, these functions satisfy the differentiation formulae

(8.16c) $$\frac{d(a^z)}{dz} = a^z \log a, \quad \frac{d(z^a)}{dz} = a z^{a-1}.$$

## Exercises 8.3

1. Consider $\int \dfrac{2z - 1}{z^2 - 1}\, dz.$

   (a) What are the values of this integral taken counterclockwise around small circles centered at 1 and at $-1$?
   (b) Describe a closed path surrounding both 1 and $-1$ about which the integral is zero.

2. Investigate the extensions of the laws of exponents,

   $$a^s a^t = a^{s+t}, \quad s^a t^a = (st)^a, \quad (a^s)^t = a^{st} = (a^t)^s,$$

   from the real to the complex domain and discuss the complications that arise from many-valuedness in the definition $z^a = \exp[a\,(\log z + 2n\pi i)]$, where log $z$ is the principal value of the logarithm.

3. (a) Show that all values of $i^i$ are real.
   (b) Find general conditions on complex $z$ $(z \neq 0)$ and $\zeta$ such that all values of $z^\zeta$ are real.
   (c) Is it possible to choose real $x$ and $\xi$, such that all the values of $x^\xi$ are real?

4. *The gamma function:* Prove that the integral

   $$\Gamma(z) = \int_0^\infty t^{z-1}\, e^{-t}\, dt,$$

   where the principal value of $t^{z-1}$ is taken, extended over all real values of the variable of integration $t$, is an analytic function of the parameter $z = x + iy$ if $x > 0$. Show directly that the expression $\Gamma(z)$ can be differentiated with respect to $z$. Prove that the gamma function thus defined for the complex variable satisfies the functional equation $\Gamma(z + 1) = z\Gamma(z)$.

5. *Riemann's zeta function:* Taking the principal value of $n^z$, form the infinite series

$$\sum_{n=1}^{\infty} \frac{1}{n^z} = \zeta(z). \qquad (z = x + iy),$$

Prove that this series converges if $x > 1$ and represents a differentiable function [$\zeta(z)$ is called Riemann's *zeta function*]. The proof can be carried out directly by a method like that for power series (cf. Volume I, p. 525).

6. (a) Apply Cauchy's theorem to the integral

$$\int \left(z + \frac{1}{z}\right)^m z^{n-1} \, dz \qquad (n > m > 0)$$

taken along a path consisting of the positive quadrant of the unit circle $|z| = 1$ and the parts of the axes between the origin and the circle, a small circular detour being made round $z = 0$; and deduce that

$$\int_0^{\pi/2} \cos^m\theta \, \cos n\theta \, d\theta = \frac{\sin[(n-m)\pi/2]}{2^{m+1}} \frac{\Gamma(m+1) \, \Gamma[(n-m)/2]}{\Gamma[(n+m)/2 + 1]}$$

(b) Prove that if $n = m$ the value of the latter integral is $\pi/2^{m+1}$. (In the complex integral the integrand may be taken as real on the positive half of the axis.)

## 8.4 Cauchy's Formula and Its Applications

### a. Cauchy's Formula

Cauchy's theorem for multiply connected regions leads to a fundamental formula, again Cauchy's, which expresses the value of an analytic function $f(z)$ at any point $z = a$ in the interior of a closed region $R$ throughout which the function is analytic, by means of the values that the function takes on the boundary $C$.

We assume that the function $f(z)$ is analytic in the simply connected region $R$ and on its boundary $C$. Then the function

$$g(z) = \frac{f(z)}{z - a}$$

is analytic everywhere in the region $R$, the boundary $C$ included, except at the point $z = a$. Out of the region $R$ we cut a circle of small radius $\rho$ about the point $z = a$, lying entirely within $R$ (Fig. 8.7), and then apply Cauchy's theorem (p. 790) to the function $g(z)$. If $K$ denotes the circumference of the circle described in the positive sense and $C$

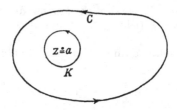

**Figure 8.7**

the boundary of $R$ described in the positive sense, Cauchy's theorem
states that [see (8.12b), p. 791]

$$\int_C g(z)\,dz = \int_K g(z)\,dz.$$

On the circle $K$ we have $z - a = \rho e^{i\theta}$, where the angle $\theta$ determines
the position of the point on the circumference. On the circle, therefore, $dz = \rho i e^{i\theta}\,d\theta$, and hence,

$$\int_K g(z)\,dz = i \int_0^{2\pi} f(a + \rho e^{i\theta})\,d\theta.$$

Since $f(z)$ is continuous at the point $a$, we have, provided $\rho$ is sufficiently small,

$$f(a + \rho e^{i\theta}) = f(a) + \eta,$$

where $|\eta|$ is less than an arbitrary prescribed positive quantity $\varepsilon$.
Hence,

$$\left| \int_0^{2\pi} f(a + \rho e^{i\theta})\,d\theta - \int_0^{2\pi} f(a)\,d\theta \right| = \left| \int_0^{2\pi} \eta\,d\theta \right| \leq 2\pi\varepsilon,$$

and therefore,

$$\int_0^{2\pi} f(a + \rho e^{i\theta})\,d\theta = 2\pi f(a) + \kappa,$$

where $|\kappa| \leq 2\pi\varepsilon$. Thus, if $\rho$ is sufficiently small,

$$\int_C g(z)\,dz = 2\pi i f(a) + \kappa i,$$

where $|\kappa i| < 2\pi\varepsilon$.

If we make ε tend to zero (by making ρ tend to zero), the right side of the equation tends to $2\pi i f(a)$, while the value of the left side, namely,

$$\int_C g(z)\, dz,$$

is unaltered. We thus obtain Cauchy's *fundamental integral formula*

(8.17a) $$f(a) = \frac{1}{2\pi i} \int_C \frac{f(z)}{z-a}\, dz.$$

If we now revert to the use of $t$ as variable of integration and then replace $a$ by $z$, the formula takes the form

(8.17b) $$f(z) = \frac{1}{2\pi i} \int_C \frac{f(t)}{t-z}\, dt.$$

This formula expresses the values of a function in the interior of a closed region in which the function is analytic by means of the values that the function takes on the boundary of the region.

In particular, if $C$ is a circle $t = z + re^{i\theta}$ with center $z$—that is, if $dt = ire^{i\theta}\, d\theta$—then

$$f(z) = \frac{1}{2\pi} \int_0^{2\pi} f(z + re^{i\theta})\, d\theta.$$

In words, *the value of a function at the center of a circular disk is equal to the mean of its values on the circumference, provided that the circle and its interior are contained in a region where the function is analytic.*

### b. Expansion of Analytic Functions in Power Series

Cauchy's formula has a number of important consequences. The chief of these is that *every analytic function can be expanded in a power series*, which connects the present theory with that in Section 8.1 (p. 772). More precisely, we have the following theorem:

*If the function $f(z)$ is analytic in the interior and on the boundary of a circle $|z - z_0| \leq R$, it can be expanded as a power series in $z - z_0$ that converges in the interior of that circle.*

In proving this we can take $z_0 = 0$ without loss of generality. (Otherwise we could merely introduce a new independent variable $z'$ by means of the transformation $z - z_0 = z'$). We now apply Cauchy's integral formula (8.17b) to the circle $C$, $|t| = R$, and write the integrand (using the geometric series) in the form

$$\frac{f(t)}{t-z} = \frac{f(t)}{t}\frac{1}{1-z/t} = \frac{f(t)}{t}\left(1 + \frac{z}{t} + \cdots \frac{z^n}{t^n}\right) + \frac{f(t)}{t}\left(\frac{z}{t}\right)^{n+1}\frac{1}{1-z/t}.$$

Since $z$ is a point in the interior of the circle, $|z/t| = q$ is a positive number less than unity, and we estimate the remainder of the geometric series,

$$r_n = \frac{1}{t}\frac{z^{n+1}}{t^{n+1}}\frac{1}{1-z/t},$$

by

$$|r_n| \leq \frac{1}{R}q^{n+1}\frac{1}{1-q}.$$

Introducing our expressions into Cauchy's formula and integrating term by term, we obtain

$$f(z) = c_0 + c_1 z + \cdots + c_n z^n + R_n,$$

where

$$c_v = \frac{1}{2\pi i}\int_C \frac{f(t)}{t^{v+1}}\,dt,$$

$$R_n = \frac{1}{2\pi i}\int_C f(t)r_n\,dt.$$

If $M$ is an upper bound of the values of $|f(t)|$ on the circumference of the circle, our estimate (8.11c) for complex integrals immediately gives

$$|R_n| \leq \frac{1}{2\pi R}\frac{q^{n+1}}{1-q}\ 2\pi RM = \frac{q^{n+1}}{1-q}M$$

for the remainder. Since $q < 1$, this remainder tends to zero as $n$ increases and we obtain the power series expansion for $f(z)$,

$$f(z) = \sum_{v=0}^{\infty} c_v z^v,$$

where

(8.18a)
$$c_\nu = \frac{1}{2\pi i} \int_C \frac{f(t)}{t^{\nu+1}} \, dt.$$

Our assertion is thus proved.

This theorem has important consequences. To begin with, we know from p. 776 that every power series can be differentiated as often as we please in the interior of its circle of convergence. Since every analytic function can be represented by a power series, it follows that *the derivative of a function in the interior of a region where the function is analytic is also differentiable* (i.e., is again an *analytic function*). In other words, *the operation of differentiation does not lead us out of the class of analytic functions*. As we already know that the same is true for the operation of integration, we see that *differentiation and integration of analytic functions can be carried out without any restrictions*. This is an agreeable state of affairs, which does not exist in the case of real functions.

Since, as we saw in Section 8.1 (p. 776), every power series is the Taylor series of the function that it represents, it now follows in general that every analytic function can be expanded in the neighborhood of a point $z = z_0$ in a region $R$ where the function is analytic in a Taylor series

(8.18b)
$$f(z) = f(z_0) + \sum_{\nu=1}^{\infty} \frac{f^{(\nu)}(z_0)}{\nu!} (z - z_0)^\nu.$$

The coefficients $c_\nu$ in (8.18a) are accordingly given by the formulae

(8.18c)
$$\frac{f^{(\nu)}(z_0)}{\nu!} = \frac{1}{2\pi i} \int_C \frac{f(z_0 + t)}{t^{\nu+1}} \, dt.$$

From this result we may also deduce an important fact about the radius of convergence of a power series. *The Taylor series of a function $f(z)$ in the neighborhood of a point $z = z_0$ converges in the interior of the largest circle whose interior lies wholly within the region where the function is defined and is analytic.*

By virtue of the theorems on differentiation and integration that we have now established as also valid for the complex variable, all the elementary functions of a real variable that we expanded in Taylor series have exactly the same Taylor series for a complex independent variable. For most of these functions we have already seen that this is true.

Here we may point out that, for example, the binomial series (cf. Volume I, p. 456).

(8.19a)
$$(1 + z)^\alpha = \sum_{v=0}^{\infty} \binom{\alpha}{v} z^v$$

is also valid for the complex variable if $|z| < 1$ and $\alpha$ is any complex exponent, provided that

(8.19b)
$$(1 + z)^\alpha = e^{\alpha \, \log(1+z)}$$

is formed from the *principal value* of log $(1 + z)$.

The fact that the radius of convergence of this series is equal to unity follows from what we have just said, together with the remark that the function $(1 + z)^\alpha$ is no longer analytic at the point $z = -1$, for if it were, all the derivatives would exist there, which is certainly not the case. The circle with radius 1 with the point $z = 0$ as center is therefore the largest circle in the interior of which the function is still analytic.

This example illustrates that the convergence behavior of power series, which real analysis leaves in mystery, becomes completely intelligible in the light of the fact that we have just proved about the radius of convergence.

For example, the failure of the geometric series representing $1/(1 + z^2)$ to converge on the unit circle is a simple consequence of the fact that the function is no longer analytic for $z = i$ and $z = -i$. We also see now that the power series

(8.20)
$$\frac{z}{e^z - 1} = \sum \frac{B_v{}^* z^v}{v!},$$

which defines *Bernoulli's numbers* (cf. Volume I, p. 562), must have the circle $|z| = 2\pi$ as its circle of convergence, for the denominator of the function vanishes for $z = 2\pi i$ but (apart from the origin) at no point interior to the circle $|z| \leq 2\pi$.

### c. The Theory of Functions and Potential Theory

Since analytic functions $f = u + iv$ may be differentiated as often as we please, it follows that the functions $u(x, y)$ and $v(x, y)$ also have continuous derivatives of any order. We may, therefore, differentiate the Cauchy-Riemann equations. If we differentiate the first equation with respect to $x$ and the second with respect to $y$ and add, we have

$$\Delta u = u_{xx} + u_{yy} = 0;$$

in the same way, the imaginary part $v$ satisfies the same equation

$$\Delta v = v_{xx} + v_{yy} = 0.$$

In other words, *the real part and the imaginary part of an analytic function are potential functions.*

If two potential functions $u$, $v$ satisfy the Cauchy-Riemann equations, $v$ is said to be *conjugate* to $u$, and $-u$ conjugate to $v$.

This suggests that the theory of functions of a complex variable and potential theory in two dimensions are essentially equivalent to one another.

### d. The Converse of Cauchy's Theorem

Cauchy's theorem has a valid converse (Morera's theorem):

*If the integral of the continuous function $\zeta = u + iv = f(z)$ around every closed curve $C$ in its region of definition $R$ vanishes, then $f(z)$ is an analytic function in $R$.*

To prove this, we note that the integral

$$F(z) = \int_{t_0}^{z} f(t)\, dt$$

taken along any path joining a fixed point $t_0$ and a variable point $z$ is independent of the path. Then by (8.11c), p. 789,

$$\frac{F(z+h) - F(z)}{h} - f(z) = \frac{1}{h}\int_{z}^{z+h} [f(t) - f(z)]\, dt \to 0 \qquad (h \to 0).$$

Hence, $F(z)$ has the derivative $F'(z) = f(z)$. $F(z)$ is therefore analytic, and by our earlier result, so is its derivative $f(z)$.

The converse of Cauchy's theorem shows that the postulate of differentiability could have been replaced by the postulate of integrability (i.e., that the line integral is independent of the path). The equivalence of these two postulates is a very characteristic feature of the theory of functions of a complex variable.

### e. Zeros, Poles, and Residues of an Analytic Function

If the function $f(z)$ vanishes at the point $z = z_0$, the constant term in the Taylor series of the function in powers of $z - z_0$

$$f(z) = f(z_0) + (z - z_0) f'(z_0) + \cdots,$$

vanishes, and possibly other terms of the series also vanish. A factor $(z - z_0)^n$ may then be taken out of the power series and we may write

$$f(z) = (z - z_0)^n \, g(z)$$

where $g(z_0) \neq 0$. A point $z_0$ for which this occurs is said to be a *zero of the function $f(z)$ of the nth order.*

The reciprocal $1/f(z) = q(z)$ of an analytic function, as we saw above, is also analytic, except at the points where $f(z)$ vanishes. If $z_0$ is a zero of $f(z)$ of the nth order, the function $q(z)$ can be represented in the neighborhood of the point $z_0$ in the form

$$q(z) = \frac{1}{(z - z_0)^n} \frac{1}{g(z)} = \frac{1}{(z - z_0)^n} h(z),$$

where $h(z)$ is analytic in the neighborhood of $z = z_0$. At the point $z = z_0$ the function $q(z)$ ceases to be analytic. We call this point a *singularity (singular point)*. In this particular case the singularity is called a *pole of the function $q(z)$ of the nth order.* If we think of the function $h(z)$ as expanded in powers of $(z - z_0)$ and then divided by $(z - z_0)^n$ term by term, in the neighborhood of the pole we obtain an expansion of the form

$$q(z) = c_{-n}(z - z_0)^{-n} + \cdots + c_{-1}(z - z_0)^{-1} + c_0 + c_1(z - z_0) + \cdots,$$

where the coefficients of the powers of $(z - z_0)$ are denoted by $c_{-n}$, $\ldots, c_{-1}, c_0, c_1, \ldots$.

If we are dealing with a pole of the first order (i.e., if $n = 1$), we obtain the coefficient $c_{-1}$ immediately from the relation

$$c_{-1} = \lim_{z \to z_0} (z - z_0)q(z).$$

Since

$$\frac{1}{q(z)(z - z_0)} = \frac{f(z)}{z - z_0} = \frac{f(z) - f(z_0)}{z - z_0},$$

we have for the coefficient of $1/(z - z_0)$ in the expansion of $q(z)$,

(8.21a)
$$c_{-1} = \frac{1}{f'(z_0)}.$$

In the same way, if $q(z) = r(z)/\phi(z)$, where $\phi(z)$ has a zero of the first order at $z = z_0$ and $r(z_0) \neq 0$, we have in the expansion of $q(z)$

(8.21b)
$$c_{-1} = \frac{r(z_0)}{\phi'(z_0)}.$$

If a function is defined and analytic everywhere in the neighborhood of a point $z_0$ but not at the point itself, its integral around a complete circle enclosing the point $z_0$ will in general not be zero. By Cauchy's theorem, however, the integral is independent of the radius of this circle and in general has the same value for all closed curves $C$ that form the boundary of a sufficiently small region enclosing the point $z_0$. The value of the integral taken around the point in the positive sense is called the *residue* at the point.

If the singularity is a pole of the $n$th order and if we integrate the expansion of the function, the integral of the series with positive indices is zero, as this power series is still analytic at the point $z_0$.

When integrated, the term $c_{-1}(z - z_0)^{-1}$ gives the value $2\pi i c_{-1}$, while the terms with higher negative indices give 0, for the indefinite integral of $(z - z_0)^{-\nu}$ for $\nu > 1$ is $(z - z_0)^{-\nu+1}/(1 - \nu)$, as in the real case, so that the integral around a closed curve vanishes.

*The residue of a function at a pole is therefore $2\pi i c_{-1}$.*

In the next section we shall become acquainted with the usefulness of this idea as expressed by the following theorem:

THEOREM OF RESIDUES. *If the function $f(z)$ is analytic in the interior of a region $R$ and on its boundary $C$ except at a finite number of interior poles, the integral of the function taken around $C$ in the positive sense is equal to the sum of the residues of the function at the poles enclosed by the boundary $C$.*

The proof follows at once from the statements above.

## Exercises 8.4

1. Prove, without using the theory of power series directly, that the derivative of an analytic function is differentiable by successive differentiation under the integral sign in Cauchy's formula and justify the validity of this process.

2. Show that the function

$$f(z) - \frac{1}{2\pi i} \int \frac{f(\zeta)}{\zeta - z} \frac{z^n}{\zeta^n} \, d\zeta,$$

where the integral is taken around a simple contour enclosing the points $\zeta = 0$ and $\zeta = z$, is a polynomial $g(z)$ of degree $n - 1$ such that

$$g^{(m)}(0) = f^{(m)}(0) \qquad (m = 0, 1, \ldots, n - 1).$$

3. Show that for every potential function $u$ it is possible to construct a conjugate function $v$ and to determine it uniquely apart from an additive constant provided the domain is simply connected.

4. What are the residues of $f(z) = (2z - 1)/(z^2 - 1)$ at its poles?

5. If $f(z)$ is bounded, $|f(z)| < M$, on the entire complex plane, show that

$$f(z) - f(0) = \frac{1}{2\pi i} \int f(\zeta) \left[ \frac{1}{\zeta - z} - \frac{1}{\zeta} \right] dt$$

can be made as small as one pleases by taking the integral over a sufficiently large circle. Consequently, $f(z) = f(0)$; that is, the function is constant.

6. Let $f(z)$ be analytic for $|z| \leq \rho$. If $M$ is the maximum of $|f(z)|$ on the circle $|z| = \rho$, then the coefficients of the power series for $f$,

$$f(z) = \sum_{v=0}^{\infty} a_v z^v,$$

satisfy the inequality

$$|a_v| \leq \frac{M}{\rho^v}.$$

Note that the conclusion of Exercise 5 follows also from this result.

7. Let $P(z) = \alpha_n z^n + \alpha_{n-1} z^{n-1} + \cdots + \alpha_0$ be a polynomial of positive degree $n$. Show that the assumption that $P(z)$ has no roots implies that $f(z) = 1/P(z)$ is bounded and, hence, constant, by Exercise 5 or Exercise 6, and, then, that $f(z)$ is identically zero. This proves the *fundamental theorem of algebra*, that every polynomial of positive degree with complex coefficients has at least one complex root.

8. Let $f(z)$ be analytic in the interior of, and on, a simple closed curve $C$ with the possible exception of a finite number of points in the interior. Consider

$$I = \frac{1}{2\pi i} \int_C \frac{f'(z)}{f(z)} dz,$$

taken in the positive sense around $C$.

   (a) Show that if $f$ has a zero of order $n$ at $\alpha$ and no other poles or zeros in the interior of or on $C$, then $I = n$.

   (b) Show that if $f$ has a pole of order $m$ at $\alpha$ and no poles or zeros at any other point in or on $C$, then $I = -m$.

   (c) Show that if $f$ has a finite number of zeros and poles in $C$, none on $C$, then $I$ is the number of zeros minus the number of poles, counting multiplicity; that is, if the zeros have multiplicities $n_1, n_2, \ldots, n_j$ and the poles, multiplicities $m_1, m_2, \ldots, m_k$, then

$$I = n_1 + n_2 + \cdots + n_j - m_1 - m_2 - \cdots - m_k.$$

9. (a) Two polynomials $P(z)$ and $Q(z)$ are such that at every point on a certain closed contour $C$

$$|Q(z)| < |P(z)|.$$

   Prove that the equations $P(z) = 0$ and $P(z) + Q(z) = 0$ have the same numbers of roots within $C$. (Consider the family of functions $P(z) + \theta Q(z)$, where the parameter $\theta$ varies from 0 to 1.)

   (b) Prove that all the roots of the equation

$$z^5 + az + 1 = 0$$

lie within the circle $|z| = r$ if

$$|a| < r^4 - \frac{1}{r}.$$

10. Use Exercise 8(b) to show that a polynomial $P(z)$ of degree $n$ has precisely $n$ roots, counting multiplicity.

11. (a) If $f(z)$ has one simple root $\alpha$ within a closed curve $C$, prove that this root is given by

$$\alpha = \frac{1}{2\pi i} \int_C z \, \frac{f'(z)}{f(z)} \, dz.$$

(b) Interpret the integral of part (a) when $f(z)$ has finitely many zeros and poles in, but not on, $C$.

12. Prove that $e^z$ cannot vanish for any value of $z$.

## 8.5 Applications to Complex Integration (Contour Integration)

Cauchy's theorem and the theorem of residues frequently enable us to evaluate real definite integrals by regarding these as integrals along the real axis of a complex plane and then simplifying the argument by suitable modification of the path of integration.[1] In this way we sometimes obtain surprisingly elegant evaluations of apparently complicated definite integrals, without necessarily being able to calculate the corresponding indefinite integrals. We shall discuss some typical examples.

### a. Proof of the Formula

$$(8.22) \qquad \int_0^\infty \frac{\sin x}{x} \, dx = \frac{\pi}{2}.$$

Here we give the following instructive proof of this important formula, which we have already discussed by other methods (Volume I, p. 589; Volume II, p. 471).

We integrate the function $e^{iz}/z$ in the complex $z$-plane along the path $C$ shown in Fig. 8.8, which consists of a semicircle $H_R$ of radius $R$ and a semicircle $H_r$ of radius $r$, both having their centers at the origin, and the two symmetrical intervals $I_1$ and $I_2$ of the real axis. Since the function $e^{iz}/z$ is regular in the circular ring enclosed by these boundaries, the value of the integral in question is zero. Combining the integrals along $I_1$ and $I_2$, we have

---

[1] It is always necessary to reduce the integral considered to one over a *closed* path in the complex plane.

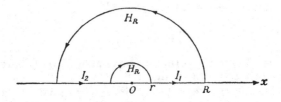

**Figure 8.8**

$$\int_{H_R} \frac{e^{iz}}{z}\, dz + \int_{H_r} \frac{e^{iz}}{z}\, dz + 2i \int_r^R \frac{\sin x}{x}\, dx = 0.$$

We now let $R$ tend to infinity. Then the integral along the semicircle $H_R$ tends to zero, for if we put $z = R(\cos \theta + i \sin \theta) = Re^{i\theta}$ for points of the semicircle, we have

$$e^{iz} = e^{iR \cos \theta}\, e^{-R \sin \theta},$$

and the integral becomes

$$i \int_0^\pi e^{iR \cos \theta}\, e^{-R \sin \theta}\, d\theta.$$

The absolute value of the factor $e^{iR \cos \theta}$ is 1, while the absolute value of the factor $e^{-R \sin \theta}$ is less than 1 and, moreover, tends uniformly to zero as $R$ tends to infinity, in every interval $\varepsilon \leq \theta \leq \pi - \varepsilon$. Hence, it follows at once that the integral along $H_R$ tends to zero as $R \to \infty$. As the reader can easily prove for himself, the integral along the semicircle $H_r$ tends to $-\pi i$ as $r \to 0$. The integral along the two symmetrical intervals $I_1$, $I_2$ of the real axis tends to

$$2i \int_0^\infty \frac{\sin x}{x}\, dx \qquad \text{as} \qquad R \to \infty \text{ and } r \to 0.$$

Combining these statements, we immediately obtain the relation (8.22).

### b. Proof of the Formula

(8.23)            $$\int_0^\infty (\cos ax)e^{-x^2}\, dx = \frac{1}{2} \sqrt{\pi}\, e^{a^2/4}$$

(Compare Section 4.12, p. 476 Exercise 9a.)

We integrate the expression $e^{-z^2}$ along a rectangle $ABB'A'$ (Fig. 8.9), in which the length of the vertical sides $AA'$, $BB'$ is $a/2$ and that

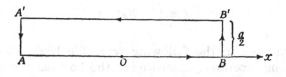

**Figure 8.9**

of the horizontal sides $AB$, $A'B'$ is $2R$. This integral has the value zero, by Cauchy's theorem. On the vertical sides we have

$$|e^{-z^2}| = |e^{-(x^2-y^2)} e^{-2ixy}| = e^{-R^2}e^{y^2} < e^{-R^2} e^{a^2/4},$$

and this expression tends uniformly to zero as $R$ tends to infinity. Thus, the portions of the whole integral that arise from the vertical sides tend to zero and if we carry out the passage to the limit $R \to \infty$ and note that $dz = d(x + \frac{1}{2}ia) = dx$, on $A'B'$ we may express the result of Cauchy's theorem as follows:

$$\int_{-\infty}^{+\infty} e^{-(x+ia/2)^2}\, dx = \int_{-\infty}^{\infty} e^{-x^2}\, dx.$$

That is, we can displace the path of integration of the infinite integral parallel to itself. By our previous result (see p. 415) the value of the integral on the right is $\sqrt{\pi}$. The integral on the left immediately becomes

$$e^{a^2/4}\int_{-\infty}^{\infty} e^{-x^2}(\cos\ ax - i\sin\ ax)dx = 2e^{a^2/4}\int_0^{\infty} \cos ax\, e^{-x^2}\, dx,$$

since $\sin ax$ is an odd function and $\cos ax$ an even function. This proves formula (8.23).

**c. *Application of the Theorem of Residues to the Integration of Rational Functions***

For the rational function

$$Q(z) = \frac{a_0 + a_1z + \cdots + a_mz^m}{b_0 + b_1z + \cdots + b_nz^n},$$

if the denominator has no real zeros and its degree exceeds that of the numerator by at least two, the integral

$$I = \int_{-\infty}^{\infty} Q(x) \, dx$$

can be evaluated in the following way: We begin by taking the integral along a contour consisting of the boundary of a semicircle $H$ of large radius $R$ (on which $z = Re^{i\theta}$, $0 \leq \theta \leq \pi$) and the real axis from $-R$ to $+R$. The radius $R$ is chosen so large that all the zeros of the denominator lie in the interior of the circle. Consequently, all the poles of the $Q(z)$ lie in the interior of the circle. On one hand, the integral is equal to the sum of the residues of $Q(z)$ within the semi-circle, while, on the other, it is equal to the integral

$$I_R = \int_{-R}^{R} Q(x) \, dx$$

plus the integral along the semicircle $H$. By our assumptions, a fixed positive constant $M$ exists such that for sufficiently large values of $R$ we have[1]

$$|Q(z)| < \frac{M}{R^2}.$$

The length of the circumference of the semicircle is $\pi R$. By our estimation formula (8.11c) on p. 789, the integral along the semicircle is therefore less in absolute value than

$$\pi R \frac{M}{R^2} = \frac{\pi M}{R}$$

and, hence, tends to zero as $R \to \infty$. This means that *the integral*

$$I = \int_{-\infty}^{\infty} Q(x) \, dx$$

*is equal to the sum of the residues of $Q(z)$ in the upper half-plane.*

We now apply this principle to some interesting special cases. We begin by taking

$$Q(z) = \frac{1}{az^2 + bz + c} = \frac{1}{f(z)},$$

---

[1]This follows immediately from the fact that $Q(z) = (1/z^2) \, R(z)$, where $R(z)$ tends to zero as $z \to \infty$ (when $n > m + 2$) or to $a_m/b_n$ (when $n = m + 2$).

where the coefficients $a$, $b$, $c$ are real and satisfy the conditions $a > 0$, $b^2 - 4ac < 0$. The function $Q(z)$ has one simple pole in the upper half-plane at the point

$$z_1 = \frac{1}{2a}\{-b + i\sqrt{4ac - b^2}\},$$

where the square root is to be taken positive, in the upper half-plane. By the general rule (8.21a), therefore, the residue is $2\pi i [1/f'(z_1)]$. Since

$$f'(z_1) = 2az_1 + b = i\sqrt{4ac - b^2},$$

we have

(8.24a)
$$\int_{-\infty}^{\infty} \frac{1}{ax^2 + bx + c}\, dx = \frac{2\pi}{\sqrt{4ac - b^2}}.$$

As a second example, we shall prove the formula (cf. Volume I, p. 290)

(8.24b)
$$\int_{-\infty}^{+\infty} \frac{dx}{1 + x^4} = \frac{1}{2}\pi\sqrt{2}.$$

Here again, we can immediately apply our general principle. In the upper half-plane the function $1/(1 + z^4) = 1/f(z)$ has the two poles $z_1 = \varepsilon = e^{(1/4)\pi i}$, $z_2 = -\varepsilon^{-1}$ (the two fourth roots of $-1$ that have a positive imaginary part). The sum of the residues is

$$2\pi i \left\{\frac{1}{f'(z_1)} + \frac{1}{f'(z_2)}\right\} = 2\pi i \frac{1}{4}\left(\frac{1}{z_1^3} + \frac{1}{z_2^3}\right) = \frac{\pi i}{2}(\varepsilon^{-3} - \varepsilon^3),$$

$$= -\pi i \cdot i \sin\frac{3\pi}{4} = \pi \sin\frac{\pi}{4} = \frac{1}{2}\pi\sqrt{2},$$

as was asserted.

The following proof of the formula

(8.24c)
$$\int_{-\infty}^{\infty} \frac{dx}{(1 + x^2)^{n+1}} = \frac{\pi}{4^n}\frac{(2n)!}{(n!)^2}$$

exemplifies the case where the residue at a pole of higher order has to be calculated.

If we replace $x$ by $z$, the denominator of the integrand is of the form $(z + i)^{n+1}(z - i)^{n+1}$, and the integrand accordingly has a pole of the $(n + 1)$-th order at the point $z = +i$. To find the residue at that point, we write

$$\frac{1}{(z^2 + 1)^{n+1}} = \frac{1}{f(z)} = \frac{1}{(z - i)^{n+1}} \frac{1}{(2i + z - i)^{n+1}}$$

$$= \frac{1}{(z - i)^{n+1}} \frac{1}{(2i)^{n+1}} \left(1 + \frac{z - i}{2i}\right)^{-n-1}.$$

If we expand the last factor by the binomial theorem, the term in $(z - i)^n$ has the coefficient

$$\frac{1}{(2i)^n} \binom{-n-1}{n} = \frac{1}{(2i)^n} (-1)^n \frac{(n + 1) \cdots 2n}{1 \cdot 2 \cdots n} = \frac{i^n}{2^n} \frac{(2n)!}{(n!)^2}.$$

The coefficient $c_{-1}$ in the series for the integrand in the neighborhood of the point $z = i$ is therefore equal to

$$\frac{1}{2^{2n+1}} \frac{1}{i} \frac{(2n)!}{(n!)^2}.$$

The residue $2\pi i c_{-1}$ is therefore

$$\frac{\pi}{2^{2n}} \frac{(2n)!}{(n!)^2},$$

which proves the formula.

As a further exercise the reader may prove for himself by the theory of residues that,

(8.24d) $$\int_0^\infty \frac{x \sin x}{x^2 + c^2} \, dx = \frac{1}{2} \pi e^{-|c|}$$

(replacing $\sin x$ by $e^{ix}$).

### d. The Theorem of Residues and Linear Differential Equations with Constant Coefficients

Let

$$a_0 + a_1 z + a_2 z^2 + \cdots + a_n z^n = P(z)$$

be a polynomial of the $n$th degree and $t$ a real parameter. We think of the integral

(8.25)
$$u(t) = \int_C \frac{e^{tz}f(z)}{P(z)}\,dz,$$

taken along any closed path $C$ in the $z$-plane, which does not pass through any of the zeros of $P(z)$, as a function $u(t)$ of the parameter $t$. Let $f(z)$ be a constant or any polynomial in $z$, of a degree that we shall assume to be less than $n$. By the rules for differentiation under the integral sign, which hold unaltered for the complex plane, we can differentiate the expression $u(t)$ once or repeatedly with respect to $t$. This differentiation with respect to $t$ under the integral sign is equivalent to multiplication of the integrand by $z$, $z^2$, $z^3$, . . . ., as the case may be. If we now form the differential expression $L[u] = a_0u + a_1u' + a_2u'' + \cdots + a_nu^{(n)}$, or, in symbolic notation, $P(D)u$, where $D$ denotes the symbol of differentiation $D = d/dt$, we have

$$P(D)u = L[u] = \int_C e^{tz}\,f(z)\,dz.$$

By Cauchy's theorem, the value of the complex integral on the right is 0; that is, the function $u(t)$ is a solution of the differential equation $L[u] = 0$. If $f(z)$ is any polynomial of the $(n-1)$-th degree, this solution contains $n$ arbitrary constants. We may accordingly expect to get in this way the most general solution of the linear differential equation with constant coefficients, $L[u] = 0$.

In fact, we do obtain the solutions in the form that we already know (cf. Chapter 6, p. 696), on evaluating the integral by the theory of residues, with the assumption that the curve $C$ encloses all the zeros $z_1, z_2, \ldots, z_n$ of the denominator $P(z) = a_n(z - z_1)(z - z_2) \cdots (z - z_n)$. If we assume to begin with that all these zeros are simple zeros, they are simple poles of the integrand, and the residue at the point $z_v$ is by formula (8.21b) given by

$$2\pi i\,\frac{f(z_v)}{P'(z_v)}\,e^{tz_v}.$$

By suitable choice of the polynomial $f(z)$ the expressions $f(z_v)/P'(z_v)$ can be made arbitrary constants; we accordingly obtain the solution in the form

$$u(t) = \sum_{v=1}^{n} c_v e^{z_v t},$$

in agreement with our previous results.

If a zero $z_v$ of the polynomial $P(z)$ is multiple, say $r$-fold, so that the corresponding pole of the integrand is of the $r$th order, the residue at the point $z_v$ must be determined by expanding the numerator $e^{tz} f(z) = e^{tz_v} e^{t(z-z_v)} f(z)$ in powers of $z - z_v$. We leave it to the reader to show that the residue at the point $z_v$ gives the solutions $te^{tz_v}$, . . ., $t^{r-1} e^{tz_v}$ as well as the solution $e^{tz_v}$.

## Exercises 8.5

1. (a) Let $f(z)$ be analytic and $g(z)$ have a pole of order $n$ at $z = \alpha$. Obtain an expression for the residue of $f(z)g(z)$ at $z = \alpha$.

   (b) In particular, if $g(z) = (z - \alpha)^{-n}$, show that the residue is

$$\frac{2\pi i}{(n-1)!} f^{(n-1)}(\alpha).$$

2. If $f(z)$ has a zero of order 2 at $\alpha$, show that the residue of $1/f(z)$ at $\alpha$ is

$$-\frac{4\pi i}{3} \frac{f'''(\alpha)}{f''(\alpha)^2}.$$

3. Evaluate, for nonnegative integers $n$, $m$ with $n > m$, the following integrals:

   (a) $\displaystyle\int_{-\infty}^{\infty} \frac{x^2}{1 + x^4}\, dx$

   (b) $\displaystyle\int_{-\infty}^{\infty} \frac{1}{(1 + x^4)^2}\, dx$

   (c) $\displaystyle\int_{-\infty}^{\infty} \frac{x^{2m}}{1 + x^{2n}}\, dx.$

4. Let $f(z)$ be a polynomial of degree $n$ with the simple roots $\alpha_1, \alpha_2, \ldots, \alpha_n$. Prove that

$$\sum_{v=1}^{n} \frac{\alpha_v{}^k}{f'(\alpha_v)} = 0 \qquad\qquad (k = 0, 1, \cdots, n-2).$$

$\left(\text{Consider } \displaystyle\int \frac{z^k}{f(z)}\, dz \text{ around a closed curve enclosing all the } \alpha_v.\right)$

5. Derive the result of (8.24d), namely,

$$\int_0^{\infty} \frac{x \sin x}{x^2 + c^2} = \frac{1}{2} \pi e^{-|c|}.$$

## 8.6  Many-Valued Functions and Analytic Extension

In defining functions both real and complex, we have hitherto always adopted the point of view that for each value of the independent variable the value of the function must be *unique*. Even Cauchy's theorem, for example, is based on the assumption that the function can be defined uniquely in the region under consideration. All the same, many-valuedness often arises of necessity in the actual construction of functions, (e.g., in finding the inverse of a unique function such as the $n$th power). In the real case, we separated different *one-valued branches* of the inverse function in inversion processes such as $\sqrt{z}$ or $\sqrt[n]{z}$. We shall see, however, that in the complex case this separation is no longer reasonable, for the various one-valued branches are now interconnected in a way that makes any separation of them rather artificial.

We must be content here with a very simple discussion based on typical examples.

For instance, we consider the inverse $\zeta = \sqrt{z}$ of the function $z = \zeta^2$. To each nonzero value of $z$ there correspond the two possible solutions $\zeta$ and $-\zeta$ of the equation $z = \zeta^2$. These two branches of the function are connected in the following way: Let $z = re^{i\theta}$. If we then put $\zeta = \sqrt{r}\, e^{i\theta/2} = f(z)$, $\zeta = f(z)$ is certainly analytic in every simply connected region $R$ excluding the origin [where $f(z)$ is no longer differentiable]. In such a region, $\zeta$ is uniquely defined, by our previous statement. If, however, we let the point $z$ move around the origin on a concentric circle $K$, say in the positive direction, $\zeta = \sqrt{r}\, e^{i\theta/2}$ will vary continuously; the angle $\theta$, however, will not return to its original value but will be increased by $2\pi$. Hence, in this continuous extension when we come back to the point $z$, we no longer have the initial value $\zeta = \sqrt{r}\, e^{i\theta/2}$, but the value $\sqrt{r}\, e^{i\theta/2}\, e^{2\pi i/2} = -\zeta$. We say that when the function $f(z)$ is continuously extended on the closed curve $K$ it is not unique.

The function $\sqrt[n]{z}$, where $n$ is an integer, exhibits exactly the same behavior. Here every revolution multiplies the value of the function by the $n$th root of unity—namely, $\varepsilon = e^{2\pi i/n}$—and the function only returns to its original value after $n$ revolutions.

In the case of the function $\log z$, we saw (p. 795) that there is a similar many-valuedness, in that, in traveling once continuously around the origin in the positive sense, the value of $\log z$ is increased by $2\pi i$.

Again, the function $z^a$ is multiplied by $e^{2\pi i a}$ per revolution.

All these functions, although in the first instance uniquely defined in a region $R$, are found to be many-valued when we extend them continuously (as analytic functions) and return to the starting point by a certain closed path. This phenomenon of many-valuedness and the associated general theory of analytic extension cannot be investigated in greater detail within the limits of this book. We merely point out that the uniqueness of the values of a function can theoretically be ensured by drawing certain lines in the $z$-plane that the path traced by $z$ is not allowed to cross, or, as we say, by making *cuts* along certain lines. These cuts are so arranged that closed paths in the plane that lead to many-valuedness are no longer possible.

For example, the function log $z$ is made one-valued by cutting the $z$-plane along the negative real axis. The same applies to the function $\sqrt{z}$. The function $\sqrt{1-z^2}$ becomes one-valued if we make a cut along the real axis between $-1$ and $+1$.

Once the plane has been cut in this way, Cauchy's theorem can at once be applied to these functions. We give a simple example by proving the formula

$$(8.27) \qquad I = \int_{-1}^{+1} \frac{1}{(x-k)\sqrt{1-x^2}}\, dx = \frac{2\pi}{\sqrt{k^2-1}},$$

where $k$ is a constant that does not lie on the real axis between $-1$ and $+1$.

We begin by noting that the function

$$\frac{1}{(z-k)\sqrt{1-z^2}}$$

is one-valued in the $z$-plane, provided we make a cut along the real axis from $-1$ to $+1$. If in the complex plane we approach this cut $S$ first from above and then from below, we obtain equal and opposite values for the square root $\sqrt{1-z^2}$, say, positive from above and negative from below. We now take the complex integral

$$\int_C \frac{dz}{(z-k)\sqrt{1-z^2}}$$

along a path $C$ as indicated in Fig. 8.10. By Cauchy's theorem we can make this path contract round the cut without altering the value of the integral. The integral is therefore equal to the limiting value obtained when this contraction is made, which is obviously equal to $2I$. On the other hand, if we take the integral of the same integrand along the circumference of a circle $K$ with radius $R$ and center at the origin, this integral, by our previous investigations, tends to zero as $R$ increases.[1] By the theorem of residues, however, the sum of the integrals along $C$ and $K$ is equal to the residue of the integrand at the

---

[1]In fact, its value is actually zero, since by Cauchy's theorem it is independent of the radius $R$, provided that the circle encloses the pole $z = k$.

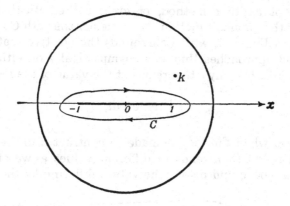

**Figure 8.10**

enclosed pole $z = k$; hence, $2I$ is equal to the residue in question. This residue is

$$2\pi i \lim_{z \to k} (z - k) \frac{1}{\sqrt{1 - z^2}} \frac{1}{(z - k)} = \frac{2\pi}{\sqrt{k^2 - 1}},$$

which proves our statement.

*Example of Analytic Extension: The Gamma Function*

In conclusion we give yet another example showing how an analytic function, originally defined in a part of the plane, can be extended beyond the original region of definition. We shall extend the gamma function, which was defined for $x > 0$ by the equation

(8.28) $$\Gamma(z) = \int_0^\infty t^{z-1} e^{-t} \, dt,$$

analytically for $x \leq 0$ also. We could do this by means of the functional equation

$$\Gamma(z) = \frac{1}{z} \Gamma(z + 1),$$

using this equation to define $\Gamma(z - 1)$ when $\Gamma(z)$ is known. By means of this equation, we can imagine $\Gamma(z)$ as extended first to the strip $-1 < x \leq 0$ and subsequently extended to the next strip $-2 < x \leq -1$, and so on.

We adopt another method, of greater theoretical interest, for extending the gamma function. We consider the path $C$ in the $t$-plane indicated in Fig. 8.11, which surrounds the positive real axis of the $t$-plane and approaches this axis asymptotically on either side. We easily see from Cauchy's theorem that the value of the loop-integral,[1]

$$\int_C t^{z-1} e^{-t}\, dt,$$

is unaltered when the loop is made to contract into the $x$-axis. The integrand $t^{z-1} e^{-t}$ then tends to different values as we approach the $x$-axis from above and below, the values differing by the factor $e^{2\pi i z}$.

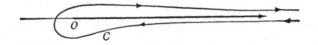

**Figure 8.11**    Loop-integral for the gamma function.

For $x > 0$, we thus obtain the formula

$$(1 - e^{2\pi i z})\, \Gamma(z) = \int_C t^{z-1} e^{-t}\, dt.$$

This formula is derived subject to the assumption that $x$, the real part of $z$, is positive. We see now, however, that the loop-integral has a meaning, no matter what the complex number $z$ is, since it avoids the origin $t = 0$. This loop-integral therefore represents a function defined throughout the $z$-plane. We then define this function by stating that it is equal to $(1 - e^{2\pi i z})\Gamma(z)$ throughout the $z$-plane. The gamma function has thus been analytically extended to the whole of the $z$-plane, except the points $x \leqq 0$ for which the factor $(1 - e^{2\pi i z})$ vanishes, that is, except the points $z = 0$, $z = -1$, $z = -2$, and so on.

For more detailed and extensive investigations the reader is referred to the literature of the theory of functions.[2]

### Miscellaneous Exercises 8

1. Write down the condition that three points $z_1$, $z_2$, $z_3$ may lie in a straight line.

---

[1]This is again an improper integral, which arises by a passage to a limit from an integral along a finite portion of $C$. The reader may satisfy himself that it exists by an argument similar to those previously employed.
[2]For example L. V. Ahlfors, *Complex Analysis*, N. Y.: McGraw-Hill, 1953.

2. Show that three distinct point $\alpha$, $\beta$, $\gamma$ of the complex plane form an isosceles triangle with vertex at $\gamma$ if and only if there exists a real positive $k$ for which

$$\frac{\gamma - \alpha}{\beta - \alpha} \frac{\gamma - \beta}{\alpha - \beta} = k.$$

3. Write down the condition that four points $z_1$, $z_2$, $z_3$, $z_4$ may lie on a cricle.

4. Let $A$, $B$, $C$, $D$ in the $z$-plane be four points in order on the circumference of a circle, with coordinates $z_1$, $z_2$, $z_3$, $z_4$. Using these complex coordinates, show that $AB \cdot CD + BC \cdot AD = AC \cdot BD$.

5. Prove that the equation $\cos z = c$ can be solved for all values of $c$.

6. For which values of $c$ has the equation $\tan z = c$ no solution?

7. For which values of $z$ is (a) $\cos z$, (b) $\sin z$ real?

8. Find the radius of convergence of the power series $\Sigma a_n z^n$, where

(a) $a_n = \dfrac{1}{n^s}$, $s$ being a complex number with a positive real part

(b) $a_n = n^n$

(c) $a_n = \log n$.

9. Evaluate the integrals

(a) $\displaystyle\int_0^\infty \frac{\cos x}{1 + x^4} dx$

(b) $\displaystyle\int_0^\infty \frac{x^2 \cos x}{1 + x^4} dx$

(c) $\displaystyle\int_0^\infty \frac{\cos x}{q^2 + x^2} dx$

(d) $\displaystyle\int_0^\infty \frac{x^{a-1}}{(x + 1)(x + 2)} dx$ for $1 < \alpha < 2$

by complex integration.

10. Find the poles and residues of the functions

$$\frac{1}{\sin z}, \quad \frac{1}{\cos z}, \quad \Gamma(z), \quad \cot z = \frac{\cos z}{\sin z}.$$

11. Show that if $x$ and $y$ are real

$$|\sinh(x + iy)| \geq A(x),$$

where $A(x)$ is independent of $y$ and tends to $\infty$ as $x \to \pm\infty$.

   By integrating $1/[(z - w) \sinh z]$ round a suitable sequence of contours, show that

$$\frac{1}{\sinh w} = \frac{1}{w} + 2w \sum_1^\infty \frac{(-1)^n}{w^2 + \pi^2 n^2}.$$

12. Find the limiting value of the integral

$$\int_{C_n} \frac{\cot \pi t}{t - z} \, dt$$

as $n \to \infty$, where the path of integration is a square $C_n$ with its sides parallel to the axes at a distance $n \pm \frac{1}{2}$ from the origin. Hence, using the theorem of residues, obtain the expression for $\cot \pi z$ in partial fractions.

13. Using the equation

$$\log(1 + z) = \int_0^z \frac{dt}{1 + t},$$

show that the power series for $\log (1 + z)$ converges everywhere on the unit circle $|z| = 1$, except at the point $z = -1$. By equating the imaginary part of the series to the imaginary part of $\log (1 + e^{i\theta})$, establishes the truth of the Fourier series (cf. Volume I, p. 592)

$$\frac{1}{2} \theta = \sin \theta - \frac{1}{2} \sin 2\theta + \frac{1}{3} \sin 3\theta - \cdots \qquad (-\pi < \theta < \pi).$$

14. Prove that if $f$ is analytic $(d^n/dx^n) f(\sqrt{x})$ is equal to the result obtained by putting $y$ and $a$ each equal to $\sqrt{x}$ in the expression for

$$2 \frac{\partial^n}{\partial y^n} \frac{y f(y)}{(y + a)^{n+1}}.$$

15. (a) Prove that the series

$$f(z) = f(x + iy) = \sum_{\nu=1}^{\infty} \frac{(-1)^{\nu+1}}{\nu^z}$$

converges for $x > 0$.

   (b) Prove that this series provides an extension of the zeta function (defined in Exercise 5, p. 797) to values of $z$ such that $0 < x \leq 1$, by means of the formula

$$f(z) = (1 - 2^{1-z})\zeta(z),$$

which is valid for $x > 1$.

   (c) Prove that the zeta function has a pole of residue 1 at $z = 1$.

# Solutions

**Exercises 1.1** (p. 10)

1. (a) Write $z = r (\cos \theta + i \sin \theta)$, in polar form with $0 < \theta < 2\pi$. Then, by De Moivre's theorem (Volume I, p. 105),

$$z^n = r^n(\cos n\theta + i \sin n\theta).$$

For $r < 1$, we have $\lim_{n \to \infty} r^n = 0$; therefore, $\lim_{n \to \infty} z^n = 0$. For $r > 1$, we have $\lim_{n \to \infty} r^n = \infty$; therefore, the distance of $z^n$ from the origin, hence from any given point, can be made arbitrarily large and the sequence diverges. For $r = 1$, there are two cases: $z = 1$ ($\theta = 0$) for which $\lim_{n \to \infty} z^n = 1$, and $z = \cos \theta + i \sin \theta$. In the latter case, we have for the distance between two successive points of the sequence

$$|z^{n+1} - z^n| = |z^n| \cdot |z - 1| = |z - 1|$$
$$= 2 - 2 \cos \theta,$$

a fixed positive value; by the Cauchy test the sequence must then diverge.

(b) The primitive $n$th root of $z$ is given in polar form by

$$z^{1/n} = r^{1/n}\left(\cos \frac{\theta}{n} + i \sin \frac{\theta}{n}\right).$$

If $z = 0$, we have $\lim_{n \to \infty} z^{1/n} = 0$. Otherwise, we have on setting $z^{1/n} = x_n + iy_n$,

$$\lim_{n \to \infty} z^{1/n} = \lim_{n \to \infty} x_n + i \lim_{n \to \infty} y_n$$

$$= \lim_{n \to \infty} r^{1/n} \cos \frac{\theta}{n} + i \lim_{n \to \infty} r^{1/n} \sin \frac{\theta}{n} = 1.$$

2. Apply the limit theorems of Volume I to the components of $P_n$ separately.

3. For a point $(a, b)$ satisfying $a^2 + b^2 < 1$, set $\alpha = \sqrt{a^2 + b^2}$. The neighborhood $(x - a)^2 + (y - b)^2 < (1 - \alpha)^2$ of $(a, b)$ is contained in the disk.

   For a point $(a, b)$ satisfying $a^2 + b^2 = 1$, every neighborhood contains points not in the disk.

4. Let $(a, b)$ be any point of $S$. Put $\gamma = b - a^2 > 0$. Consider an $\varepsilon$-neighborhood of $(a, b)$,

$$(x - a)^2 + (y - b)^2 < \varepsilon^2.$$

For all points of the neighborhood, we have $|x - a| < \varepsilon$, $|y - b| < \varepsilon$. Using

$$a^2 = x^2 - 2(x-a)a - (x-a)^2,$$

we obtain

$$y > b - \varepsilon = a^2 + \gamma - \varepsilon$$
$$= x^2 - 2(x-a)\,a - (x-a)^2 + \gamma - \varepsilon$$
$$> x^2 + \gamma - 2\varepsilon|a| - \varepsilon^2 - \varepsilon > x^2$$

provided $\varepsilon$ is taken as the smaller of 1 or $\gamma/(2|a|+2)$. Thus the $\varepsilon$-neighborhood is in $S$.

5. The segment (together with its end points if these are not considered as points of the segment).

## Problems 1.1 (p. 11)

1. By definition, every neighborhood of the boundary point $P$ contains points of $S$. Choose $P_1$ in $S$ so that $\overline{P_1P} < 1/2$. Since $P$ is not in $S$, $P_1 \neq P$, and therefore, $\overline{P_1P} > 0$. Now proceed by induction: given $P_n$ choose $P_{n+1}$ in $S$ so that $\overline{P_{n+1}P} < \frac{1}{2}\,\overline{P_nP}$. Clearly, the $P_n$ are distinct and $\overline{P_nP} < 1/2^n$.

2. Let $S$ be the given set; $S_c$, the closure of $S$; and $S_{cc}$, the closure of $S_c$. Every point of $S_{cc}$ is either in $S_c$ or the boundary of $S_c$. If $P$ is in the boundary of $S_c$, then every neighborhood of $P$ contains at least one point $Q$ of $S_c$ and one point $R$ not in $S_c$. Since $R$ is not in $S_c$, it is not in $S$. Since a neighborhood is open, the neighborhood of $P$ contains a neighborhood of $Q$ that must contain a point of $S$. Thus $P$ is in $S_c$.

3. Let $X$ be any point of $S$ on $\overline{PQ}$. The set of values of $\overline{PX}$ is bounded, since $\overline{PX} \leqslant \overline{PQ}$. Let $R$ be the point on $\overline{PQ}$ at distance equal to lub $\overline{PX}$ from $P$. Any neighborhood of $R$ contains points of $\overline{PQ}$ that are in $S$ and points that are not in $S$.

4. All points of $G$ are interior points.

## Exercises 1.2 (p. 16)

1. (a) $\dfrac{27}{8}$

 (c) $\dfrac{1}{(\log \pi)^e}$

 (e) 5.

2. The domain is the set of points $(x, y)$ and the range, the set of values $u$, where

 (a) $y \geq -x,\, u \geq 0$         (j) $x = y = 0,\, u = 0$

 (c) $y > -x,\, u > 0$         (k) $|y| < |x|,\, u$ real

 (e) $y > -\dfrac{x}{5},\, u$ real      (l) $(x, y) \neq (0, 0),\, 0 \leq u \leq \dfrac{\pi}{4}$

(g) $x^2 + y^2 + z^2 \leq a^2, 0 \leq w \leq a$    (m) $y \neq -x, -\dfrac{\pi}{2} < u < \dfrac{\pi}{2}$

(h) $y \neq -x, u$ real                (n) $x \neq 0, 0 < u \leq 1$

(i) $x^2 + 2y^2 \leq 3, 0 \leq u \leq \sqrt{3}$    (o) $\dfrac{1}{e} < x + y < e, 0 \leq u \leq \pi$

(p) $2n\pi - \dfrac{\pi}{2} \leq x \leq 2n\pi + \dfrac{\pi}{2}$ and $y \geq 0$, or

$2n\pi + \dfrac{\pi}{2} \leq x \leq 2n\pi + \dfrac{3\pi}{2}$ and $y \leq 0. u \geq 0.$

3. For $k$ variables,

$$\frac{1}{k!}(n+1)(n+2)\cdots(n+k).$$

(Compare Volume I, Chapter 1, p. 117, Exercise 11.)

## Exercises 1.3 (p. 24)

2. Discontinuous at $x = y = 0$.
3. (a) Set $x = \rho \cos \theta, y = \rho \sin \theta$. Then

$$|f(x,y)| = \rho^3|\cos 3\theta - 3\cos\theta\sin^2\theta| < 4\rho^3.$$

Take $\delta(\varepsilon) = \sqrt[3]{\varepsilon/4}$. $f(x,y)$ has at least the order of $\rho^3$.
4. As in the theory of functions of one real variable, sums and products of continuous functions and continuous functions of continuous functions are continuous.
   (a) Continuous.
   (b) Discontinuity possible only at $(0,0)$. Note with $x = \rho \cos \theta, y = \rho \sin \theta$ from $|\sin \alpha| < |\alpha|$, that

$$\left|\frac{\sin xy}{\sqrt{x^2+y^2}}\right| < \rho;$$

hence, the limit at $(0,0)$ exists and is 0.
5. Use the mean value theorem of the differential calculus to obtain for $z \geq 0, z + h > 0$

$$\left|\sqrt{1+(z+h)} - \sqrt{1+z}\right| = \frac{|h|}{2\sqrt{1+(x+\theta h)}} \leq \frac{|h|}{2};$$

hence, it is sufficient with appropriate choice of $z$ in each case to require $|h| < 2\varepsilon$. Set $\Delta x = \rho \cos \theta, \Delta y = \rho \sin \theta$, where $\rho < \delta(\varepsilon, x, y)$
   (a) With $z = x^2 + 2y^2$ and $h = \Delta z$ note that

$$|\Delta z| = \rho|2x\cos\theta + 4y\sin\theta + \rho(\cos^2\theta + 2\sin^2\theta)|$$
$$\leq \rho(2|x| + 4|y| + 3\rho) \leq \rho(2|x| + 4|y| + 3),$$

where we impose $\delta < 1$. For $|\Delta z| < 2\varepsilon$, it is sufficient to require

$$\delta < \min \left( \frac{2\varepsilon}{2|x| + 4|y| + 3}, 1 \right).$$

6. On the lines $y = \pm x$.

7. On the lines $x = n + \frac{1}{2}$, $y = n + \frac{1}{2}$.

8. For all values. (By definition, a function is continuous in the exterior of its domain.)

9. Set $z = 1/u$ where $u = 1 - x^2 - y^2$. $|\Delta z| = |\Delta u|/(u + \theta \Delta u)^2$. For $u > 0$, choose $|\Delta u| < 2/u$. Then $u + \theta \Delta u > u/2$ and

$$|\Delta z| < \frac{4|\Delta u|}{u^2}.$$

Now, with $\Delta x = \rho \cos \theta$, $\Delta y = \rho \sin \theta$, $\rho < \delta \leq 1$ and $|x|, |y| < 1$,

$$|\Delta u| = |\rho (2x \cos \theta + 2y \sin \theta) + \rho^2|$$
$$< \rho (2|x| + 2|y| + 1) < 5\delta.$$

Therefore, to enforce $|z| < \varepsilon$, take

$$\delta = \min \left[ \frac{\varepsilon}{20} (1 - x^2 - y^2)^2, 1 \right].$$

11. With $x = \rho \cos \theta$, $y = \rho \sin \theta$, we have

$$P = \rho^2 (a \cos^2\theta + 2b \cos \theta \sin \theta + c \sin^2\theta)$$
$$= \rho^2 f(\theta).$$

The expression $f(\theta)$ must not vanish for any value of $\theta$. Thus we must have $ac - b^2 > 0$.

12. All discontiuous, (a) on line $x = 0$, (c) on line $y = -x$.

13. For the approach along a straight line set $x = \rho \cos \theta$, $y = \rho \sin \theta$ with $\theta$ fixed. To show discontinuity for $f(x, y)$, approach along the parabola, $x = ay^2$ with arbitrary $a$, for $g(x, y)$, along the circle $(x - \frac{1}{2})^2 + y^2 = \frac{1}{4}$.

14. For (e) and (g) limits exist. For (h), set $y = e^{-a/|x|}$ with arbitrary positive $\alpha$ and show for

$$f(x, y) = \frac{y^{|x|} \sqrt{x^2 + y^2}}{\sqrt{x^2 + y^2} + \left| \frac{y}{x} \right|}$$

that $\lim\limits_{x \to 0} f(x, e^{-a/|x|}) = e^{-a}$.

15. For Exercise 14(e),

$$\delta(\varepsilon) = -\frac{1}{\sqrt{2} \log \varepsilon}.$$

For Exercise 14(g),

$$\delta = \min \left( -\frac{\log 2}{\log \varepsilon}, \frac{1}{2} \right).$$

16. First set $x = y = 0$, then set $z = 0$.

17. Follows since $R(x, y)$ is not defined at the origin and the origin is a boundary point of the domain of $R$.

18. (a) 1

    (b) 0

    (c) 0.

19. Set $y = mx$. Then $\lim\limits_{x \to 0} z = 3(1 - m)/(1 + m)$.

20. Compare Exercise 13.

21. Approach along straight lines other than $x = 0$ yields the limiting value 0. Approach along the curve $y = a/\log x$ yields the arbitrary limiting value $a$.

23. $\phi$ maps the part of its domain within any circle of sufficiently small radius $\rho$ about the origin into an interval of radius $C\rho$ centered at 0, where the constant $C$ may be fixed independently of $\rho$.

## Problems 1.3 (p. 26)

1. Let S be the domain of $f$, $S^*$ the domain of $f^*$. If $Q$ is an interior point of $S$, then there exists a neighborhood of $Q$ entirely within $S$ and continuity for $f^*$ is identical with continuity for $f$. If $Q$ in $S^*$ is a boundary point of $S$, then whether or not $Q$ is in $S$, there exists a $\delta$-neighborhood of $Q$ wherein $|f(P) - f^*(Q)| < \varepsilon/2$. For any point $\hat{Q}$ of $S^*$ in the $\delta$-neighborhood of $Q$ but not in $S$, there are points $P$ in $S$ for which $f(P)$ is arbitrarily close to $f^*(\hat{Q})$, say $|f(P) - f^*(\hat{Q})| < \varepsilon/2$. It follows that $|f^*(\hat{Q}) - f^*(Q)| < \varepsilon$.

2. If $\lim\limits_{(x,y) \to (\xi, \eta)} f(x, y) = L$ and $\lim\limits_{n \to \infty} (x_n, y_n) = (\xi, \eta)$, then for any positive $\varepsilon$ there is a $\delta$ such that $|f(x, y) - L| < \varepsilon$ whenever $(x, y)$ lies within the $\delta$-neighborhood of $(\xi, \eta)$. Furthermore, there is an $N$ such that $(x_n, y_n)$ lies within the $\delta$-neighborhood of $(\xi, \eta)$ for $n > N$. For $n > N$, then, $|f(x_n, y_n) - L| < \varepsilon$.

    Conversely, suppose for every sequence of points $(x_n, y_n)$ in the domain of $f$ with limit $(\xi, \eta)$, we have $\lim\limits_{n \to \infty} f(x_n, y_n) = L$. If $f$ did not have the limit $L$ at $(\xi, \eta)$, then for some $\varepsilon > 0$ and for all $\delta > 0$, there exists a point $(x, y) \neq (\xi, \eta)$ in the $\delta$-neighborhood of $(\xi, \eta)$ for which $|f(x, y) - L| > \varepsilon$. Set $\delta_1 = 1$ and choose $(x_1, y_1)$ in the $\delta_1$-neighborhood of $(\xi, \eta)$ so that $|f(x_1, y_1) - L| \geq \varepsilon$. Define $\delta_n$ and $(x_n, y_n)$ sequentially by $\delta_n = \frac{1}{2}\sqrt{(x_{n-1} - \xi)^2 + (y_{n-1} - \eta)^2}$, and $\sqrt{(x_n - \xi)^2 + (y_n - \eta)^2} < \delta_n$ with $|f(x_n, y_n) - L| \geq \varepsilon$. In this way, a sequence $(x_n, y_n)$ is constructed that violates the hypotheses if $f$ does not have the limit $L$ at $(\xi, \eta)$.

## Exercises 1.4a (p. 30)

1. (a) $\dfrac{\partial z}{\partial x} = nax^{n-1}$; $\dfrac{\partial z}{\partial y} = mby^{m-1}$.

    (c) $\dfrac{\partial z}{\partial x} = \dfrac{2x^2 - 3y^2}{x^2 y}$; $\dfrac{\partial z}{\partial y} = \dfrac{3y^2 - 2x^2}{xy^2}$.

(e) $\dfrac{\partial z}{\partial x} = 2xy^{3/2}$;  $\dfrac{\partial z}{\partial y} = \dfrac{3}{2}x^2 y^{1/2}$.

(g) $\dfrac{\partial z}{\partial x} = \dfrac{y^{3/4}}{2x^{1/2}}$;  $\dfrac{\partial z}{\partial y} = \dfrac{3x^{1/2}}{4y^{1/4}}$.

(j) $\dfrac{\partial z}{\partial x} = -2x \sin (x^2 + y)$;  $\dfrac{\partial z}{\partial y} = -\sin (x^2 + y)$.

(l) $\dfrac{\partial z}{\partial x} = -\dfrac{\sin x}{\sin y}$;  $\dfrac{\partial z}{\partial y} = -\dfrac{\cos x \cos y}{\sin^2 y}$.

(n) $\dfrac{\partial z}{\partial x} = \dfrac{2x^2 + y^2}{\sqrt{x^2 + y^2}}$;  $\dfrac{\partial z}{\partial y} = \dfrac{xy}{\sqrt{x^2 + y^2}}$.

2. (a) $\dfrac{\partial f}{\partial x} = \dfrac{2x}{3(x^2 + y^2)^{2/3}}$ ;  $\dfrac{\partial f}{\partial y} = \dfrac{2y}{3(x^2 + y^2)^{2/3}}$

(c) $\dfrac{\partial f}{\partial x} = e^{x-y}$;  $\dfrac{\partial f}{\partial y} = -e^{x-y}$

(e) $\dfrac{\partial f}{\partial x} = yz \cos xz$;  $\dfrac{\partial f}{\partial y} = \sin xz$;  $\dfrac{\partial f}{\partial z} = xy \cos xz$.

3. (a) $\dfrac{\partial f}{\partial x} = y$;  $\dfrac{\partial f}{\partial y} = x$.  $\dfrac{\partial^2 f}{\partial x^2} = \dfrac{\partial^2 f}{\partial y^2} = 0$;  $\dfrac{\partial^2 f}{\partial x \, \partial y} = 1$.

(c) Use $f(x, y) = \dfrac{x + y}{1 - xy}$;

$\dfrac{\partial f}{\partial x} = \dfrac{1 + y^2}{(1 - xy)^2}$;  $\dfrac{\partial f}{\partial y} = \dfrac{1 + x^2}{(1 - xy)^2}$.

$\dfrac{\partial^2 f}{\partial x^2} = \dfrac{2(y + y^3)}{(1 - xy)^3}$;  $\dfrac{\partial^2 f}{\partial x \, \partial y} = \dfrac{2 (x + y)}{(1 - xy)^3}$;  $\dfrac{\partial^2 f}{\partial y^2} = \dfrac{2(x + x^3)}{(1 - xy)^3}$.

(e) $\dfrac{\partial f}{\partial x} = yx^{y-1} e^{(x^y)}$;  $\dfrac{\partial f}{\partial y} = x^y e^{(x^y)} \log x$.

$\dfrac{\partial^2 f}{\partial x^2} = yx^{y-2} e^{(x^y)} (y - 1 + yx^y)$;

$\dfrac{\partial^2 f}{\partial x \, \partial y} = x^{y-1} e^{(x^y)} (1 + y \log x + yx^y \log x)$;

$\dfrac{\partial^2 f}{\partial y^2} = x^y (\log x)^2 e^{(x^y)} (1 + x^y)$.

4. $f_x = 0$, $f_y = 0$, $f_z = -3$.

5. 1.

8. $(2/r)$.

9. $a = -3$.

## Problems 1.4a (p. 31)

1. $\binom{n+k}{k}$. (Compare Exercises 1.2, number 3.)

2. Consider a function of the form $f(x, y) = \alpha(x)\beta(y)$ where $\alpha$ is differentiable and $\beta$ is not.

3. Differentiate with respect to $x$ and $y$ to obtain for all $x$ and $y$,

$$\phi'(x^2 + y^2) = \frac{\psi'(x)}{2x}\psi(y) = \frac{\psi'(y)}{2y}\psi(x);$$

whence, $\psi'(x)/2x\psi(x)$ is constant. $f(x, y) = ce^{a(x^2 + y^2)}$.

## Exercises 1.4c (p. 36)

2. (a) Observe that the first partial derivatives,

$$\frac{\partial f}{\partial x} = \begin{cases} \dfrac{2x}{(x^2 + y^2)^2} \exp\left[-1/(x^2 + y^2)\right], & x, y \neq 0 \\ 0, & x = y = 0 \end{cases}$$

$$\frac{\partial f}{\partial y} = \begin{cases} \dfrac{2y}{(x^2 + y^2)^2} \exp\left[-1/(x^2 + y^2)\right], & x, y \neq 0 \\ 0, & x = y = 0, \end{cases}$$

are bounded.

(b) The origin is the only point in question. Consider

$$\frac{\partial f}{\partial x} = \begin{cases} 2x\,\dfrac{x^4 + y^4}{x^2 + y^2} + 4x^3 \log\,(x^2 + y^2), & x, y \neq 0 \\ 0, & x = y = 0, \end{cases}$$

in the neighborhood $x^2 + y^2 < \delta^2$. Then

$$\frac{\partial f}{\partial x} < 2\delta^3 + 8\delta^2|\delta \log \delta|$$

$$< 10\delta^2,$$

for $\delta < 1$, where we have used $|\delta \log \delta| < 1$, for $\delta < 1$.

## Exercises 1.4d (p. 39)

1. (a) $2ab$

   (c) $ab\,f''(ax + by)$

   (e) $-\dfrac{1}{(x + y)^2}$.

2. (b) $f_x = y \sinh xy$, $f_y = x \sinh xy$, $f_{xx} = y^2 \cosh xy$,

   $f_{xy} = xy \cosh xy + \sinh xy$, $f_{yy} = x^2 \cosh xy$,

$$f_{xxx} = y^3 \sinh xy, \quad f_{xxy} = xy^2 \sinh xy + 2y \cosh xy,$$
$$f_{xyy} = x^2 y \sinh xy + 2x \cosh xy, \quad f_{yyy} = x^3 \sinh xy.$$

(d) $f_x = 1/y - y/x^2, \quad f_y = 1/x - x/y^2, \quad f_{xx} = 2y/x^3,$

$\quad f_{xy} = (-1/x^2) - 1/y^2, \quad f_{yy} = 2x/y^3, \quad f_{xxx} = -6y/x^4, \quad f_{xxy} = 2/x^3,$

$\quad f_{xyy} = 2/y^3, \quad f_{yyy} = -6x/y^4.$

## Problems 1.4d (p. 39)

1. (b) Set $z = \log u$. Then $z_{xy} = 0$. Thus $z_x$ does not depend on $y$. Set $z_x = \alpha(x)$; then,

$$z = \int \alpha(x)\,dx + \psi(y) = \phi(x) + \psi(y);$$

whence,

$$u = e^z = e^{\phi(x)}\,e_{\psi}(y).$$

## Exercises 1.5a (p. 42)

1. (a), (b)  $f_x(0, 0)$ does not exist.

(c) Set $h = \rho \cos \theta$, $k = \rho \sin \theta$. For differentiability it would be necessary that

$$f(h, k) - f(0, 0) = \rho \sin 2\theta = f_x(0, 0)h + f_y(0, 0)k + o(\rho),$$

but $f_x(0, 0) = f_y(0, 0) = 0$, a contradiction.

2. For $s$ between $x$ and $x + \delta_1$, $t$ between $y$ and $y + \delta_2$, we have $|g(s) - g(x)| < \varepsilon_1(\delta_1)$, $|h(t) - h(y)| < \varepsilon_2(\delta_2)$ where $\lim_{\delta_1 \to 0} \varepsilon_1(\delta_1) = \lim_{\delta_2 \to 0} \varepsilon_2(\delta_2) = 0$. Consequently, by the mean value theorem of integral calculus,

$$\int_{x_0}^{x+\delta_1} g(s)\,ds = \int_{x_0}^{x} g(s)\,ds + \delta_1 g(\xi)$$

where $|g(\xi) - g(x)| < \varepsilon_1(\delta_1)$; a similar result holds for $h(t)$. It follows that

$$f(x + \delta_1, y + \delta_2) = \left[ \int_{x_0}^{x} g(s)\,ds + \delta_1 g(x) + o(\delta_1) \right]$$

$$\cdot \left[ \int_{y_0}^{y} h(t)\,dt + \delta_2 h(y) + o(\delta_2) \right]$$

$$= f(x, y) + \delta_1 g(x) + \delta_2 h(y) + o(\sqrt{\delta_1^2 + \delta_2^2}).$$

## Problems 1.5a (p. 43)

1. Set $\rho = \sqrt{h^2 + k^2}$. Then

$$|f(x, y) - f(a, b)| \le \rho(|f_x(a, b)| + |f_y(a, b)| + \varepsilon),$$

where $\lim_{\rho \to 0} \varepsilon = 0$. Thus, $f$ is not only continuous, but Lipschitz continuous: for $P = (x, y)$, $A = (a, b)$, we have in some neighborhood of $A$, $|f(P) - f(A)| \le M |P - A|$, where $M$ is constant.

## Exercises 1.5b (p. 45)

1. The slope of the section of the surface $z = f(x, y)$ with the plane arc $\tan[(y - y_0)/(x - x_0)] = \alpha$; that is, the slope in the $z$, $\rho$-plane of the curve $z = \phi(\rho) = f(x + \rho \cos \alpha, y + \rho \sin \alpha)$.

2. (a) $a$, $\dfrac{a\sqrt{3} + b}{2}$, $\dfrac{a + b\sqrt{3}}{2}$, $b$.

   (c) $2$, $\sqrt{3} - 2$, $1 - 2\sqrt{3}$, $-4$.

   (e) $-1$, $-\dfrac{\sqrt{3}}{2}$, $-\dfrac{1}{2}$, $0$.

   (g) $0, 0, 0, 0$.

3. (a) $-8/5$

   (b) $-1$

   (c) $-2/\sqrt{3}$.

4. $f(x, y) = xy/(x^2 + y^2)$.

6. $\partial^2 f/\partial r^2 = \sin 2\theta$.

## Exercises 1.5c (p. 48)

1. (a) $z = 8y - 4$

   (c) $3x + 3y - 4z + 5 - 3 \log 2 = 0$

   (e) $z = [\exp(1/\sqrt{2})/\sqrt{2}] (x - y + \sqrt{2} + \pi/4)$

   (g) $z = 2e^{-2}(x + y + \frac{1}{2} e^2 \int_0^2 e^{-t^2} dt - 2)$.

2. The common point is the origin.

3. The equation of the plane through the three points can be put in the form

$$z - z_0 = $$
$$\frac{(x - x_0) [k_1(z_2 - z_0) - k_2(z_1 - z_0)] + (y - y_0) [h_2(z_1 - z_0) - h_1(z_2 - z_0)]}{h_2 k_1 - h_1 k_2},$$

where $h_i = x_i - x_0$, $k_i = y - y_0$, for $i = 1, 2$. Set $h_i = \rho_i \cos \alpha_i$, $k_i = \rho_i \sin \alpha_i$. Then $z_i - z_0 = \rho_i[(\cos \alpha_i)(\partial z/\partial x) + (\sin \alpha_i)(\partial z/\partial y)] + o(\rho_i)$. Enter this in the equation of the plane with $\sin (\alpha_1 - \alpha_2) \ne 0$, and $(x, y)$ fixed to obtain the desired result,

$$z - z_0 = (x - x_0) \frac{\partial z}{\partial x} + (y - y_0) \frac{\partial z}{\partial y} + \frac{o(\rho_2)}{\rho_2} + \frac{o(\rho_1)}{\rho_1}.$$

4. We may suppose not all coefficients vanish, say $c \neq 0$. Then $(x_0, y_0, z_0)$ lies on one of the surfaces

$$z = \pm \sqrt{\frac{1 - ax^2 - by^2}{c}}.$$

The tangent plane has the equation

$$z - z_0 = (x - x_0) z_x (x_0, y_0) + (y - y_0) z_y(x_0, y_0).$$

Differentiate the equation for the quadric surface to obtain

$$2ax_0 + 2cz_0 \frac{\partial z}{\partial x} = 0$$

$$2by_0 + 2cz_0 \frac{\partial z}{\partial y} = 0$$

and insert the values for $\frac{\partial z}{\partial x}$ and $\frac{\partial z}{\partial y}$ in the equation for the tangent plane to obtain (if $z_0 \neq 0$),

$$z - z_0 = -\frac{ax_0}{cz_0} (x - x_0) - \frac{by_0}{cz_0} (y - y_0),$$

whence

$$ax_0 x + by_0 y + cz_0 z = ax_0^2 + by_0^2 + cz_0^2 = 1.$$

## Exercises 1.5d (p. 51)

1. (a) $(2xy^2 + 3y^3) \, dx + (2x^2y + 9xy^2 - 8y^3) \, dy.$

   (c) $4x^3 \, dx - 3y^2 \, dy/(x^4 - y^4).$

   (e) $-(dx + y^{-1} \, dy) \sin (x + \log y).$

   (g) $dx + dy/(1 + (x + y)^2).$

   (i) $(dx + dy - dz) \sinh (x + y - z).$

2. $(-2/10) + (7 \sqrt[3]{5}/25)$

3. $e^{x^2+y^2}[(8x^3 + 12x) \, dx^3 + (8x^2y + 4y) \, dx^2 \, dy + (8xy^2 + 4x) \, dx \, dy^2 + (8y^3 + 12y) \, dy^3].$

## Exercises 1.5e (p. 53)

1. $z$ varies from $-3$ to $-3.5$.

2. $-\dfrac{1}{600}.$

3. $1/2 \, (y|h| + x|k|).$

4. From $dz = y \, dx + x \, dy, \, dz/z = dx/x + dy/y.$

5. From $dg = 2dx/t^2 - 4x \, dt/t^3$, the relative error in $g$ is $dg/g = dx/x - 2dt/t.$

Thus a given relative error in the measurement of $t$ will have twice the effect of the same relative error in the measurement of $x$.

## Exercises 1.6a (p. 57)

1. (a) $z_x = -2x \log (1 + y)$, $z_y = -\dfrac{x^2}{1 + y}$, $z_{xx} = -2 \log (1 + y)$,

   $z_{xy} = -\dfrac{2x}{(1 + y)}$, $z_{yy} = \dfrac{x^2}{(1 + y)^2}$.

   (e) Set $u = x$, $v = \arctan y$, $z_x = v \sec^2(uv)$, $z_y = [\sec^2(uv)]/(1 + y^2)$,

   $z_{xx} = 2v^2 \sec^2(uv) \tan (uv)$, $z_{xy} = [\sec^2 (uv)/(1 + y^2)] [1 + 2v \tan (uv)]$,

   $z_{yy} = x \sec^2(uv)/(1 + y^2)^2 [x \tan(uv) - 2y]$.

2. (a) $w_x = \dfrac{-x - y \cos z}{(x^2 + y^2 + 2xy \cos z)^{3/2}}$,

   $w_y = \dfrac{-y - x \cos z}{(x^2 + y^2 + 2xy \cos z)^{3/2}}$,

   $w_z = \dfrac{xy \sin z}{(x^2 + y^2 + 2xy \cos z)^{3/2}}$.

   (b) $w_x = \dfrac{1}{\sqrt{z^2 + 2zy^2 + y^4 - x^2}}$,

   $w_y = \dfrac{-2xy}{(z + y^2)\sqrt{z^2 + 2zy^2 + y^4 - x^2}}$,

   $w_z = \dfrac{-x}{(z + y^2)\sqrt{z^2 + 2zy^2 + y^4 - x^2}}$.

   (c) $w_x = 2x + \dfrac{2xy}{1 + x^2 + y^2 + z^2}$,

   $w_y = \log (1 + x^2 + y^2 + z^2) + \dfrac{2y^2}{1 + x^2 + y^2 + z^2}$,

   $w_z = \dfrac{2yz}{1 + x^2 + y^2 + z^2}$.

   (d) $w_x = \dfrac{1}{2(1 + x + yz)\sqrt{x + yz}}$,

   $w_y = \dfrac{z}{2(1 + x + yz)\sqrt{x + yz}}$,

   $w_z = \dfrac{y}{2(1 + x + yz)\sqrt{x + yz}}$.

3. (a) Consider the derivative of $z = u^v$ where $u$ and $v$ are functions of $x$:

$$\frac{dz}{dx} = vu^{v-1} \frac{du}{dx} + u^v \log u \frac{dv}{dx}.$$

Employ this formula for $u = x$, $v = x$ to obtain

$$\frac{d}{dx}(x^x) = x^x(1 + \log x).$$

Now employ the formula again for $u = x$, $v = x^x$ to obtain

$$\frac{d}{dx}(x^{(x^x)}) = x^{(x^x)} x^x \left[\frac{1}{x} + \log x + (\log x)^2\right].$$

(b) Set $y = 1/x$. Then

$$\frac{dz}{dx} = -\frac{1}{x^2}\frac{dz}{dy}.$$

Use $z = (y^y)^y = u^v$, where $u = y$, $v = y^2$ to obtain

$$\frac{dz}{dy} = y^{(y^2+1)}(1 + 2\log y) = yz(1 + 2\log y),$$

whence,

$$\frac{dz}{dx} = \frac{2\log x - 1}{x^{3+1/x^2}}.$$

4. See Problem 1.
5. Use the symmetry in the several variables and calculate in each case:

(a) $f_{xx} = \dfrac{y^2 - x^2}{(x^2 + y^2)^2}$,

(b) $g_{xx} = \dfrac{2x^2 - y^2 - z^2}{(x^2 + y^2)^2}$,

(c) $h_{xx} = \dfrac{6x^2 - 2y^2 - 2z^2 - 2w^2}{(x^2 + y^2)^3}$.

## Problems 1.6a (p. 58)

1. Use the Cauchy-Riemann equations in

$$\phi_{xx} + \phi_{yy} = (u_x^2 + u_y^2)f_{uu} + 2(u_x v_x + u_y v_y)f_{uv} + (v_x^2 + v_y^2)f_{vv}$$
$$+ (u_{xx} + u_{yy})f_u + (v_{xx} + v_{yy})f_v,$$

and note that $u$ and $v$ are also solutions of Laplace's equation.

2. Let the vertex of the cone be located at the origin (no loss of generality is entailed since a translation of axes will not affect the derivatives of $f$). If a point $(x, y, z)$ lies on the cone, then so also does the point $(\lambda x, \lambda y, \lambda z)$ where $\lambda$ is any real number. We therefore have

$$\frac{z}{x} = f\left(\frac{x}{x}, \frac{y}{x}\right) = f\left(1, \frac{y}{x}\right) = \phi\left(\frac{y}{x}\right);$$

thus the equation of the cone can be written in terms of a function $\phi$ of one real variable:

$$z = x\phi\left(\frac{y}{x}\right).$$

The result follows on differentiation.

3. (a) $g_{rr} + \dfrac{2}{r}\, g_r$.

  (b) From $g_{rr}/g_r = -2/r$, obtain $\log g_r = -2 \log r +$ constant, etc.

4. (a) $g_{rr} + \dfrac{n-1}{r}\, g_r$.

  (b) If $n = 1$, $ar + b$.

   If $n = 2$, $a \log r + b$.

   If $n > 2$, $a/r^{n-2} + b$ (compare Problem 3).

## Exercises 1.6c (p. 63)

1. $\sqrt{u_r{}^2 + (1/r^2)\, u_\theta{}^2}$.

2. Set $u = f(x, y)$ and introduce new variables by $\xi = x \cos \theta + y \sin \theta$, $\eta = y \cos \theta - x \sin \theta$. Obtain $u_{xx} = \cos^2 \theta\, u_{\xi\xi} - 2 \cos \theta \sin \theta\, u_{\xi\eta} + \sin^2\theta\, u_{\eta\eta}$, $u_{yy} = \sin^2\theta\, u_{\xi\xi} + 2 \cos \theta \sin \theta\, u_{\xi\eta} + \cos^2 \theta\, u_{\eta\eta}$.

4. $z_x = 3$, $z_y = 1$, $z_r = z_x \cos \theta + z_y \sin \theta$, $z_\theta = - z_x r \sin \theta + z_y r \cos \theta$.

5. Note that the derivatives do not depend on $a$ and $b$. The transformation is essentially a rotation and translation of the $x$, $y$-axes. Compare Exercise 2 and 3. Use

$$u_{xx} = \alpha^2 U_{\xi\xi} - 2\alpha\beta U_{\xi\eta} + \beta^2 U_{\eta\eta},$$

$$u_{xy} = \alpha\beta U_{\xi\xi} + (\alpha^2 - \beta^2)\, U_{\xi\eta} - \alpha\beta U_{\eta\eta},$$

$$u_{yy} = \beta^2 U_{\xi\xi} + 2\alpha\beta U_{\xi\eta} + \alpha^2 U_{\eta\eta}.$$

For a geometrical interpretation see 1.6 a, Problem 2.

6. $\dfrac{z^3}{2x^2}\, T_z + T_{xx} + \dfrac{z}{x}\, T_{xz} + \dfrac{z^2}{x^2}\, T_{zz}$.

## Problems 1.6c (p. 64)

1. $\dfrac{1}{r^2} \dfrac{\partial}{\partial r}\left(r^2 \dfrac{\partial u}{\partial r}\right) + \dfrac{1}{\sin \theta} \dfrac{\partial^2 u}{\partial \phi^2} + \dfrac{\partial}{\partial u}\left(\sin \theta \dfrac{\partial u}{\partial \theta}\right)$.

   To compare with 1.6 a, Problem 3, let derivatives of $u$ with respect to $\theta$ and $\phi$ vanish.

2. Under the given transformation, the equation $Af_{xx} + 2Bf_{xy} + Cf_{yy} = 0$ is transformed into $A^*f_{\xi\xi} + 2B^*f_{\xi\eta} + C^*f_{\eta\eta} = 0$, where

$$A^* = a^2A + 2abB + b^2C$$

$$B^* = acA + (ad + bc)B + bdC$$

$$C^* = c^2A + 2cdB + d^2C$$

(compare Exercise 3). Observe that

$$B^{*2} - A^*C^* = (ad - bc)^2 (B^2 - AC).$$

Thus, the sign of $B^{*2} - A^*C^*$ is independent of the linear transformation. It follows that no such transformation exists for (a) if $B^2 - AC \geq 0$ or for (b) if $B^2 - AC < 0$.

(a) Assume $B^2 - AC < 0$, and set $A^* = 1$, $B^* = 0$, $C^* = 1$ above. Observe from $AC > B^2 \geq 0$ that $A$ and $C$ have the same nonzero sign, which we may assume to be positive. If $B = 0$, take $b = c = 0$, $a = 1/\sqrt{A}$, $d = 1/\sqrt{C}$. If $B \neq 0$, first reduce to the case $B = 0$, for example, by taking

$$b = 0, \quad a = \frac{1}{\sqrt{A}}, \quad c = \frac{B}{\sqrt{A(AC - B^2)}}, \quad d = \frac{-A}{\sqrt{A(AC - B^2)}}.$$

(b) Assume $B^2 - AC > 0$ and set $A^* = C^* = 0$, $B^* = 1$ above. If $B = 0$, then $A$ and $C$ have opposite signs. In that case, satisfy the equations

$$\frac{a}{b} = \sqrt{-\frac{C}{A}}, \quad \frac{d}{c} = \sqrt{-\frac{C}{A}}, \quad bc\sqrt{-AC} = 1;$$

for example, take

$$a = 1, \quad b = \sqrt{-\frac{A}{C}}, \quad c = \frac{1}{2}, \quad d = \frac{1}{2}\sqrt{-\frac{C}{A}}.$$

If $B \neq 0$ and at least one of $A$ or $C$ is nonvanishing, say $A > 0$, first reduce to the case $B = 0$, for example, by taking $A^* = A$, $C^* = -1/A$, $b = 0$, then

$$a = 1, \quad d = \frac{1}{\sqrt{B^2 - AC}}, \quad c = -\frac{B}{\sqrt{A(B^2 - AC)}}.$$

## Exercises 1.7a (p. 66)

1. (a) $(h + k) \cos (x + h + y + k)$.

   (b) $-\dfrac{h(y + k)}{(x + h)^2} + \dfrac{k}{x + h}$.

2. (a) $-\dfrac{1}{8}$.

   (b) $\dfrac{5}{8} e^{5/16}$.

   (c) $\dfrac{\pi}{8}$.

## Exercises 1.7b (p. 68)

1. For a curve defined by the intersection with the surface $z = f(x, y)$ of a vertical plane $h(\eta - y) - k(\xi - x) = 0$ through the point $(x, y)$, there exists

a tangent at some interior point of any arc that is parallel to the chord joining the end points.

2. (a) $\frac{1}{2}$.

(b) $\frac{8}{3\pi}$ arc sin $\frac{8 - 4\sqrt{2} - \sqrt{2}}{3\pi}$.

3. Take $x = 0$, $y = -\frac{1}{2}$, $h = k = \frac{1}{2}$.

5. (a) $\frac{3}{7}$.

(b) $\frac{23}{54}$.

**Problems 1.7b** (p. 68)

1. It is sufficient to prove that $f$ has the same value for any two points that can be connected by a segment within the domain.

**Exercises 1.7c** (p. 70)

1. $xy$.
2. Observe that $df$ vanishes at $(2, 3)$ for $h = 0.1$, $k = -0.1$. Thus, approximately, $f(2.1, 2.9) = f(2, 3) + \frac{1}{2}d^2f(2, 3) = 79.9$.
3. The approximation is exact. The error is zero to all orders.
4. (a) $x^3 - 2x^2y + y^2 + h(3x^2 - 4xy) + k(2y - 2x^2) + h^2(3x - 2y) - hk4x$
   $+ k^2 + 6h^3 - 2h^2k$.

   (b) $\sum\limits_{n=1}^{\infty} \frac{(-1)^n(h + 2k)^{2n-1}}{(2n - 1)!}$.

   (c) The cases $x + h > 0$, $x + h < 0$ must be taken separately; the two cases yield different first order terms in $h$:

   $x^4y - 2y^2x - \sqrt{3}|x| + h(4x^3y - 2y^2 - \sqrt{3}\ \text{sgn}(x + h)$
   $+ k(x^4 - 4yx) + h^26x^2y + hk4x^3 - k^22x + h^34xy$
   $+ h^2k6x^2 - 2hk^2 + h^4y + 4h^3k + h^4k$.

5. $x + x(y - 1) - 2x(z + 1) - 2x(y - 1)(z + 1) + 2x(z + 1)^2$
   $+ x(y - 1)(z + 1)^2$.

6. (a) $y - x^2 - \frac{y^3}{3} + x^4y - x^2y^3 + \frac{y^5}{5} + \cdots$

   (b) $y + \frac{x^2y}{2} + \frac{y^3}{6} + \frac{x^2y^3}{12} + \frac{x^4y}{24} + \frac{y^5}{120} \cdots$

(c) $1 + y + \dfrac{y^2}{2} - \dfrac{x^4}{6} - \dfrac{x^3 y}{3} + \dfrac{xy^3}{6} + \dfrac{y^4}{24} + \cdots$

(d) $1 + x + \dfrac{x^2}{2} - \dfrac{y^2}{2} + \dfrac{x^3}{6} - \dfrac{xy^2}{2} + \cdots$

(e) $x - \dfrac{x^3}{6} + \dfrac{xy^2}{2} + \dfrac{x^5}{120} - \dfrac{x^3 y^2}{12} + \dfrac{5xy^4}{24} + \cdots$

(f) $xy + \dfrac{x^2 y}{2} + \dfrac{xy^2}{2} + \dfrac{x^3 y}{3} + \dfrac{x^2 y^2}{4} + \dfrac{xy^3}{3} + \cdots$

(g) $1 + x^2 - y^2 + \dfrac{x^4}{2} - x^2 y^2 + \dfrac{y^4}{2} + \cdots$

(h) $1 - \dfrac{3x^2}{2} - xy - \dfrac{y^2}{2} + \cdots$

(i) $1 - \dfrac{x^2}{2} - \dfrac{x^2 y^2}{2} + \dfrac{x^4}{24} - \dfrac{x^6}{120} - \dfrac{x^4 y^2}{12} + \dfrac{x^2 y^4}{3} + \cdots$

(j) $x^2 + y^2 - \dfrac{x^6}{6} - \dfrac{x^4 y^2}{2} - \dfrac{x^2 y^4}{2} - \dfrac{y^6}{6} + \cdots$

7. Observe that the error is fourth order. To fourth order

$$\frac{\cos x}{\cos y} = 1 - \frac{x^2 - y^2}{2} + \frac{x^4 - 6x^2 y^2 + 5y^4}{24} + \cdots;$$

for the fourth-order term we have

$$\frac{x^4 - 6x^2 y^2 + 5y^4}{24} = \frac{(y^2 - x^2)(5y^2 - x^2)}{24}.$$

For $|x| \le \pi/6$, $|y| \le \pi/6$ the two factors reach their maxima at $x = 0$, $y = \pi/6$. Thus, we estimate the error as about

$$\frac{5}{24}\left(\frac{\pi}{6}\right)^4 \approx .016$$

## Problems 1.7c (p. 70)

1. (a) $\displaystyle\sum_{n=0}^{\infty}\sum_{r=0}^{n}\binom{n}{r}x^r y^{n-r} = \sum_{n=0}^{\infty}\sum_{m=0}^{\infty}\binom{m+n}{n}x^m y^n;$

   converges in the strip $|x + y| < 1$.

   (b) $\displaystyle\sum_{n=0}^{\infty}\sum_{r=1}^{n}\frac{x^r}{r!}\frac{y^{n-r}}{(n-r)!} = \sum_{n=0}^{\infty}\sum_{m=0}^{\infty}\frac{x^m}{m!}\frac{y^n}{n!};$

   converges for all values of $x$ and $y$.

2. Expand both sides of the spherical formula to second order in $x$, $y$, and $z$.

3. Expand $f(2h, e^{-1/2h})$ and $f(0, 0)$ to second order in the neighborhood of $(h, e^{-1/h})$; add and divide by $h^2$.

4. Convergence follows by convergence of the expansion of the exponential function for one variable. Differentiate with respect to $x$ to obtain

$$2yf(x, y) = \sum_{n=0}^{\infty} \frac{H_n'(x)y^n}{n!} = \sum_{n=1}^{\infty} \frac{2H_{n-1}(x)y^n}{(n-1)!}$$

whence (b) follows on equating coefficients. From (b) and $H_0(x) = 1$, (a) follows inductively. To obtain (c), differentiate with respect to $y$ and equate coefficients. To obtain (d), use (b) to replace $2nH_{n-1}$ in (c) by $H_n'$ and then differentiate to obtain

$$H_{n+1}' - 2xH_n' + 2H_n' + H_n'' = 0.$$

Next use (b) in this result to replace $H_{n+1}'$ by $2(n+1)H_n$.

### Exercises 1.8b (p. 80)

1. Use the uniform continuity of $\beta_k(x, k)$ for $x$ in the closed interval $a \le x \le b$ and $k$ restricted to any closed subinterval of $k_0 < k < k_1$.

2. (a) For $\varepsilon = k^{-2/3}$ and $1 - \varepsilon < x < 1$, we have for large $k$

$$k \log x = k(x-1) + 0(k^{-1/3})$$

$$\frac{x-1}{\log x} = 1 + 0(k^{-2/3}),$$

   hence

$$\frac{x^k(x-1)}{\log x} = e^{k(x-1)} (1 + 0(k^{-1/3})),$$

   while for $0 < x < 1 - \varepsilon$

$$\frac{x^k(x-1)}{\log x} = 0\left(\frac{x-1}{\log x} e^{-k^{1/3}}\right).$$

   It follows that

$$F(k) = \int_{1-\varepsilon}^{1} + \int_{0}^{1-\varepsilon} = \frac{1}{k} + 0(k^{-4/3}).$$

   (b) By Ex. 1,

$$F'(k) = \int_{0}^{1} x^k(x-1)\, dx = \frac{1}{k+2} - \frac{1}{k+1}.$$

   Hence $F(k) = \log \frac{2+k}{1+k} + c$, where the value of the constant $c$ turns out to be 0 from (a).

### Exercises 1.9b (p. 92)

1. (a) $\int_{0}^{2\pi}(-t \sin t + \cos^2 t + \sin t)\, dt = 3\pi$

   (b) $\int_{-1}^{1}(-2t^2x_0 - 2tx_0y_0(1-t^2) + y_0(1-t^2))\, dt = -\frac{4}{3}(x_0 - y_0).$

**Exercises 2.1** (p. 141)

1. If $X = (x, y, z)$ is an arbitrary point of the line, then

$$\overrightarrow{PX} = \lambda \mathbf{A},$$

where $\lambda$ may be any real number. Thus,

$$(x + 2, y, z - 4) = \lambda(2, 1, 3),$$

or

$$\frac{x + 2}{2} = y = \frac{z - 4}{3}.$$

2. Set $\overrightarrow{PQ} = \mathbf{A}$. Any point $X$ of the line satisfies $\overrightarrow{PX} = \lambda \mathbf{A}$. Let $\mathbf{B}$, $\mathbf{C}$, and $\mathbf{V}$ be the position vectors of $P$, $Q$, and $X$, respectively. Then,

$$\overrightarrow{PX} = \mathbf{V} - \mathbf{B} = \lambda \mathbf{A} = \lambda(\mathbf{C} - \mathbf{B});$$

or

$$\mathbf{V} = (1 - \lambda)\mathbf{B} + \lambda \mathbf{C}$$

In particular, if $P = (3, -2, 2)$ and $Q = (6, -5, 4)$, as given in (a),

$$(x, y, z) = \lambda(3, -3, 2),$$

or

$$\frac{x}{3} = -\frac{y}{3} = \frac{z}{2}.$$

3. If $\mathbf{V}$ is the position vector of any point $X$ on the line joining $P$ to $Q$, then, by the solution to Exercise 2,

$$\mathbf{V} = (1 - \lambda) \mathbf{A} + \lambda \mathbf{B}.$$

for some real $\lambda$. Thus,

$$(1 - \lambda) (\mathbf{V} - \mathbf{A}) = \lambda (\mathbf{B} - \mathbf{V}) = (1 - \lambda) \lambda (\mathbf{B} - \mathbf{A}).$$

If $0 < \lambda < 1$, it follows that $\mathbf{V} - \mathbf{A}$, $\mathbf{B} - \mathbf{V}$ and $\mathbf{B} - \mathbf{A}$ have the same direction and $|\mathbf{V} - \mathbf{A}|/|\mathbf{B} - \mathbf{V}| = \lambda/(1 - \lambda)$

4. Write the position vector in the form

$$\mathbf{V} = \mathbf{A} + \lambda(\mathbf{B} - \mathbf{A}),$$

where $\mathbf{B} - \mathbf{A}$ is represented by $\overrightarrow{PQ}$, to see that $\lambda > 0$.

5. Let $\mathbf{A}$, $\mathbf{B}$, $\mathbf{C}$, $\mathbf{D}$, $\mathbf{E}$ be the position vectors of the points $P$, $Q$, $R$, $S$, $M$, respectively. Take the origin $O$ at the point dividing $MS$ in the ratio $1/3$. Thus, $\mathbf{D} = -3\mathbf{E}$. Since $\mathbf{E} = 1/3 (\mathbf{A} + \mathbf{B} + \mathbf{C})$, it follows that

$$\frac{1}{4}(\mathbf{A} + \mathbf{B} + \mathbf{C} + \mathbf{D}) = 0.$$

Hence, $O$ is the center of mass by the general definition and clearly does not depend on the order of the vertices.

6. Let the edges be $PQ$ and $RS$; in the notation of the preceding solution their midpoints have position vectors $\frac{1}{2}(A + B)$ and $\frac{1}{2}(C + D)$, respectively. From the solution to Exercise 5, $\frac{1}{2}(A + B) = -\frac{1}{2}(C + D)$; hence, the midpoints are collinear with the center of mass $O$ and equidistant from it.

7. If $P_k = (x_k, y_k, z_k)$, for $k = 1, 2, \ldots, n$, then

$$G = (x_0, y_0, z_0) = \left( \frac{\Sigma m_k x_k}{\Sigma m_k}, \frac{\Sigma m_k y_k}{\Sigma m_k}, \frac{\Sigma m_k z_k}{\Sigma m_k} \right)$$

$$\Sigma m_k \, A_k = (\Sigma m_k(x_k - x_0), \Sigma m_k(y_k - y_0), \Sigma m_k(z_k - z_0)) = (0, 0, 0).$$

8. The zero vector is the real number 1. "Multiplication" of the "vector" $a$ by the scalar $\lambda$ means raising $a$ to power $\lambda$. Thus, if vector "addition" is denoted by $\oplus$, scalar multiplication by $\odot$,

$$\lambda \odot (a \oplus b) = (ab)^\lambda = a^\lambda b^\lambda = (\lambda \odot a) \oplus (\lambda \odot b).$$

9. The complex number $a + ib$ corresponds to the vector $(a, b)$.

10. Take the origin as center of the sphere and let $A$, $B$, $R$ be the position vectors of $P$, $Q$, $R$, respectively. If the radius of the sphere is $\rho$,

$$|A|^2 = |B|^2 = |R|^2 = \rho^2$$

and $B = -A$. Consequently, from (15c)

$$(R - A) \cdot (R - B) = (R - A) \cdot (R + A) = |R^2| - |A|^2 = 0.$$

11. (a) From $(X - P) \cdot A = 0$, an equation of the plane is

$$x + 2y - 2z = -1.$$

With the unit normal $B = (-1/3, -2/3, 2/3)$, obtain the normal form

$$-\frac{1}{3}x - \frac{2}{3}y + \frac{2}{3}z = \frac{1}{3}.$$

(b) 2/3.

(c) Same.

12. (a) Set $P = (y_1, y_2, \ldots, y_n)$ and let $B$ be the position vector of $P$. If $Q = (x_1, x_2, \ldots, x_n)$ with position vector $X$ is the foot of the perpendicular, then

$$A \cdot X = c \quad \text{and} \quad B - X = \lambda A.$$

Thus $A \cdot (B - \lambda A) = c$, hence $\lambda = (A \cdot B - c)/|A|^2$ and

$$X = B + A (c - A \cdot B)/|A|^2.$$

(b) $(-1/9, 2/9, 2/9)$ and $(7/9, -13/9, -5/9)$, respectively.

13. Observe first that $C \neq O$; otherwise,

$$A = \frac{A \cdot B}{|B|^2} B,$$

violating the condition that $A$ and $B$ are nonparallel. $B \cdot C = 0$.

14. The angle between the line and the plane is the complement of the angle between the line and the normal; that is,

$$\sin \phi = \frac{\alpha A + \beta B + \gamma C}{\sqrt{\alpha^2 + \beta^2 + \gamma^2} \sqrt{A^2 + B^2 + C^2}}.$$

### Exercises 2.2 (p. 158)

1. (a) The line $x = -1 + 4\lambda$, $y = 2$, $z = 1 + 3\lambda$.
   (b) The plane $x = 2 + 3\mu + \nu$, $y = 1 - 2\mu$, $z = -4 + \mu - \nu$; or $x + 2y + z = 0$.
   (c) The two-dimensional linear space of points $(x, y, z, w)$ satisfying $x + 2y + z = 0$ and $2y + 2z + w = -4$.

2. (a) $A_1 = \sqrt{2}\, E_1 + 2E_3$.

3. For $E_1$, only $E_1 = A_1/|A_1|$ is possible. Suppose such vectors up to index $k - 1$ have been found. Take $E_k = V_k/|V_k|$ where

$$V_k = A_k - \sum_{\mu=1}^{k-1}(A_\mu \cdot E_\mu)\, E_\mu.$$

Observe that if $E_\mu$ depends on $A_1$, $A_2$, ..., $A_\mu$, for $\mu = 1, 2, \ldots,$ $k - 1$, then $E_k$ depends on $A_1$, $A_2$, ..., $A_k$.

4. Let $A_k$, $k = 1, 2, \ldots, n + 1$ be any set of $n + 1$ vectors. If $A_1, \ldots,$ $A_n$ are dependent so is the full set of $n + 1$ vectors; if not, the vectors $E_1, \ldots, E_n$ are dependent on $A_1, \ldots, A_n$ by Exercise 3. Since $E_k$, $k = 1, 2, \ldots, n$ may be taken as coordinate vectors, $A_{n+1}$ depends on $E, \ldots, E_n$; hence, a fortiori, it depends on $A_1$, $A_2$, ..., $A_n$.

5. In the vector form the line has the equation

$$Z = At + B$$

where $B = (b, d, f)$ and $A = (a, c, e)$. Let $Q$ be the foot of the perpendicular from $P$ to the line and $X_0 = (x_0, y_0, z_0)$, $X_1 = (x_1, y_1, z_1)$ the position vectors of $P$ and $Q$, respectively. Since $Q$ is on the line, for some number $\tau$, $X_1 = A\tau + B$. But, from $(X_1 - X_0) \cdot A = 0$ the desired distance $d$ is given by

$$d^2 = |X_1 - X_0|^2 = (X_1 - X_0) \cdot (A\tau + B - X_0) = (X_1 - X_0) \cdot (B - X_0)$$

$$= (x_1 - x_0)(b - x_0) + (y_1 - y_0)(d - y_0) + (z_1 - z_0)(f - z_0),$$

where

$$(x_1, y_1, z_1) = (a\tau + b, c\tau + d, e\tau + f)$$

and

$$\tau = \frac{(X_0 - B) \cdot A}{|A|^2} = \frac{a(x_0 - b) + c(y_0 - d) + e(z_0 - f)}{a^2 + c^2 + e^2}.$$

6. No. To prove this, show that the coefficient vectors $(1, 2, 3)$, $(2, 3, 1)$, $(3, 1, 2)$ are linearly independent. For example, use the method of

solution of Exercise 3 to construct a set of three mutually perpendicular vectors that depend on the coefficient vectors.

7. This is equivalent to solving the system of linear equations in Exercise 6 with constants $a_1, a_2, a_3$ instead of 0, 0, 0 on the right

$$x_1 = \frac{1}{18}(-5a_1 + a_2 + 7a_3), \quad x_2 = \frac{1}{18}(a_1 + 7a_2 - 5a_3),$$

$$x_3 = \frac{1}{18}(7a_1 - 5a_2 + a_3).$$

8. From the solution to Exercise 7

$$\frac{1}{18}\begin{pmatrix} -5 & 1 & 7 \\ 1 & 7 & -5 \\ 7 & -5 & 1 \end{pmatrix}.$$

9. If **a** is singular, the column vectors $A_1, A_2, \ldots, A_n$ are dependent. If a solution $X = (x_1, x_2, \ldots, x_n)$ existed for every **Y**, then every **Y** would have a representation

$$Y = x_1A_1 + x_2A_2 + \cdots + x_nA_n,$$

but the $A_k$ do not span the space.

10.
$$ab = \begin{pmatrix} -2 & 3 & 4 \\ 1 & 0 & 1 \\ -4 & 3 & 2 \end{pmatrix}, \quad ba = \begin{pmatrix} -2 & -4 & 1 \\ -4 & -2 & 1 \\ 3 & 3 & 0 \end{pmatrix}.$$

11. $\Delta = ad - bc \neq 0.$

$$\frac{1}{\Delta}\begin{pmatrix} d & -b \\ -c & a \end{pmatrix}.$$

12. Suppose that $ae = ea = a$ and $a'e = e'a = a$ for all square matrices **a**. Then $e'e = ee' = e = e'$.

13. $b^{-1} a^{-1}$.

14. From our definition, a matrix is singular if and only if the column vectors are dependent. Thus, at least one of the column vectors can be expressed as a linear combination of the others. It follows that any image vector in the mapping can be expressed as a linear combination of no more than $n - 1$ given vectors. Conversely, if the dimension of the image space is less than $n$, the column vectors of the matrix must be linearly dependent, for if they were independent, their linear combinations would span $n$-dimensional space.

15. Express **X** in the form $(r\cos\theta, r\sin\theta)$. Then, for

$$a = \begin{pmatrix} \cos\gamma & -\sin\gamma \\ \sin\gamma & \cos\gamma \end{pmatrix},$$

$$aX = (r\cos(\theta+\gamma), r\sin(\theta+\gamma));$$

hence, **a** may be interpreted as a rotation of vectors through the angle $\gamma$ or a rotation of axes through the angle $-\gamma$. For

$$\mathbf{b} = \begin{pmatrix} \cos\gamma & \sin\gamma \\ \sin\gamma & -\cos\gamma \end{pmatrix},$$

$$\mathbf{bX} = (r\cos(\gamma+\theta),\ r\sin(\gamma-\theta)\,);$$

a reflection of vectors in the line inclined at angle $\frac{1}{2}\gamma$ with respect to the $x$-axis or a reversal of sense of the $y$-axis followed by a rotation of axes through the angle $-\gamma$.

16. The condition is necessary for orthogonality by (49a). It is also sufficient, for if $\mathbf{A}_k$ is the $k$th column vector of **a**, it is the $k$th row vector of $\mathbf{a}^T$. By the definition of matrix multiplication $\mathbf{aa}^T = \mathbf{e}$ implies

$$\mathbf{A}_j \cdot \mathbf{A}_k = \begin{cases} 0, & \text{if } j \neq k \\ 1, & \text{if } j = k. \end{cases}$$

17. Set $\mathbf{c} = \mathbf{ab}$. If $\mathbf{c} = (c_{ij})$, then $\mathbf{c}^T = (c_{ij}{}^T)$, where

$$c_{ij}{}^T = c_{ji} = \sum_{k=1}^{n} a_{jk}\, b_{ki} = \sum_{k=1}^{n} b_{ik}{}^T\, a_{kj}{}^T = \mathbf{b}^T\mathbf{a}^T.$$

18. From Exercises 13, 17, and 16, if **a** and **b** are orthogonal,

$$(\mathbf{ab})^T = \mathbf{b}^T\mathbf{a}^T = \mathbf{b}^{-1}\,\mathbf{a}^{-1} = (\mathbf{ab})^{-1}.$$

which is sufficient for the orthogonality of **ab**.

19. If $\mathbf{X} = (x_1, x_2, \ldots, x_n)$ and $\mathbf{Y} = (y_1, y_2, \ldots, y_n)$, then by (47),

$$(\mathbf{aX}) \cdot (\mathbf{aY}) = (x_1\mathbf{A}_1 + x_2\mathbf{A}_2 + \cdots + x_n\mathbf{A}_n) \cdot (y_1\mathbf{A}_1 + y_2\mathbf{A}_2 + \cdots + y_n\mathbf{A}_n)$$

$$= x_1 y_1 + x_2 y_2 + \cdots + x_n y_n.$$

20. A length-preserving matrix **a** must also preserve scalar products; for

$$|\mathbf{aX} + \mathbf{aY}|^2 = |\mathbf{aX}|^2 + |\mathbf{aY}|^2 + 2(\mathbf{aX}) \cdot (\mathbf{aY})$$

$$= |\mathbf{X}|^2 + |\mathbf{Y}|^2 + 2(\mathbf{aX}) \cdot (\mathbf{aY}) = |\mathbf{a}(\mathbf{X} + \mathbf{Y})|^2 = |\mathbf{X} + \mathbf{Y}|^2$$

$$= |\mathbf{X}|^2 + |\mathbf{Y}|^2 + 2\mathbf{X} \cdot \mathbf{Y}$$

(compare the answer to Exercise 18). Condition (47) follows since each coordinate vector $\mathbf{E}_k$ is mapped on to the column vector $\mathbf{A}_k$ of **a**.

21. Let the particles be $\mathbf{X}_1, \mathbf{X}_2, \ldots, \mathbf{X}_k$ and their masses $m_1, m_2, \ldots, m_k$, respectively. Assume the affine transformation is given in the form $\mathbf{X}' = \mathbf{aX} + \mathbf{A}$. Let the centers of mass before and after transformation be $\mathbf{X}_0 = \left(\sum_{j=1}^{k} m_j\mathbf{X}_j\right)\Big/ \sum_{j=1}^{k} m_j$, $\mathbf{Y}_0 = \left(\sum_{j=1}^{k} m_j\mathbf{X}_j'\right)\Big/ \sum_{j=1}^{m} m_j$, respectively. Observe that $\mathbf{X}_0' = \mathbf{aX}_0 + \mathbf{A} = \mathbf{Y}_0$.

## Exercises 2.3 (p. 177)

1. (a) 0.
   (b) 2.

(c) 12.

(d) $(x - y)(y - z)(z - x)(x + y + z)$.

2. $a + c = 2b$.

3. (a) Use det $(\mathbf{ea})$ = det $(\mathbf{a})$.

 (b) Use det $(\mathbf{e})$ = det $(\mathbf{aa}^{-1})$.

4. (a) $-1$.

 (b) 1.

 (c) $-1$.

 (d) 1.

5. If all the elements of the determinant vanish, the result is immediate. Otherwise, we may suppose $a_{11} \neq 0$, for if $a_{ij} \neq 0$, we may interchange the first and $i$th rows and the first and $j$th columns to place $a_{ij}$ in the first row and column, with perhaps a change of sign in the determinant. Multiply the first column by $a_{1j}/a_{11}$ and subtract from the $j$th column to make the first element in the $j$th column vanish. Proceed similarly to make the first element in any row vanish. By means of this operation and a multiplication of the first row by $-1$ if necessary, the determinant is put in the form

$$\begin{vmatrix} \alpha & 0 & 0 \\ 0 & b_{11} & b_{12} \\ 0 & b_{21} & b_{22} \end{vmatrix}.$$

The same procedures applied to the subdeterminant $\begin{vmatrix} b_{11} & b_{12} \\ b_{21} & b_{22} \end{vmatrix}$ put it in the form $\begin{vmatrix} \beta & 0 \\ 0 & \gamma \end{vmatrix}$. Since the operations on the subdeterminant can be extended to the rows and columns of the original determinant without affecting the zero elements in the first row and column, the desired form has been attained.

6. In (66a) the only possible nonzero term is that for which $j_1 = 1$, $j_2 = 2$, $\dots, j_n = n$.

7. In $a_{j_1 1} a_{j_2 2} \cdots a_{j_n n}$, let $k$ be the least index for which $j_k \neq k$. If $j_k < k$, the product vanishes. If $j_k > k$, then $k$ must appear as a row index for a factor $a_{km}$, where $k < m$; hence, again the product vanishes. Thus, $a_{11} a_{22} \cdots a_{nn}$ is the only possible nonzero term in (66a).

8. (a) $(x - y)(y - z)(z - x)$.

 (b) $-12$.

 (c) $2! 2! 3! 4!$.

9. $x = 3$, $y = 2$, $z = 1$.

10. Apply det $(\mathbf{a}) \cdot$ det $(\mathbf{b})$ = det $(\mathbf{a}^T \mathbf{b})$.

11. Use $D = (A + 2B)(A - B)^2$
$$= [(x + y + z)(x^2 + y^2 + z^2 - xy - yz - xz)]^2.$$

12. Since the determinant is an alternating form in the column vectors, it is immediate that $\Delta = A + Bx$. For $x = -a$, the matrix is lower-tri-

angular and for $x = -b$, upper-triangular. Hence, from Exercise 7, $A + Ba = f(a)$ and $A + Bb = f(b)$.

13. From (57a), with $\mathbf{c} = (c_{jk})$

$$f(\mathbf{A}, \mathbf{B}) = \sum_{j, k=1}^{n} c_{jk} a_j b_k$$

$$= \sum_{j=1}^{n} a_j \sum_{k=1}^{n} c_{jk} b_k$$

$$= \mathbf{A} \cdot (\mathbf{cB})$$

$$= \sum_{k=1}^{n} b_k \sum_{j=1}^{n} c_{jk} a_j$$

$$= \mathbf{B} \cdot (\mathbf{c}^T \mathbf{A}).$$

14. Set $\mathbf{X} = (x, y, z)$, $\mathbf{A} = (g, h, i)$, and

$$\mathbf{a} = \begin{pmatrix} a & \frac{1}{2}d & \frac{1}{2}l \\ \frac{1}{2}d & b & \frac{1}{2}f \\ \frac{1}{2}l & \frac{1}{2}f & c \end{pmatrix}$$

and rewrite the equation of the quadric in the form

$$\mathbf{X} \cdot (\mathbf{aX}) + \mathbf{A} \cdot \mathbf{X} + j = 0.$$

If the affine transformation is given in the form

$$\mathbf{X}' = \mathbf{bX} + \mathbf{B},$$

its inverse is

$$\mathbf{X} = \mathbf{cX}' + \mathbf{C}$$

where $\mathbf{c} = \mathbf{b}^{-1}$ and $\mathbf{C} = -\mathbf{b}^{-1} \mathbf{B}$. Thus the equation of the quadric in the new coordinate system is

$$\mathbf{cX}' \cdot (\mathbf{ac} \, \mathbf{X}') + \mathbf{C} \cdot (\mathbf{ac} \, \mathbf{X}') + \mathbf{cX}' \cdot (\mathbf{aB})$$
$$+ \mathbf{A} \cdot \mathbf{cX}' + \mathbf{C} \cdot (\mathbf{aC}) + \mathbf{A} \cdot \mathbf{B} + j = 0.$$

Apply the result of the preceding exercise to put this in the form

$$\mathbf{X}' \cdot (\mathbf{a'X}') + \mathbf{A}' \cdot \mathbf{X}' + j' = 0,$$

where

$$\mathbf{a}' = \mathbf{c}^T \mathbf{ac},$$

$$\mathbf{A}' = \mathbf{c}^T (\mathbf{a}^T \mathbf{C} + \mathbf{aB} + \mathbf{A}),$$

$$j' = \mathbf{C} \cdot \mathbf{aC} + \mathbf{A} \cdot \mathbf{B} + j.$$

15. Compare with the homogeneous linear system

$$a_1 x + a_2 y + dz = 0$$

$$b_1 x + b_2 y + ez = 0$$

$$c_1 x + c_2 y + fz = 0.$$

If this system has a solution with $z = -1$, and hence a nontrivial solution, the determinant D must vanish. Conversely, if the determinant vanishes, the column vectors are dependent.

Thus, there exist constants $x$, $y$, $z$, not all zero, such that

$$x\mathbf{A}_1 + y\mathbf{A}_2 + z\mathbf{B} = 0$$

where $\mathbf{A}_i = (a_i, b_i, c_i)$ and $\mathbf{B} = (d, e, f)$. It is not possible that $z = 0$, for then $\mathbf{A}_1$ and $\mathbf{A}_2$ would be dependent and all three of the given $2 \times 2$ determinants would vanish. We may therefore divide by $-z$ to make $-1$ the coefficient of $\mathbf{B}$; hence, the desired solution exists.

16. In vector form the lines may be written as

$$\mathbf{X} = \mathbf{A}t + \mathbf{B}, \quad \mathbf{X} = \mathbf{C}t + \mathbf{D}.$$

The lines are parallel if and only if $\mathbf{A}$ and $\mathbf{C}$ are parallel (this includes the case that the lines are the same). They intersect if and only if there exist numbers $t_1$, and $t_2$ for which $\mathbf{A}t_1 + \mathbf{B} = \mathbf{C}t_2 + \mathbf{D}$. Thus, by the solution of the preceding exercise, the condition is that the matrix with column vectors $\mathbf{A}$, $\mathbf{C}$, $\mathbf{B} - \mathbf{D}$ have a vanishing determinant; that is,

$$\begin{vmatrix} a_1 & c_1 & b_1 - d_1 \\ a_2 & c_2 & b_2 - d_2 \\ a_3 & c_3 & b_3 - d_3 \end{vmatrix} = 0$$

17. A set of interchanges that permutes $j_1, j_2, \ldots, j_n$ into $1, 2, \ldots, n$, also permutes $1, 2, \ldots, n$ into $k_1, k_2, \ldots, k_n$. Consequently, $j_1, j_2, \ldots, j_n$ and $k_1, k_2, \ldots, k_n$ are either both even or both odd permutations of $1, 2, \ldots, n$.

18. In vector form this states that the vector equation

$$\mathbf{aX} = \lambda \mathbf{X}$$

must have at least one nontrivial solution. Rewrite the equation in the form of a homogeneous equation:

$$(\mathbf{a} - \lambda \mathbf{e})\,\mathbf{X} = O,$$

where $\mathbf{e}$ is the unit matrix. This equation has a nontrivial solution if and only if

$$\det(\mathbf{a} - \lambda \mathbf{e}) = 0.$$

In $n$-dimensional space this is a polynomial equation in $\lambda$ of $n$th degree with leading term $(-1)^n \lambda^n$. Thus, a solution always exists if $n$ is odd.

## Exercises 2.4 (p. 202)

1. Let $\mathbf{X}_0$ be the position vector of $P$ and express the line in the vector form $\mathbf{X} = \mathbf{A}t + \mathbf{B}$. The distance $r$ from $P$ to $l$ is $|\mathbf{X}_0 - \mathbf{B}| \sin \theta$, where

$\theta$ is the angle between $\mathbf{P} - \mathbf{B}$ and $\mathbf{A}$; hence,

$$r = |(\mathbf{X}_0 - \mathbf{B}) \times \mathbf{A}| / |\mathbf{A}|.$$

2. The velocity is $r\omega$, where $r$ is the distance of the point from the axis of rotation. From the solution of the preceding, with $\mathbf{B}$ representing the origin $\mathbf{X}_0 = (x, y, z)$ and $\mathbf{A} = (\alpha, \beta, \gamma)$.

$$r\omega = \omega \left[ (y\gamma - z\beta)^2 + (z\alpha - x\gamma)^2 + (x\beta - y\alpha)^2 \right]^{1/2}.$$

3. Name the position vectors of the three points $\mathbf{X}_1, \mathbf{X}_2, \mathbf{X}_3$, respectively. If $\mathbf{X} = (x, y, z)$ represents any point of the plane, the three vectors $\mathbf{X}_1 - \mathbf{X}, \mathbf{X}_2 - \mathbf{X}, \mathbf{X}_3 - \mathbf{X}$ lie in a two-dimensional space and, hence, are dependent. Consequently,

$$\det (\mathbf{X}_1 - \mathbf{X}, \mathbf{X}_2 - \mathbf{X}, \mathbf{X}_3 - \mathbf{X}) = 0.$$

4. Let the equations of the lines be given in vector form by $l:\mathbf{X} = \mathbf{A}t + \mathbf{B}$ and $l':\mathbf{X}' = \mathbf{A}'t' + \mathbf{B}'$. The shortest segment $PP'$ with one end point on each line must be perpendicular to both. For, say, $PP'$ is not perpendicular to $l'$ at $P'$; then the perpendicular from $P$ to $l'$ would be shorter. If $\mathbf{X}$ and $\mathbf{X}'$ are the position vectors of $P$ and $P'$, respectively,

$$\mathbf{X} - \mathbf{X}' = \mathbf{A}t + \mathbf{B} - \mathbf{A}'t' + \mathbf{B}'$$

$$= k(\mathbf{A} \times \mathbf{A}').$$

To determine $k$, take the dot product with $(\mathbf{A} \times \mathbf{A}')$ in this equation, which yields

$$k = \frac{(\mathbf{B} - \mathbf{B}') \cdot (\mathbf{A} \times \mathbf{A}')}{|\mathbf{A} \times \mathbf{A}'|},$$

which yields the desired distance $d$ through

$$d^2 = |\mathbf{X} - \mathbf{X}'|^2 = k^2 |\mathbf{A} \times \mathbf{A}'|^2$$

or

$$d = \frac{|(\mathbf{B} - \mathbf{B}') \cdot (\mathbf{A} \times \mathbf{A}')|}{|\mathbf{A} \times \mathbf{A}'|}.$$

5. The sum does not depend on the choice of origin, since a different choice of origin $(a, b)$ amounts to replacing each determinant

$$\Delta_k = \begin{vmatrix} x_k & x_{k+1} \\ y_k & y_{k+1} \end{vmatrix} \quad \text{by} \quad \Delta_k' = \begin{vmatrix} x_k - a & x_{k+1} - a \\ y_k - b & y_{k+1} - b \end{vmatrix}$$

Because

$$\Delta_k' = \Delta_k - \begin{vmatrix} x_k & a \\ y_k & b \end{vmatrix} + \begin{vmatrix} x_{k+1} & a \\ y_{k+1} & b \end{vmatrix},$$

each aditional determinant $\begin{vmatrix} x_k & a \\ y_k & b \end{vmatrix}$ appears twice in the total, but with opposite signs. Thus, we may choose the origin in the interior of the polygon. The polygon is the sum of the areas of the triangles $OP_kP_{k+1}$, $k = 1, \ldots, n$ (where $P_{n+1} = P_1$), but the area of $OP_kP_{k+1}$ is precisely

$$\frac{1}{2}\begin{vmatrix} x_k & x_{k+1} \\ y_k & y_{k+1} \end{vmatrix}.$$

6. Subtract the third row from the first two to show that the determinant equals $\frac{1}{2}\mathbf{X}_1 \times \mathbf{X}_2$, where $\mathbf{X}_1 = (x_1 - x_3, \ y_1 - y_3)$ and $\mathbf{X}_2 = (x_2 - x_3, \ y_2 - y_3)$.

7. If the coordinates of the vertices are rational, the area of the triangle as defined by the determinant is clearly rational. But, for an equilateral triangle with side length $s$, the area is $\frac{1}{4}s^2\sqrt{3}$, where

$$s^2 = (x_i - x_j)^2 + (y_i - y_j)^2 \qquad\qquad (i \neq j).$$

is plainly rational.

8. (a) In vector form, this states

$$\mathbf{A} \cdot (\mathbf{A}' \times \mathbf{A}'') \leq |\mathbf{A}| \cdot |\mathbf{A}'| \cdot |\mathbf{A}''|,$$

which is obviously true, since

$$|\mathbf{A}' \times \mathbf{A}''| \leq |\mathbf{A}'| \cdot |\mathbf{A}''|$$

and

$$|D| = |\mathbf{A} \cdot (\mathbf{A}' \times \mathbf{A}'')| \leq |\mathbf{A}| \cdot |\mathbf{A}' \times \mathbf{A}''|.$$

(b) Equality can hold only if it holds in both the preceding inequalities. Thus $\mathbf{A}$, $\mathbf{A}'$, and $\mathbf{A}''$ must be mutually perpendicular.

9. (a) If $\mathbf{B}$ and $\mathbf{C}$ are dependent, say, $\mathbf{C} = \lambda\mathbf{B}$, the identity is trivially true. Otherwise, form the orthonormal basis $\mathbf{E}_1$, $\mathbf{E}_2$, $\mathbf{E}_3$, where the respective vectors are unit vectors in the directions of $\mathbf{B}$, $\mathbf{B} \times \mathbf{C}$, $\mathbf{B} \times (\mathbf{B} \times \mathbf{C})$. Write $\mathbf{A}$, $\mathbf{B}$, and $\mathbf{C}$ in terms of this basis:

$$\mathbf{A} = a_1\mathbf{E}_1 + a_2\mathbf{E}_2 + a_3\mathbf{E}_3$$

$$\mathbf{B} = b\mathbf{E}_1, \quad \mathbf{C} = c_1\mathbf{E}_1 + c_3\mathbf{E}_3$$

to obtain $\mathbf{B} \times \mathbf{C} = -bc_3\mathbf{E}_2$ and

$$\mathbf{A} \times (\mathbf{B} \times \mathbf{C}) = bc_3(a_3\mathbf{E}_1 - a_1\mathbf{E}_3).$$

Employ $\mathbf{E}_1 = (1/b)\,\mathbf{B}$ and $\mathbf{E}_3 = 1/c_3\,[\mathbf{C} - (c_1/b)\mathbf{B}]$ to obtain

$$\mathbf{A} \times (\mathbf{B} \times \mathbf{C}) = (a_1c_1 + a_3c_3)\mathbf{B} - (a_1b)\mathbf{C}.$$

(b) Observe that

$$\begin{aligned} Z = (\mathbf{X} \times \mathbf{Y}) \cdot (\mathbf{X}' \times \mathbf{Y}') &= \det(\mathbf{X}, \mathbf{Y}, \mathbf{X}' \times \mathbf{Y}') \\ &= \det(\mathbf{Y}, \mathbf{X}' \times \mathbf{Y}', \mathbf{X}) \\ &= [\mathbf{Y} \times (\mathbf{X}' \times \mathbf{Y}')\,] \cdot \mathbf{X}. \end{aligned}$$

Apply Exercise 9a to obtain

$$Z = [(\mathbf{Y} \cdot \mathbf{Y}')\mathbf{X}' - (\mathbf{Y} \cdot \mathbf{X}')\mathbf{Y}'] \cdot \mathbf{X}$$

(c) Apply Exercise 9a to rewrite the expression on the left as

$$U = [\,(\mathbf{X} \cdot \mathbf{Z})\mathbf{Y} - (\mathbf{X} \cdot \mathbf{Y})\mathbf{Z}] \cdot \mathbf{V},$$

where

$$\mathbf{V} = [\,(\mathbf{Y} \cdot \mathbf{X})\mathbf{Z} - (\mathbf{Y} \cdot \mathbf{Z})\mathbf{X}\,] \times [\,(\mathbf{Z} \cdot \mathbf{Y})\mathbf{X} - (\mathbf{Z} \cdot \mathbf{X})\mathbf{Y}\,]$$
$$= (\mathbf{Y} \cdot \mathbf{X})\,(\mathbf{Y} \cdot \mathbf{Z})\,(\mathbf{Z} \times \mathbf{X}) + (\mathbf{X} \cdot \mathbf{Y})\,(\mathbf{X} \cdot \mathbf{Z})\,(\mathbf{Y} \times \mathbf{Z})$$
$$+ (\mathbf{Z} \cdot \mathbf{Y})\,(\mathbf{Z} \cdot \mathbf{X})\,(\mathbf{X} \times \mathbf{Y}).$$

Thus,

$$U = (\mathbf{X} \cdot \mathbf{Z})\,(\mathbf{Y} \cdot \mathbf{X})\,(\mathbf{Y} \cdot \mathbf{Z})\,[\,\mathbf{Y} \cdot (\mathbf{Z} \times \mathbf{X})\,]$$
$$- (\mathbf{X} \cdot \mathbf{Y})\,(\mathbf{Z} \cdot \mathbf{Y})\,(\mathbf{Z} \cdot \mathbf{X})\,[\,\mathbf{Z} \cdot (\mathbf{X} \times \mathbf{Y})\,] = 0.$$

10. Let $\mathbf{E}$ be the unit vector in the direction of $(-1, 0, 1)$; thus, $\mathbf{E} = (-\frac{1}{2}\sqrt{2}, 0, \frac{1}{2}\sqrt{2})$. Let $\mathbf{X} = (x, y, z)$ be the position vector of any point and $\mathbf{A}$ the foot of the perpendicular from the point to the axis of rotation:

$$\mathbf{A} = (\mathbf{X} \cdot \mathbf{E})\mathbf{E} = \left(\frac{1}{2}(x - z),\ 0,\ \frac{1}{2}(z - x)\right).$$

Note that $\mathbf{X} - \mathbf{A}$ is perpendicular to $\mathbf{A}$ and introduce the mutual perpendicular $\mathbf{E} \times (\mathbf{X} - \mathbf{A})$ to these two. If $\mathbf{X}'$ is the position vector of the image of $(x, y, z)$ in the rotation, then $\mathbf{X}' - \mathbf{A}$ is perpendicular to $\mathbf{A}$ and the given orientation condition yields

$$(\mathbf{X} - \mathbf{A}) \times (\mathbf{X}' - \mathbf{A}) = r^2 \sin \phi\ \mathbf{E},$$

where $r = |\mathbf{X} - \mathbf{A}| = |\mathbf{X}' - \mathbf{A}|$ is the distance of $\mathbf{X}$ from the axis. Set

$$\mathbf{X}' = \lambda \mathbf{A} + \mu(\mathbf{X} - \mathbf{A}) + \nu[\mathbf{E} \times (\mathbf{X} - \mathbf{A})]$$

as we may, since the vectors appearing in the linear combination are mutually perpendicular. From $(\mathbf{X}' - \mathbf{A}) \cdot \mathbf{A} = 0$, it follows that $\lambda = 1$; from $(\mathbf{X}' - \mathbf{A}) \cdot (\mathbf{X} - \mathbf{A}) = r^2 \cos \phi$, we have $\mu = \cos \phi$. Finally, from Exercise 9a

$$r^2 \sin \phi\ \mathbf{E} = (\mathbf{X} - \mathbf{A}) \times (\mathbf{X}' - \mathbf{A})$$
$$= \nu(\mathbf{X} - \mathbf{A}) \times [\mathbf{E} \times (\mathbf{X} - \mathbf{A})]$$
$$= \nu r^2 \mathbf{E};$$

thus, $\nu = \sin \phi$. Employ

$$\mathbf{X} - \mathbf{A} = \left(\frac{1}{2}(x + z),\ y,\ \frac{1}{2}(x + z)\right)$$

$$\mathbf{E} \times (\mathbf{X} - \mathbf{A}) = \mathbf{E} \times \mathbf{X} = \frac{1}{2}\sqrt{2}\,(-y,\ x + z,\ -y)$$

to obtain $\mathbf{X}' = \mathbf{a}\mathbf{X}$, where

$$\mathbf{a} = \begin{pmatrix} \frac{1}{2}(\cos \phi + 1) & -\frac{1}{2}\sqrt{2}\sin \phi & \frac{1}{2}(\cos \phi - 1) \\ \frac{1}{2}\sqrt{2}\sin \phi & \cos \phi & \frac{1}{2}\sqrt{2}\sin \phi \\ \frac{1}{2}(\cos \phi - 1) & -\frac{1}{2}\sqrt{2}\sin \phi & \frac{1}{2}(\cos \phi + 1) \end{pmatrix}.$$

11. From Exercise 9a,

$$\mathbf{X} = [\,(\mathbf{A} \times \mathbf{B}) \cdot \mathbf{D}]\mathbf{C} - [\,(\mathbf{A} \times \mathbf{B}) \cdot \mathbf{C}]\mathbf{D}$$
$$= [\,(\mathbf{C} \times \mathbf{D}) \cdot \mathbf{A}]\mathbf{B} - [\,(\mathbf{C} \times \mathbf{D}) \cdot \mathbf{B}]\mathbf{A}.$$

Since **A**, **B**, **C** are independent, $(\mathbf{A} \times \mathbf{B}) \cdot \mathbf{C} \neq 0$ and we may solve for **D**.

12. Let $\mathbf{E}_1'$, $\mathbf{E}_2'$, $\mathbf{E}_3'$ be the unit coordinate vectors in the new coordinate system. We are given $\mathbf{E}_3 \cdot \mathbf{E}_3' = \cos\theta$, $\mathbf{E}_1 \times (\mathbf{E}_3 \times \mathbf{E}_3') = \sin\theta \sin\phi\, \mathbf{E}_3$, and $\mathbf{E}_1' \times (\mathbf{E}_3 \times \mathbf{E}_3') = -\sin\theta \sin\psi\, \mathbf{E}_3'$. Furthermore, $\mathbf{E}_1 \cdot (\mathbf{E}_3 \times \mathbf{E}_3') = \sin\theta \cos\phi$ and $\mathbf{E}_1' \cdot (\mathbf{E}_3 \times \mathbf{E}_3') = \sin\theta \cos\psi$. Thus, from Exercise 9a, $(\mathbf{E}_1 \cdot \mathbf{E}_3') = \sin\theta \sin\phi$ and $\mathbf{E}_1' \cdot \mathbf{E}_3 = \sin\theta \sin\psi$. Now, set

$$\mathbf{E}_i = \sum_{j=1}^{3} a_{ij}\, \mathbf{E}_j'$$

where

$$(a_{ij}) = (\mathbf{E}_i \cdot \mathbf{E}_j')$$

is the matrix we seek. The information we already have yields

$$a_{13} = \sin\theta \sin\phi, \quad a_{31} = \sin\theta \sin\psi, \quad a_{33} = \cos\theta.$$

Form $\mathbf{E}_3 \times \mathbf{E}_3' = \sin\theta \sin\psi\, \mathbf{E}_2' + a_{32}\, \mathbf{E}_1'$ and take the scalar product with $\mathbf{E}_1'$ to find

$$\mathbf{E}_1' \cdot (\mathbf{E}_3 \times \mathbf{E}_3') = \sin\theta \cos\psi = a_{32}.$$

Thus,

$$\mathbf{E}_3 = -\sin\theta \sin\psi\, \mathbf{E}_1' + \sin\theta \cos\psi\, \mathbf{E}_2' + \cos\theta\, \mathbf{E}_3'.$$

Using this expression for $\mathbf{E}_3$, solve for $a_{11}$ and $a_{12}$ in the equations

$$\mathbf{E}_1 \cdot \mathbf{E}_3 = 0, \quad |\mathbf{E}_1|^2 = 1,$$

to obtain

$$a_{11} = -\cos\theta \sin\phi \sin\psi \pm \cos\phi \cos\psi,$$
$$a_{12} = -\cos\theta \sin\phi \cos\psi \pm \cos\phi \sin\psi.$$

The undetermined signs in these expressions for $a_{11}$ and $a_{12}$ are fixed by the condition $\mathbf{E}_1 \cdot (\mathbf{E}_3 \times \mathbf{E}_3') = \sin\theta \cos\phi$, which yields the plus sign in the expression for $a_{11}$ and the minus sign for $a_{12}$. Set $\mathbf{E}_2 = \mathbf{E}_3 \times \mathbf{E}_1$ to obtain, finally,

$$(a_{ij}) = \begin{pmatrix} -\cos\theta \sin\phi \sin\psi & -\cos\theta \sin\phi \cos\psi & \sin\theta \sin\phi \\ \quad + \cos\phi \cos\psi & \quad -\cos\phi \sin\psi & \\ \cos\theta \cos\phi \cos\psi & \cos\theta \cos\phi \cos\psi & -\sin\theta \cos\phi \\ \quad + \sin\phi \cos\psi & \quad -\sin\phi \sin\psi & \\ \sin\theta \sin\psi & \sin\theta \cos\psi & \cos\theta \end{pmatrix}.$$

Note that this result holds also for $\theta = 0$ or $\pi$, when $\phi$ and $\psi$ become indeterminate with $\phi + \psi = x0x'$ or $\phi - \psi = x0x'$, respectively. The angles $\phi$, $\psi$, $\theta$, are so-called Eulerian angles, and our result shows that the most general orthogonal matrix with determinant $\Delta$ of value $+1$

may be expressed "parametrically" by means of the three variables $\phi$, $\psi$, $\theta$, subject to the inequalities

$$0 \leqq \theta \leqq \pi, \quad 0 \leqq \phi < 2\pi, \quad 0 \leqq \psi < 2\pi.$$

13. Let $\mathbf{A} = a_1\mathbf{E}_1 + a_2\mathbf{E}_2 + \cdots + a_m\mathbf{E}_m$ be a nonzero vector of $\pi$ perpendicular to all the vectors of $\pi'$ with, say, $a_1 \neq 0$. Using $\mathbf{E}_1 = 1/a_1(\mathbf{A} - a_2\mathbf{E}_2 - \cdots - a_m\mathbf{E}_m)$, we obtain from (85a)

$$\mu = \frac{1}{a_1} [\mathbf{A} - a_2\mathbf{E}_2 - \cdots - a_m\mathbf{E}_m, \mathbf{E}_2, \ldots, \mathbf{E}_m; \mathbf{E}_1', \ldots, \mathbf{E}_m')$$

$$= \frac{1}{a_1} [\mathbf{A}, \mathbf{E}_2, \ldots, \mathbf{E}_m; \mathbf{E}_1', \mathbf{E}_2', \ldots, \mathbf{E}_m'] = 0.$$

Conversely, if $\mu = 0$, the column vectors in the determinant representation (85a) of $\mu$ are dependent: for some nontrivial set of coefficients,

$$\lambda_1\mathbf{E}_k \cdot \mathbf{E}_1' + \lambda_2\mathbf{E}_k \cdot \mathbf{E}_2' + \cdots + \lambda_m\mathbf{E}_k \cdot \mathbf{E}_m' = 0 \quad (k = 1, 2, \ldots, m).$$

Then

$$\mathbf{E}_k \cdot (\lambda_1\mathbf{E}_1' + \lambda_2\mathbf{E}_2' + \cdots + \lambda_m\mathbf{E}_m') = 0$$

and we have a vector of $\pi'$ orthogonal to every basis vector and, hence, every vector of $\pi$.

## Exercises 2.5 (p. 215)

1. Let the coordinates of $P$ be $(x_1', x_2', x_3')$; of $Q$, $(x_1'', x_2'', x_3'')$. Thus $\overrightarrow{PQ}$ represents the vector $\mathbf{U}$, where $u_i = x_i'' - x_i'$. The coordinates of $P$ and $Q$ in the new system are given by (89a) with appropriate primes and $\overrightarrow{PQ}$ represents the vector $v_i = y_i'' - y_i'$ whose components clearly satisfy (89a).

5. Let the curve be expressed vectorially by $\mathbf{X}(t)$, and let the three values of the parameter be given by $t$, $t_1$, $t_2$, and the corresponding points by $\mathbf{X} = \mathbf{X}(t)$, $\mathbf{X}_1 = \mathbf{X}(t_1)$, $\mathbf{X}_2 = \mathbf{X}(t_2)$. The normal to the plane through the three points is parallel to

$$(\mathbf{X}_1 - \mathbf{X}) \times (\mathbf{X}_2 - \mathbf{X}).$$

Setting $t_1 - t = h_1$, $t_2 - t = h_2$ and using Taylor's theorem, obtain

$$\mathbf{X}_i = \mathbf{X} + \frac{d\mathbf{X}}{dt}h_i + \frac{1}{2}\frac{d^2\mathbf{X}}{dt^2}h_i^2 + \cdots.$$

Thus, to lowest order,

$$(\mathbf{X}_1 - \mathbf{X}) \times (\mathbf{X}_2 - \mathbf{X}) = \frac{1}{2}\frac{d\mathbf{X}}{dt}\frac{d^2\mathbf{X}}{dt^2}(hk^2 - kh^2).$$

In the limit as $h$ and $k$ approach 0 and as $t$ approaches $t_0$, the normal to the osculating plane takes the direction of $d\mathbf{X}/dt \times d^2\mathbf{X}/dt^2$ at $\mathbf{X}_0 =$

$\mathbf{X}(t_0)$. Thus, the position vector $\mathbf{Y}$ of a point of the osculating plane satisfies

$$(\mathbf{Y} - \mathbf{X}_0) \cdot \left( \frac{d\mathbf{X}}{dt} \times \frac{d^2\mathbf{X}}{dt^2} \right) = 0.$$

6. From the result of the preceding exercise, we must show that $d\mathbf{X}/ds$ and $d^2\mathbf{X}/ds^2$ are both perpendicular to $d\mathbf{X}/dt \times d^2\mathbf{X}/dt^2$. This is immediate from

$$\frac{d\mathbf{X}}{ds} = \frac{d\mathbf{X}}{dt} \frac{dt}{ds} \quad \text{and} \quad \frac{d^2\mathbf{X}}{ds^2} = \frac{d\mathbf{X}}{dt} \frac{d^2t}{ds^2} + \frac{d^2\mathbf{X}}{dt^2} \left( \frac{dt}{ds} \right)^2.$$

7. Let the curve be given by $\mathbf{X}(s)$, where $s$ is arc length, and expand $\mathbf{X}$ by Taylor's theorem:

$$\mathbf{X}(s) = \mathbf{X}(s_0) + \mathbf{X}'(s_0)l + \mathbf{Y}O(l^2),$$

where $l = s - s_0$ and $\mathbf{Y}$ is bounded. Thus, since $|\mathbf{X}'(s_0)| = 1$,

$$d - l = |\mathbf{X}(s) - \mathbf{X}(s_0)| - l$$
$$= |\mathbf{X}'(s_0)l + \mathbf{Y}O(l^2)| - l$$
$$\leq |\mathbf{X}'(s_0)| l + O(l^2) - l;$$

that is, $d - l = O(l^2) = o(l)$.

8. From the solution to the preceding problem 6.

$$k = \left| \frac{d^2\mathbf{X}}{ds^2} \right| = \left| \mathbf{X}' \frac{d^2t}{ds^2} + \mathbf{X}'' \left( \frac{dt}{ds} \right)^2 \right|.$$

Note that

$$\frac{dt}{ds} = \frac{1}{|\mathbf{X}'|};$$

hence,

$$\frac{d^2t}{ds^2} = -\frac{\mathbf{X}' \cdot \mathbf{X}''}{|\mathbf{X}'|^4}.$$

Thus,

$$k^2 = \frac{|\mathbf{X}'|^2 |\mathbf{X}''|^2 - (\mathbf{X}' \cdot \mathbf{X}'')^2}{|\mathbf{X}'|^6}$$

9. From the solution to Exercise 6, $d^2\mathbf{X}/dt^2$ is a linear combination of $d\mathbf{X}/ds$ and $d^2\mathbf{X}/ds^2$.

10. Let $C$ be represented by $\mathbf{X}(t)$ and assume that the position vector $\mathbf{X}(t_0)$ of $B$ is not an end point of $C$. Let $\mathbf{Y}$ be the position vector of $A$. $|\mathbf{Y} - \mathbf{X}(t_0)|$ is a minimum if

$$\frac{d}{dt}|\mathbf{Y} - \mathbf{X}(t)|^2 \bigg|_{t = t_0} = 0;$$

that is,

$$[\mathbf{Y} - \mathbf{X}(t_0)] \cdot \mathbf{X}'(t_0) = 0.$$

11. Let the curve be given parametrically by $\mathbf{X}(\theta)$ where $x = a \cos \theta$, $y = a \sin \theta$. The tangent plane depends only on $x$ and $y$, not $z$, and it makes the angle $\theta$ with the $y$-axis. The $z$-component of the tangent vector $\mathbf{X}'$ to the curve satisfies

$$\frac{z'}{\sqrt{x'^2 + y'^2 + z'^2}} = \cos \theta.$$

or

$$\frac{z'}{\sqrt{a^2 + z'^2}} = \cot \theta.$$

Thus,

$$z' = \pm a \cot \theta;$$

whence,

$$z = c \pm a \log \sin \theta.$$

For the curvature, see Exercise 8.

12. From $d\mathbf{X}/d\theta = (-\sin \theta, \cos \theta, \sinh A\theta)$, we have

$$\frac{d^2\mathbf{X}}{d\theta^2} = (-\cos \theta, -\sin \theta, A\cosh A\theta),$$

the solution yields the equation for any point $\mathbf{Y}$ of the osculating plane

$$0 = (\mathbf{Y} - \mathbf{X}) \cdot \left( \frac{d\mathbf{X}}{d\theta} \times \frac{d^2\mathbf{X}}{d\theta^2} \right),$$

where the normal vector is given by

$$\frac{d\mathbf{X}}{d\theta} \times \frac{d^2\mathbf{X}}{d\theta^2} = (N_1, N_2, N_3)$$

and

$$N_1 = A \cos \theta \cosh A\theta + \sin \theta \sinh A\theta.$$
$$N_2 = A \sin \theta \cosh A\theta - \cos \theta \sinh A\theta$$
$$N_3 = 1.$$

The distance of the plane from the origin is $|\mathbf{X} \cdot \mathbf{N}|/|\mathbf{N}|$, and, since $\mathbf{X} \cdot \mathbf{N} = (A + 1/A) \cosh A\theta$ and $|\mathbf{N}|^2 = (A^2 + 1) \cosh^2 A\theta$, the result follows.

13. (a) Let $\mathbf{X}(t)$ be the parametric representation of the curve and set $\mathbf{X}_i = \mathbf{X}(t_i)$. The plane through the three points, by Exercise 3 of Section 2.4, is

$$(\mathbf{X}_1 - \mathbf{X}) \cdot [ (\mathbf{X}_2 - \mathbf{X}) \times (\mathbf{X}_3 - \mathbf{X}) ] = 0$$

or

$$\mathbf{X} \cdot [\mathbf{X}_1 \times \mathbf{X}_2 + \mathbf{X}_2 \times \mathbf{X}_3 + \mathbf{X}_3 \times \mathbf{X}_1] = \mathbf{X}_1 \cdot (\mathbf{X}_2 \times \mathbf{X}_3),$$

from which the result follows.

(b) The three osculating planes have the equations

$$(\mathbf{X} - \mathbf{X}_i) \cdot (\mathbf{X}_i' \times \mathbf{X}_i'') = 0$$

(from Exercise 6) or, in terms of coordinates,

$$\frac{3x}{a} - \frac{6t_i}{b}y + \frac{3t_i^2}{c}z - t_i^3 = 0.$$

Thus, if $(x, y, z)$ is a point common to the three osculating planes, $t_1, t_2, t_3$ are the three roots of the above equation with coefficients:

$$t_1 + t_2 + t_3 = \frac{3z}{c},$$

$$t_1t_2 + t_2t_3 + t_3t_1 = \frac{6y}{b},$$

$$t_1t_2t_3 = \frac{3x}{a}.$$

14. Since a sphere is determined by any four of its noncoplanar points, we may impose four conditions on the sphere of closest contact: that the contact of curve and sphere be of third order. Let $\mathbf{X}(s)$ be the representation of the curve in terms of arc length and $\mathbf{A}$ the center of the sphere. Require that $|\mathbf{X} - \mathbf{A}|^2$ vanish to third order; thus, from $|\dot{\mathbf{X}}|^2 = 1$ and $\dot{\mathbf{X}} \cdot \ddot{\mathbf{X}} = 0$,

$$(\mathbf{X} - \mathbf{A}) \cdot \dot{\mathbf{X}} = 0,$$

$$(\mathbf{X} - \mathbf{A}) \cdot \ddot{\mathbf{X}} + 1 = 0$$

$$(\mathbf{X} - \mathbf{A}) \cdot \dddot{\mathbf{X}} = 0.$$

From the first and last of these equations, $\mathbf{X} - \mathbf{A} = \lambda(\dot{\mathbf{X}} \times \ddot{\mathbf{X}})$, where $\lambda$ is given by the second equation. Hence,

$$\mathbf{A} = \mathbf{X} + \frac{\dot{\mathbf{X}} \times \ddot{\mathbf{X}}}{\ddot{\mathbf{X}} \cdot [\dot{\mathbf{X}} \times \dddot{\mathbf{X}}]}.$$

15. Set $|\mathbf{X} - \mathbf{A}| = 1$ in the solution of the preceding exercise.

16. Since, by Exercise 6, $\boldsymbol{\xi}_3$ is normal to the osculating plane, $\frac{1}{\tau} = |\dot{\boldsymbol{\xi}}_3|$.

Furthermore, since $\dot{\boldsymbol{\xi}}_i$ and $\boldsymbol{\xi}_i$ are perpendicular

$$\dot{\boldsymbol{\xi}}_2 = a\boldsymbol{\xi}_1 + b\boldsymbol{\xi}_3 \text{ and } \dot{\boldsymbol{\xi}}_3 = c\boldsymbol{\xi}_1 + d\boldsymbol{\xi}_2.$$

Differentiate $\boldsymbol{\xi}_1 = \boldsymbol{\xi}_2 \times \boldsymbol{\xi}_3$ to obtain

$$\frac{1}{\rho}\boldsymbol{\xi}_2 = (\boldsymbol{\xi}_2 \times \dot{\boldsymbol{\xi}}_3) + (\dot{\boldsymbol{\xi}}_2 \times \boldsymbol{\xi}_3)$$

$$= -a\dot{\boldsymbol{\xi}}_2 - c\boldsymbol{\xi}_3;$$

hence $a = -1/\rho$ and $c = 0$. From $\dot{\boldsymbol{\xi}}_3 = d\boldsymbol{\xi}_2$, $d = \pm 1/\tau$; choose the minus sign. To determine $b$, differentiate $\boldsymbol{\xi}_3 = (\boldsymbol{\xi}_1 \times \boldsymbol{\xi}_2)$:

$$\dot{\boldsymbol{\xi}}_3 = -\frac{1}{\tau}\boldsymbol{\xi}_2 = (\boldsymbol{\xi}_1 \times \dot{\boldsymbol{\xi}}_2) - (\boldsymbol{\xi}_2 \times \dot{\boldsymbol{\xi}}_1)$$

$$= -b\,\boldsymbol{\xi}_2;$$

whence $b = 1/\tau$.

17. (a) Differentiate $\ddot{\mathbf{X}} = \dot{\xi}_1 = k\xi_2$ to obtain

$$\dddot{\mathbf{X}} = \dot{k}\xi_2 + k\dot{\xi}_2$$

$$= -k^2\xi_1 + \dot{k}\xi_2 + \frac{k}{\tau}\xi_3.$$

(b) From the result of Exercise 14,

$$\frac{\xi_2}{\tau} + \frac{\dot{k}}{k^2\tau}\xi_3.$$

18. Since $1/\tau = |\dot{\xi}_3| = 0$, then $\dot{\xi}_3 = 0$ and, therefore, $\xi_3$ must be a constant vector. From $0 = \xi_1 \cdot \xi_3 = \dot{\mathbf{X}} \cdot \xi_3 = \dfrac{d}{ds}(\mathbf{X} \cdot \xi_3)$, it follows that $\mathbf{X} \cdot \xi_3$ = constant.

19. Let $\mathbf{A}$ and $\mathbf{P}$ be the position vectors of $A$ and $P$ respectively. Set $\mathbf{X} = \mathbf{A} - \mathbf{P}$, hence $\dot{\mathbf{X}} = -\dot{\mathbf{P}}$. The equation states

$$\frac{d}{dt}|\mathbf{X}| = -\mathbf{a} \cdot \dot{\mathbf{P}},$$

which follows directly from the differentiation formula

$$\frac{d}{dt}|\mathbf{X}| = \frac{d}{dt}\sqrt{\mathbf{X} \cdot \mathbf{X}} = \frac{\mathbf{X} \cdot \dot{\mathbf{X}}}{|\mathbf{X}|}$$

with $a = \mathbf{X}/|\mathbf{X}|$.

20. (a) Set $\mathbf{X} = \mathbf{A} - \mathbf{P}$ as in the preceding solution. From that solution,

$$-\dot{\mathbf{P}} = \dot{\mathbf{X}} = \frac{d}{dt}(|\mathbf{X}|\mathbf{a}) = -(\mathbf{a} \cdot \dot{\mathbf{P}})\mathbf{a} + |\mathbf{X}|\dot{\mathbf{a}}.$$

and the desired result is immediate.

(b) Introduce the expression for $\dot{\mathbf{a}}$ and the similar expressions for $\dot{\mathbf{b}}$ in

$$\ddot{\mathbf{P}} = u\dot{\mathbf{a}} + v\dot{\mathbf{b}} + w\dot{\mathbf{c}} + \dot{u}\mathbf{a} + \dot{v}\mathbf{b} + \dot{w}\mathbf{c}.$$

21. (a) Let the curve be given by $\mathbf{X}(t)$. The surface then has the parametric equation

$$\mathbf{y} = \mathbf{X}(t) + \lambda\dot{\mathbf{X}}(t)$$

The vector $\partial\mathbf{y}/\partial\lambda \times \partial\mathbf{y}/\partial t$ is normal to the surface, but

$$\frac{\partial\mathbf{y}}{\partial\lambda} \times \frac{\partial\mathbf{y}}{\partial t} = \dot{\mathbf{X}}(t) \times [\dot{\mathbf{X}}(t) + \lambda\ddot{\mathbf{X}}(t)] = \lambda\dot{\mathbf{X}}(t) \times \ddot{\mathbf{X}}(t)$$

is also normal to the osculating plane.

(b) Set $\mathbf{Y} = (x, y, z)$ and $\mathbf{X}(t) = (\alpha(t), \beta(t), \gamma(t))$. Thus, $x$ and $y$ are functions of $t$ and $\lambda$ satisfying

$$x = \alpha(t) + \lambda\dot{\alpha}(t)$$

$$y = \beta(t) + \lambda\dot{\beta}(t).$$

Use

$$u(x, y) = \gamma(t) + \lambda\dot{\gamma}(t)$$

to calculate $u_{xx}$, $u_{yy}$, and $u_{xy}$ in terms of derivatives with respect to $t$ and $\lambda$.

Differentiate $\mathbf{Y} = \mathbf{X}(t) + \lambda\dot{\mathbf{X}}(t)$ with respect to $x$ to obtain, $(\lambda = s)$

$$\mathbf{Y}_x = (1, 0, u_x) = (\dot{\mathbf{X}} + \lambda\ddot{\mathbf{X}})t_x + \dot{\mathbf{X}}s_x.$$

Form $\dot{\mathbf{X}} \times \mathbf{Y}_x$ and equate components in the $x$ and $z$ directions to obtain

$$\dot{\beta}u_x = st_x(\beta, \gamma), \quad \dot{\beta} = -st_x(\alpha, \beta),$$

where $(u, v)$ is defined by

$$(u, v) = \dot{u}\dot{v} - \dot{v}\ddot{u}.$$

Thus,

$$u_x = -\frac{(\beta, \gamma)}{(\alpha, \beta)}, \quad t_x = -\frac{\dot{\beta}}{s(\alpha, \beta)}.$$

Similarly, from $\dot{\mathbf{X}} \times \mathbf{Y}_y$ obtain

$$u_y = -\frac{(\gamma, \alpha)}{(\alpha, \beta)}, \quad t_y = \frac{\dot{\alpha}}{s(\alpha, \beta)}$$

Note that $u_x$ and $u_y$ do not depend on $\lambda$. Consequently,

$$u_{xx} = t_x\frac{d}{dt}u_x = \frac{\dot{\beta}}{s(\alpha, \beta)}\frac{d}{dt}\frac{(\beta, \gamma)}{(\alpha, \beta)}$$

$$u_{yy} = t_y\frac{d}{dt}u_y = \frac{\dot{\alpha}}{s(\alpha, \beta)}\frac{d}{dt}\frac{(\alpha, \gamma)}{(\alpha, \beta)}$$

and

$$u_{xy} = t_y\frac{d}{dt}u_x = -\frac{\dot{\alpha}}{s(\alpha, \beta)}\frac{d}{dt}\frac{(\beta, \gamma)}{(\alpha, \beta)}$$

$$= t_x\frac{d}{dt}u_y = -\frac{\dot{\beta}}{s(\alpha, \beta)}\frac{d}{dt}\frac{(\alpha, \gamma)}{(\alpha, \beta)},$$

from which the result is immediate.

## Exercises 3.1a (p. 219)

1. Set $y_{n+1} = y_n + cf(a, y_n)$, where $c$ is constant, and apply the methods of Volume 1, Sections 6.3c and d, with $\varphi(y) = y + cf(a, y)$. To guarantee convergence, we require $|\varphi'(y)| \leqq q < 1$ on some interval containing $b$, and the smaller the $q$, the better. Consequently, we attempt to fix $c$ so that $\varphi'(y)$ is nearly zero, or

$$c \approx -\frac{1}{f_y(a, b)}.$$

Thus we begin with the assumption $f_y(a, b) \neq 0$.

In practice, we choose $c = -1/f_y(a, y_0)$, where $y_0$ is close to the sought-for solution $b$. The condition for convergence then becomes

$$|\varphi'(y)| = \left|\frac{f_y(a, y_0) - f_y(a, y)}{f_y(a, y_0)}\right| \leq q < 1$$

for all $y$ in some neighborhood of $b$. Suppose $f_y$ satisfies a Lipschitz condition

$$|f_y(a, \eta_2) - f_y(a, \eta_1)| < K|\eta_2 - \eta_1|$$

on some neighborhood of $b$. Within this neighborhood, let $\varepsilon$ be the radius of some perhaps smaller neighborhood where $\partial f/\partial y$ is bounded away from 0,

$$f_y(a, y) > m > 0;$$

such a neighborhood exists by virtue of the Lipschitz condition and $f_y(a, b) \neq 0$. For an initial choice $y_0$ satisfying

$$|y_0 - b| < \max\left\{\varepsilon, \frac{qm}{2K}\right\},$$

the iteration scheme converges to $b$ through

$$|y_n - b| \leq \frac{1}{2}q^n|y_0 - b|.$$

## Exercises 3.1b (p. 221)

1. (a) The tangent plane is horizontal. The surface intersects the tangent plane in the pair of lines $y = x$ and $y = -x$; hence, $y$ cannot be expressed as a function of $x$ in the neighborhood of $(x_0, y_0)$.

   (b) The surface is a cylinder with generators parallel to the vector $\mathbf{i} - \mathbf{j}$. Thus, the line $y = 1 - x$, $z = 0$ lies on the surface and yields the desired solution $y = 1 - x$.

   (c) The surface is a cylinder with generators parallel to $\mathbf{i} - \mathbf{j}$. The solution is $y = 1/2 - x$.

   (d) The tangent plane $y + z = 0$ is not horizontal. Thus, the curve $f(x, y) = 0$ is tangent to the line $y = 0$ at the origin.

## Exercises 3.1c (p. 225)

1. By subtracting the constant on the right from both sides, we may put each of these equations in the form $F(x, y) = 0$. The conditions of the theorem are satisfied. In particular, each given point is an initial solution $F(x_0, y_0) = 0$; and $F_y(x_0, y_0)$ has nonzero values, namely, (a) 4, (b) $-1$, (c) 2, (d) 6.

2. (a) $-\dfrac{2x + y}{x + 2y}$;  $-\dfrac{5}{4}$.

   (b) Explicitly, $y = \pi/2x$; hence, $y' = -\pi/2x^2$. Implicitly,

$$y' = \frac{\cot xy - xy}{x^2}; \quad -\frac{\pi}{2}.$$

(c) Explicitly, $y = 1/x$; hence, $y' = -1/x^2$. Implicitly, $y' = -y/x$; $-1$.

(d) $y' = -\dfrac{y + 5x^4}{x + 5y^4}; \quad -1.$

3. (a) $y'' = \dfrac{-6(x^2 + xy + y^2)}{(x + 2y)^3} = \dfrac{-42}{(x + 2y)^3}; \quad -\dfrac{21}{32}.$

(b) $y'' = \dfrac{\pi}{x^3}; \quad \pi.$

(c) $y'' = \dfrac{2y}{x^2} = \dfrac{2}{x^3}; \quad 2.$

(d) $y'' = -\dfrac{[150\ x^3 y^3(10 - xy) + 20(x^6 + y^6) + 8xy - 30]}{(x + 5y^4)^3}; \quad -\dfrac{19}{3}.$

4. From the positive sign of their second derivatives, $b$ and $c$.

5. Assume that the equation defines $y$ as a differentiable function of $x$ in a neighborhood of each extreme value. Then at an extremum $F_x(x, y) = 0$. Maximum, $y = 6$; minimum, $y = -6$.

6. Set $F\ (x, y) = y - y_0 - \int_{x_0}^{x} f_y(\xi, y)d\xi$ and note that

$$F_y(x,\ y) = 1 - \int_{x_0}^{x} f_y(\xi, y)d\xi > 0$$

for $x$ sufficiently close to $x_0$.

## Exercises 3.1d (p. 228)

1. $f(x, y) = y^3 + x$ near $(0, 0)$.
2. Same as for Exercise 1.
3. Since $F_y(x, y) = (3y^2 - 2y + 1) + x^2$ is the sum of a positive quadratic expression in $y$ and a square, it follows that $F_y(x, y) > 0$ for each $x$ and all $y$. Consequently, for each $x$, $F(x, y)$ is strictly increasing in $y$. Thus, $F(x, y) = 0$ can have no more than one solution $y$ corresponding to each fixed $x$. Such a solution must exist because for each $x$, $y^3 - y^2 + (1 + x^2)y = G(x, y)$ takes on arbitrarily large values of both signs, positive and negative, for appropriate values of $y$. It follows by the intermediate value theorem that $G(x,y)$ takes on all real values. In particular, for some value of $y$, $G(x, y) = \phi(x)$; hence, for each $x$ and this value of $y$, $F(x, y) = G(x, y) - \phi(x) = 0$.

## Exercises 3.1e (p. 230)

1. Set $F(x, y, z) = x + y + z - \sin xyz$. $F_z(0, 0, 0) = 1 \neq 0$.

$$\frac{\partial z}{\partial x} = \frac{yz \cos xyz - 1}{1 - xy \cos xyz}, \qquad \frac{\partial z}{\partial y} = \frac{xz \cos xyz - 1}{1 - xy \cos xyz}.$$

2. Since each equation can be put in the form $F(z, x, y, \ldots) = 0$, where $F$ is formed by rational operations and application of continuously differentiable functions of one variable, it is only necessary to test that the derivative $F_z$ at the point is nonzero.

(a) $F_z = 1$

(b) $F_z = -6$

(c) For $F(x, y, z) = 1 + x + y - \cosh(x + z) - \sinh(y + z)$, $F_z = 1$.

3. For $f(x, y, z) = x + y + z + xyz^3$, $f_z(0, 0, 0) = 1 \neq 0$. Second- through fourth-order terms vanish; $z = -x - y + \cdots$.

## Exercises 3.2a (p. 235)

1. (a) Equation satisfied only by point $(0, 0)$; tangent and normal do not exist.

(b) $(\xi - x) [e^x \sin y - e^y \sin x] + (\eta - y) [e^x \cos y + e^y \cos x] = 0$;

$(\eta - x) [e^x \cos y + e^y \cos x] - (\eta - y) [e^x \sin y - e^y \sin x] = 0$.

(c) Equation satisfied only by points $(-1, \pi/2 + 2k\pi)$; tangent and normal do not exist.

(d) $(\xi - x) (2x + \cos x) + (\eta - y) (2y - 1) = 0$;

$(\xi - x) (2y - 1) - (\eta - y) (2x + \cos x) = 0$.

(e) $(\xi - x) (3x^2) + (\eta - y) (4y^3 - \sinh y) = 0$;

$(\xi - x) (4y^3 - \sinh y) - (\eta - y) (3x^2) = 0$.

(f) Equation satisfied only on positive $x$- and $y$-axes. For $x = 0$, $y > 0$, tangent is $x = 0$, and normal, $\eta = y$; for $y = 0$, $x > 0$, tangent is $y = 0$, and normal $\xi = x$.

2. $-1$.

3. From Volume I, p. 437, Problem 5 of 4.1h,

$$k = \frac{r^2 + 2r'^2 - rr''}{(r^2 + r'^2)^{3/2}},$$

where the primes indicate derivatives with respect to $\theta$. Enter the expressions for $r'$ and $r''$ in terms of the partial derivatives of $f$ in the formula for $k$ to obtain

$$k = \frac{r^2 f_r^3 + r(f_r^2 f_{\theta\theta} - 2f_\theta f_r f_{r\theta} + f_\theta^2 f_{rr}) + 2f_\theta^2 f_r}{(f_\theta^2 + r^2 f^2)^{3/2}}.$$

4. Observe that $F_{xx} = F_{yy} = 6(x + y - a) = 0$ when $x + y = a$. Apply (13):

$$F_y{}^2 F_{xx} - 2F_x F_y F_{xy} + F_x{}^2 F_{yy} = -54axy\, F_{xy} = 0,$$

since $xy = 0$ at an intersection.

5. $a = \pm 1$, $b = -\frac{1}{2}$.

6. The circles $K$, $K'$, $K''$ may be denoted by the equations

$$K = x^2 + y^2 + ax + by + c = 0,$$
$$K' = x^2 + y^2 + a'x + b'y + c' = 0,$$
$$K'' = x^2 + y^2 + a''x + b''y + c'' = 0.$$

Then any circle passing through $A$ and $B$ is given by $K' + \lambda K'' = 0$. The conditions that the circle $K$ should be orthogonal to $K'$ and $K''$ are $aa' + bb' - 2(c + c') = 0$, $aa'' + bb'' - 2(c + c'') = 0$. From these conditions the corresponding relation expressing the orthogonality of $K$ and $K' + \lambda K''$ readily follows.

### Exercises 3.2b (p. 237)

1. (a) Double point
   (b) Two branches tangent to $x$-axis
   (c) A corner: for $x = 0^+$ the slope is 0, for $x = 0^-$ the slope is 1
   (d) Cusp
   (e) Cusp.
2. The coordinate axes.
3. $y = x^2(1 \pm x^{1/2})$. The two branches of the curve forming the cusp at the origin lie on the same side of their common tangent.
4. The curves are obtained by rotation through the angle $\alpha$ from the curve $(x - b)^3 = cy^2$.
5. Differentiate the equation $F = 0$ twice with respect to $x$ and use the fact that $F_y = 0$.

$$\varphi = \text{arc tan} \frac{2\sqrt{F_{xy}{}^2 - F_{xx}F_{yy}}}{F_{xx} + F_{yy}};$$

thus,

   (a) $\pi/2$;
   (b) $\pi/2$.
6. Note that the tangents at the origin are $y = 0$ and $ax + by = 0$. In the respective cases, expand $y$ to second order:

$$y = \frac{1}{2}y_0'' x^2 + \cdots \quad \text{and} \quad y = -\frac{a}{b}x + \frac{1}{2}y_0''x^2 + \cdots.$$

Enter these expressions in the original equation to obtain $y_0''$.

$$k = \frac{2c}{a}, \quad k = \frac{2(a^3g - a^2bf - ab^2e - b^3c)}{a(a^2 + b^2)^{3/2}}.$$

### Exercises 3.2c (p. 240)

1. (a) $5x + 7y - 21z + 9 = 0$
   (b) $20x + 13y + 3z = 36$
   (c) $x - y - z + \pi/6 = 0$

(d) $x + 2z - 2 = 0$

(e) The surface has no tangent plane at the point.

(f) $z = 0$.

2. Each equation is in the form $F(x, y, z) =$ constant. The vectors $(F_x, F_y, F_z)$ perpendicular to the respective surfaces are given by

$$\left(\frac{y}{z}, \frac{x}{z}, -\frac{xy}{z^2}\right), \quad \left(\frac{x}{\sqrt{x^2 + z^2}}, \frac{y}{\sqrt{y^2 + z^2}}, \frac{z}{\sqrt{x^2 + z^2}} + \frac{z}{\sqrt{y^2 + z^2}}\right),$$

$$\left(\frac{x}{\sqrt{x^2 + z^2}}, -\frac{y}{\sqrt{y^2 + z^2}}, \frac{z}{\sqrt{x^2 + z^2}} - \frac{z}{\sqrt{y^2 + z^2}}\right).$$

The scalar product of any two of these vectors vanishes.

3. $x(y + z) = ay$.

4. Since this is a surface of revolution, we may assume $y = 0$, Let $(a, 0, c)$ be a point of the surface, that is, $a^2 - c^2 = 1$. The tangent plane at the point is $ax - cz = 1$. The intersection lines are $(z - c)c = (x - a)a = \pm acy$.

5. From Euler's relation the equation

$$(\xi - x)F_x + (\eta - y)F_y + (\zeta - z)F_z = 0$$

for the tangent plane can be put in the form

$$\xi F_x + \eta F_y + \zeta F_z = xF_x + yF_y + zF_z = hF(x, y, z) = h.$$

6. $z_x = \dfrac{yz - x^2}{z^2 - xy}$, $z_y = \dfrac{xz - y^2}{z^2 - xy}$.

7. (a) 0

(b) arc cos $1/\sqrt{6}$

(c) arc cos $4/5$

(d) $\pi/2$

(e) Not defined.

## Exercises 3.3a (p. 246)

1. (a) Circles $\xi^2 + \eta^2 = e^{2x}$; lines through origin $\xi \sin y - \eta \cos y = 0$.

(b) Parabolic arcs, $\eta = \sqrt{x^2 - 2\xi x}$, $\eta = \sqrt{y^2 + 2\xi y}$.

(c) $\eta = \cos x(1 + 1/\xi^2)$, $\eta = \cos y(1 + \xi^2)$.

(d) Parabolas $\xi = \eta^2 - 2\eta(x^2 + 1) + x^4 + 3x + 1$, $\eta = \xi^2 - 2\xi y + y^4 + y + 1$.

(e) $\xi = x^{\eta^{1/x}}$, $\eta = y^{\xi^{1/y}}$.

(f) Lines $\xi =$ constant, $\eta =$ constant $(\eta \geq 1)$.

(g) Elliptical arcs $\xi^2 - 2\xi\eta \sin 2x + \eta^2 = \cos^2 2x$, $\xi^2 - 2\xi\eta \sin 2x + \eta^2 = \cos^2 2y$.

(h) Segments $\xi = e^{\cos x}$, $(e^{-1} \leq \eta \leq e)$, $\eta = e^{\cos y}$, $(e^{-1} \leq \xi \leq e)$.

2. The equation admits only the values $x = y = 0$. Hence, the region is the plane. Its image is the open first quadrant in the $\xi$, $\eta$-plane.

3. The region bounded by the two circles $\xi^2 + \eta^2 = 8$, $\xi^2 + \eta^2 = 32$ and the hyperbolas $\xi^2 - \eta^2 = 2$, $\xi^2 - \eta^2 = 6$.

4. No. The origin of the $\xi, \eta$-plane is the image of any point $(0, y)$.

## Exercises 3.3b (p. 248)

1. For this, it is only necessary to show that at a given point with Cartesian coordinates $(a, b)$ the curves $\xi = \alpha$, $\eta = \beta$, where $\alpha = (\sin b)/(a - 1)$ and $\beta = a \tan b$, have different directions. For $\xi = \alpha$,

$$\frac{dx}{dy} = \frac{(a - 1) \cos b}{\sin b} \; ;$$

for $\eta = \beta$,

$$\frac{dx}{dy} = \frac{-a}{\cos^2 b \sin b} .$$

Thus, curvilinear coordinates are defined for all points except those that satisfy $\cos^3 b = a/(1 - a)$.

2. $(\xi - 1)^{2/3} + \eta^2(\xi - 1)^{-2/3} = 1$.

3. As in the solution of Exercise 1, those points with Cartesian coordinates $(a, b)$ for which the curves $\xi = \alpha$ and $\eta = \beta$ have the same direction, in this case, the points on the 45°-lines $b = \pm a$.

## Exercises 3.3c (p. 251)

1. Use

$$\xi^2 + \eta^2 + \zeta^2 = (x^2 + y^2 + z^2)^{-1}$$

to obtain

$$x = \frac{\xi}{\xi^2 + \eta^2 + \zeta^2}, \quad y = \frac{\eta}{\xi^2 + \eta^2 + \zeta^2}, \quad z = \frac{\zeta}{\xi^2 + \eta^2 + \zeta^2} .$$

2.
$$r = \sqrt{x^2 + y^2 + z^2 + w^2}$$

$$\phi = \text{arc tan} \frac{\sqrt{x^2 + y^2 + z^2}}{w}, \quad \psi = \text{arc tan} \frac{\sqrt{y^2 + z^2}}{x},$$

$\theta = \text{arc tan } z/y$. Here $r = $ constant, is a three-sphere of radius $r$ centered at the origin; $\phi = $ constant, is the hypercone generated by all lines through 0 making the angle $\phi$ with the $w$-axis; the set $\psi = $ constant is the union of all planes through the $w$-axis that meet the $x$ axis at the angle $\psi$. The set $\theta = $ constant is the union of all three-spaces containing the $x$- and $w$-axes that meet the $y$-axis at angle $\theta$.

## Exercises 3.3d (p. 255)

1. (a) $ad - bc$   (d) $\dfrac{1}{x^2 + y^2}$

(b) $1/\sqrt{x^2 + y^2}$      (e) $-3x^2y^2$

(c) $4xy$      (f) $9x^2y^2 + 1$.

2. If $ad - bc = 0$, all points; if $ad - bc \neq 0$, none.

   (b) None. (The transformation is not defined for $x = y = 0$.)

   (c) The coordinates axes.

   (d) None. Note, however, that there is no over-all inverse because the points( $x, y + 2n\pi$) all have the same image.

   (e) The coordinate axes.

   (f) None.

3. (a) $D = e^{2x}$; $x_\xi = y_\eta = \xi/(\xi^2 + \eta^2)$; $x_\eta = -y_\xi = \eta/(\xi^2 + \eta^2)$; $x_{\xi\xi} = y_{\xi\eta} = -x_{\eta\eta} = (\xi^2 - \eta^2)/(\xi^2 + \eta^2)^2$; $y_{\xi\xi} = -x_{\xi\eta} = -y_{\eta\eta} = -2\xi\eta/(\xi^2 + \eta^2)^2$.

  (b) $D = 4(x^2 + y^2)$; with $r = \sqrt{\xi^2 + \eta^2}$, $\theta = $ arc tan $\eta/\xi$; $x_\xi = y_\eta = \frac{1}{2}\sqrt{r} \cos \frac{1}{2} \theta$; $y_\xi = -x_\eta = -\frac{1}{2} \sqrt{r} \sin \frac{1}{2}\theta$; $x_{\xi\xi} = y_{\xi\eta} = -x_{\eta\eta} = -\frac{1}{4} r^{3/2} \cos 3\theta/2$; $y_{\xi\xi} = -x_{\xi\eta} = -y_{\eta\eta} = \frac{1}{4} r^{3/2} \sin 3\theta/2$.

  (c) $D = 2 \sin(x - y)/\cos^2(x + y)$. $x_\xi = y_\xi = 1/2(1 + \xi^2)$; $x_\eta = y_\eta = 1/2\sqrt{1 - \eta^2}$; $x_{\xi\xi} = y_{\xi\xi} = - \xi/(1 + \xi^2)^2$; $x_{\xi\eta} = y_{\xi\eta} = 0$; $x_{\eta\eta} = -y_{\eta\eta} = \eta/2(1 - \eta^2)^{3/2}$.

  (d) $D = \cosh(x + y); x_\xi = (\cosh y)/D; x_\eta = -(\sinh y)/D; y_\xi = (\sinh x)/D; y_\eta = (\cosh x)/D$.

    $x_{\xi\xi} = - [\cosh^2 y \sinh(x + y) + \sinh^2 x]/D^3$;

    $x_{\xi\eta} = \frac{1}{2}[\sinh 2y \sinh(x + y) - \sinh 2x]/D^3$;

    $x_{\eta\eta} = -[\sinh^2 y \sin(x + y) + \cosh^2 x]/D^3$;

    $y_{\xi\xi} = [\cosh^2 y - \sinh^2 x \sinh(x + y)]/D^3$;

    $y_{\xi\eta} = -\frac{1}{2}[\sinh 2y + \sinh 2x \sinh(x + y)]/D^3$;

    $y_{\eta\eta} = [\sinh^2 y - \cosh^2 x \sinh(x + y)]/D^3$.

  (e) $D = 6x^3y - 3y^4$. $x_\xi = 2x/3(2x^3 - y^3)$

    $x_\eta = -y/(2x^3 - y^3)$, $y_\xi = -y/3(2x^3 - y^3)$;

    $y_\eta = x^2/y(2x^3 - y^3)$. $x_{\xi\xi} = - \frac{2}{3}x(8x^3 + 5y^3)/(2x^3 - y^3)^3$;

    $x_{\xi\eta} = 2y(7x^3 + y^3)/3(2x^3 - y^3)^3$;

    $x_{\eta\eta} = -2x^2(x^3 + 4y^3)/y(2x^3 - y^3)^3$;

    $y_{\xi\xi} = 2y(7x^3 + y^3)/3(2x^3 - y^3)^3$

    $y_{\xi\eta} = -2x^2(x^3 + 4y^3)/3y(2x^3 - y^3)^3$

    $y_{\eta\eta} = 2x(y^6 + 3x^3y^3 - x^6)/y^3(2x^3 - y^3)^3$.

  (a) Let $m_1$ and $m_2$ be the slopes of two curves passing through the point $(a, b)$ of the $x, y$-plane. Let $\mu_1$ and $\mu_2$ be the corresponding slopes at the corresponding point in the $\xi$, $\eta$-plane. Use

$$\mu = \frac{d\eta}{d\xi} = \frac{d\eta/dx}{d\xi/dx} = \frac{(\partial\eta/\partial x) + m(\partial\eta/\partial y)}{(\partial\xi/\partial x) + m(\partial\xi/\partial y)} = \frac{m(a^2 - b^2) - 2ab}{b^2 - a^2 - 2mab}$$

to obtain

$$\frac{\mu_2 - \mu_1}{1 + \mu_1\mu_2} = \frac{m_1 - m_2}{1 + m_1 m_2}.$$

Thus, the angle between the two curves is preserved in magnitude but reversed in orientation.

(b) Observe that $\xi^2 + \eta^2 = 1/(x^2 + y^2)$. Express the circle $(x - a)^2 + (y - b)^2 = r^2$ in the form $x^2 + y^2 - 2ax - 2by = r^2 - a^2 - b^2$. This transforms into the curve

$$\frac{1}{\xi^2 + \eta^2} - \frac{2a\xi}{\xi^2 + \eta^2} - \frac{2by}{\xi^2 + \eta^2} = r^2 - a^2 - b^2$$

or

$$(\xi^2 + \eta^2)(r^2 - a^2 - b^2) + 2a\xi + 2b\eta = 1.$$

This is a circle in the $\xi$, $\eta$-plane unless the original circle passes through the origin; then $r^2 - a^2 - b^2 = 0$ and the image is a straight line.

(c) $-1/(x^2 + y^2)^2$.

5. By the solution of Exercise 4(b), an inversion maps $P_1 P_2 P_3$ into an ordinary triangle with the same angles.

6. Let $m_1$, $m_2$ be the slopes of curves passing through the point $(a, b)$ and $\mu_1$, $\mu_2$ the corresponding slopes of their images. From

$$\mu = \frac{dv/dx}{du/dx} = \frac{\psi_x + m\psi_y}{\phi_x + m\phi_y} = \frac{\psi_x + m\psi_y}{\psi_y - m\psi_c},$$

it follows that

$$\frac{\mu_2 - \mu_1}{1 + \mu_2\mu_1} = \frac{m_2 - m_1}{1 + m_1 m_2}.$$

7. The normal is given by

$$\frac{\xi - x}{u_x} = \frac{\eta - y}{u_y} = u - z.$$

It passes through the $z$-axis if and only if $xu_y - yu_x = 0$. The surface is a surface of revolution if and only if $z = f(w)$ where $w = x^2 + y^2$. Thus, the curves $z = $ constant and $w = $ constant are the same and the mapping $(x, y) \to (w, z)$ must have a vanishing Jacobian, that is,

$$\frac{d(w, z)}{d(x, y)} = 2 \begin{vmatrix} x & y \\ u_x & u_y \end{vmatrix} = 0.$$

8. (a) If either $t < b$ (ellipse) or $b < t < a$ (hyperbola), the foci are $(0, \pm c)$, where $c = \sqrt{a - b}$.

(b) If we denote the left-hand side of the equation defining $t_1$ and $t_2$ by $F(x, y, t)$, two curves $t_1 = $ constant and $t_2 = $ constant are given implicitly by the equations $F(x, y, t_1) = 1$ and $F(x, y, t_2) = 1$, respectively. The condition that these should be orthogonal is therefore

$$0 = F_x(x, y, t_1) \, F_x(x, y, t_2) + F_y(x, y, t_1) \, F_y(x, y, t_2)$$

$$= \frac{4x^2}{(a - t_1)(a - t_2)} + \frac{4y^2}{(b - t_1)(b - t_2)} \, ;$$

but this relation is an immediate consequence of $F(x, y, t_1) - F(x, y, t_2) = 0$.

(c) The coefficients of the quadratic equation defining $t_1$ and $t_2$ are equal to $t_1$, $t_2$, and $-(t_1 + t_2)$, respectively. We thus obtain two linear equations in $x^2$ and $y^2$, whence

$$x = \pm \sqrt{\frac{(a - t_1)(a - t_2)}{a - b}}, \; y = \pm \sqrt{\frac{(b - t_1)(b - t_2)}{b - a}}.$$

(d) $\dfrac{d(t_1, t_2)}{d(x, y)} = \dfrac{4xy(a - b)}{\sqrt{\{(a + b)^2 - 2(a - b)(x^2 - y^2) + (x^2 + y^2)^2\}}}.$

(e) $\dfrac{f_1' g_1'}{(a - t_1)(b - t_1)} = \dfrac{f_2' g_2'}{(a - t_2)(b - t_2)}.$

9. (a) Let $F(t)$ be the left-hand side of the equation defining $t$. $F$ is a continuous function of $t$ in $-\infty < t < c$, for which $F(-\infty) = 0$, $F(c - 0) = +\infty$; hence, $F = 1$ at one point at least of that interval. Similar conclusions apply to the other intervals.

(b) Cf. Exercise 8 (b).

(c) Cf. Exercise 8 (c). $x = \pm \sqrt{\dfrac{(a - t_1)(a - t_2)(a - t_3)}{(a - b)(a - c)}}$,

with similar formulae for $y$ and $z$.

10. (a) Apply the result of Exercise 6.

(b) Let $x = r \cos\theta$, $y = r \sin\theta$. Then the straight line $\theta = $ constant is transformed into the conic $t_1 = \frac{1}{2} - \cos^2\theta$ and the circle $r = $ constant. into the conic $t_2 = -\frac{1}{4}[r^2 + (1/r^2)]$.

11. (b) Use (24d) as follows

$$x_{\xi\eta} = \frac{\partial}{\partial\xi}\left(\frac{-\xi_y}{D}\right) = x_{\eta\xi} = \frac{\partial}{\partial\eta}\left(\frac{\eta_y}{D}\right),$$

or apply the result of part (a).

## Exercises 3.3e (p. 260)

1. (a) 1.    (b) $4x^3$.    (c) $\dfrac{\exp[2x/(x^2 + y^2)]}{(x^2 + y^2)^2}.$

2. (a), (c). In part (b), $u_0 = v_0 = 1$ is not in the range of the composite transformation.
3. Apply (31b).
4. The inverse transformation

$$x = p(\xi, \eta), \quad y = q(\xi, \eta)$$

exists. The first result is obtained by forming the composition of the given mapping with

$$z = f(p(\xi), q(\eta)) = \alpha(\xi, \eta)$$

$$\eta = \eta = \beta(\xi, \eta),$$

whence

$$\frac{d(z, \eta)}{d(\xi, \eta)} = \frac{d(z, \eta)}{d(x, y)} \frac{d(x, y)}{d(\xi, \eta)} = \frac{d(z, \eta)/d(x, y)}{d(\xi, \eta)/d(x, y)}.$$

But

$$\frac{d(z, \eta)}{d(\xi, \eta)} = \begin{vmatrix} \dfrac{\partial z}{\partial \xi} & \dfrac{\partial z}{\partial \eta} \\ 0 & 1 \end{vmatrix} = \frac{\partial z}{\partial \xi}.$$

## Exercises 3.3f (p. 266)

1. (a), (b). In part (c), the given values do not satisfy the equations.

## Exercises 3.3g (p. 273)

1. With $w = v - 1$,

$$x_2 = 1 + \frac{1}{2}(u + w) + \frac{1}{8}(u^2 - 2uw - w^2),$$

$$y_2 = 1 - \frac{1}{2}(u - w) + \frac{1}{8}(u^2 + 2uw - w^2).$$

2. The same.

## Exercises 3.3h (p. 275)

1. $\xi = x^2 + x|x|, \quad \eta = y.$
2. If the functions are dependent, $\partial(\xi, \eta)/\partial(x, y) = a\beta - b\alpha = 0.$

## Exercises 3.3i (p. 277)

1. (a) $-e^{3z} \cos y$
   (b) 0.

(c) $-\left[\dfrac{y^z \log y \sinh x}{\cosh^2 y} - \dfrac{\cosh z}{y} - (\cosh z)y^{z-1} \sinh x\right].$

(d) $-x^2 \sin z.$

(e) $x.$

2. There exists a region on which some function of $\xi$, $\eta$, $\zeta$ vanishes. The condition for this is $\partial(\xi, \eta, \zeta)/\partial(x, y, z) = 0.$

3. The triple of Exercise 1(b) is dependent:

$$(\eta^2 + \rho^2)\,[(\eta + \rho - \xi)^2 + \xi^2] = 2(\eta + \rho)^2.$$

4. $\dfrac{\partial(\xi, \eta, \zeta)}{\partial(x,y,z)} = \begin{vmatrix} 1 & 1 & 1 \\ 2x & 2y & 2z \\ y+z & x+z & y+x \end{vmatrix} \equiv 0; \quad \xi^2 - \eta - 2\zeta = 0.$

5. (a) Since the angle between two surfaces is the angle between their normals, we need show only that the angle between any two directions is unchanged. Let $s$ be arc length on any curve in $x$, $y$, $z$-space and $\mathbf{t} = (\dot{x}, \dot{y}, \dot{z}) = \dot{\mathbf{X}}$ the unit tangent vector, where the dot denotes differentiation with respect to $s$. The direction of $\mathbf{t}$ maps into the direction of $\tau = \dfrac{(\dot{\xi}, \dot{\eta}, \dot{\zeta})}{(\dot{\xi}^2 + \dot{\eta}^2 + \dot{\zeta}^2)^{1/2}} = \dot{\mathbf{Y}}/|\dot{\mathbf{Y}}|$. The image direction $\tau$ is given in terms of $\mathbf{t}$ and $\mathbf{X}$ by

$$\tau = \mathbf{t} - \frac{2(\mathbf{t} \cdot \mathbf{X})\mathbf{X}}{|\mathbf{X}|^2}.$$

From this it follows easily that the cosine of the angle between two curves meeting at $\mathbf{X}$ is given by $\tau_1 \cdot \tau_2 = \mathbf{t}_1 \cdot \mathbf{t}_2.$

(b) Follows as does the solution of Exercise 4(b), p. 256

(c) $-1/(x^2 + y^2 + z^2)^3.$

## Exercises 3.4a (p. 286)

1. (a) $ds^2 = \sin^2 v\, du^2 + dv^2$

   (b) $ds^2 = \cosh^2 v\, du^2 + (1 + 2 \sinh^2 v)dv^2$

   (c) $ds^2 = (1 + f'^2)dz^2 + f^2\, d\theta^2$

   (d) $ds^2 = \dfrac{(t_1 - t_2)\,(t_1 - t_3)}{4(a - t_1)\,(b - t_1)\,(c - t_1)}\, dt_1^2 + \dfrac{(t_2 - t_1)\,(t_2 - t_3)}{4(a - t_2)\,(b - t_2)\,(c - t_2)}\, dt_2^2.$

2. $E = G = \cosh^2(t/a)$, $F = 0.$

3. $\mathbf{X}_u = (\cos v, \sin v, \alpha)$; $\mathbf{X}_v = (-u \sin v, u \cos v, 0)$; hence, $\mathbf{X}_u \cdot \mathbf{X}_v = 0.$

4. $ds^2 = (1 + z_x{}^2)dx^2 + 2z_x z_y\, dx\, dy + (1 + z_y)^2\, dy^2.$

5. $EG - F^2 = \begin{vmatrix} y_u & z_u \\ y_v & z_v \end{vmatrix}^2 + \begin{vmatrix} z_u & x_u \\ z_v & x_v \end{vmatrix}^2 + \begin{vmatrix} x_u & y_u \\ x_v & y_v \end{vmatrix}^2$; use the

   transformation formula for Jacobians.

6. Introduce coordinates $x, y, z$ such that $P$ becomes the origin; the tangent plane at $P$, the $x, y$-plane; and $t$, the $x$-axis. The equation of $S$ then takes the form $z = f(x, y)$, where $f(0, 0) = f_x(0, 0) = 0$. A plane $\Sigma$ through $t$ is given by the equation $z = \alpha y$. We now introduce $r = \sqrt{y^2 + z^2}$ and $x$ as coordinates in $\Sigma$; then the intersection of $\Sigma$ and $S$ is given implicitly by the equation

$$\frac{r\alpha}{\sqrt{1 + \alpha^2}} = f\left\{x, \frac{r}{\sqrt{1 + \alpha^2}}\right\}.$$

The curvature of the curve of intersection at the point $x = 0$, $r = 0$ is therefore (cf. p. 232) given by

$$k = f_{xx} \frac{\sqrt{1 + \alpha^2}}{\alpha}$$

Thus, the center of curvature of this section has the coordinates

$$x = 0, y = \frac{1}{k\sqrt{1 + \alpha^2}} = \frac{\alpha}{f_{xx}(1 + \alpha^2)}, \ z = \frac{\alpha}{k\sqrt{1 + \alpha^2}} = \frac{\alpha^2}{f_{xx}(1 + \alpha^2)};$$

that is, it lies on the circle

$$f_{xx}(y^2 + z^2) - z = 0.$$

7. Take the tangent plane at $P$ as the $x, y$-plane. Then the equation of $S$ may be taken to be $z = f(x, y)$. A normal plane is given by the equation $x = \alpha y$. Take $r = \sqrt{x^2 + y^2}$ and $z$ as coordinates in the plane;

$$z = f\left\{\frac{\alpha r}{\sqrt{1 + \alpha^2}}, \frac{r}{\sqrt{1 + \alpha^2}}\right\},$$

and its a curvature at $r = 0$ by

$$k = f_{xx}(0, 0)\frac{\alpha^2}{1 + \alpha^2} + 2f_{xy}(0, 0)\frac{\alpha}{1 + \alpha^2} + f_{yy}(0, 0)\frac{1}{1 + \alpha^2};$$

the final point of the vector of length $1/\sqrt{k}$ along the line $t$ then has the coordinates

$$x = \frac{\alpha}{\sqrt{1 + \alpha^2}}\frac{1}{\sqrt{k}}, y = \frac{1}{\sqrt{1 + \alpha^2}}\frac{1}{\sqrt{k}}, z = 0;$$

that is, it lies on the conic

$$x^2 f_{xx} + 2xy f_{xy} + y^2 f_{yy} = 1.$$

8. (a) By differentiating the two equations with respect to a parameter $t$ of the curve, we obtain

$$xx' + yy' + zz' = 0, \quad axx' + byy' + czz' = 0.$$

From these relations we can find the ratio $x':y':z'$, that is, the direction of the tangent. If $(\xi, \eta, \zeta)$ are current coordinates, the equations of the tangent are

$$(\xi - x) : (\eta - y) : (\zeta - z) = \frac{c - b}{x} : \frac{a - c}{y} : \frac{b - a}{z}.$$

(b) By differentiating the equations of the curve a second time and using the result of (a), we obtain

$$xx'' + yy'' + zz'' = -(x'^2 + y'^2 + z'^2)$$

$$= \lambda \left\{ \frac{(c-b)^2}{x^2} + \frac{(a-c)^2}{y^2} + \frac{(b-a)^2}{z^2} \right\}$$

and

$$axx'' + byy'' + czz'' = \lambda \left\{ \frac{a(c-b)^2}{x^2} + \frac{b(a-c)^2}{y^2} + \frac{c(b-a)^2}{z^2} \right\},$$

where $\lambda$ is a factor of proportionality. Eliminating $\lambda$, we have

$$(xx'' + yy'' + zz'') \left\{ \frac{a(c-b)^2}{x^2} + \frac{b(a-c)^2}{y^2} + \frac{c(b-a)^2}{z^2} \right\}$$

$$= (axx'' + byy'' + czz'') \left\{ \frac{(c-b)^2}{x^2} + \frac{(a-c)^2}{y^2} + \frac{(b-a)^2}{z^2} \right\}.$$

This linear equation in $x''$, $y''$, $z''$ remains valid if we substitute $x'$, $y'$, $z'$ for $x''$, $y''$, $z''$. Hence, it is still satisfied if we replace $x''$, $y''$, $z''$ by some linear combination $\lambda x' + \mu x''$, $\lambda y' + \mu y''$, $\lambda z' + \mu z''$, respectively. Now if $(\xi, \eta, \zeta)$ is in the plane, $\xi - x, \eta - y, \zeta - z$ are just such a linear combination (cf. Exercise 6, p. 215).

The equation of the osculating plane is hence found to be

$$\frac{ax^3}{c-b}(\xi - x) + \frac{by^3}{a-c}(\eta - y) + \frac{cz^3}{b-a}(\zeta - z) = 0.$$

9. Take $\theta$ as parameter for both curves. Then with $u = \theta$, $v = \phi$, set $du/dt = dv/d\tau = 1$, $dv/dt = -1$, $dv/d\tau = 1$, $E = a^2$, $G = a^2 \sin^2\theta$ in (48). The tangents of the curves are given in coordinate vectors $\mathbf{i}, \mathbf{j}, \mathbf{k}$ by

$$\dot{\mathbf{X}} = \mathbf{X}_\theta \pm \mathbf{X}_\phi$$

$$= a(\cos \theta \cos \phi \pm \sin \theta \sin \phi)\mathbf{i}$$

$$+ a(\cos \theta \sin \phi \mp \sin \theta \cos \phi)\mathbf{j} - a \sin \theta \, \mathbf{k},$$

and $|\dot{\mathbf{X}}|^2 = a^2(1 + \sin^2\theta)$ in both cases.

$$\ddot{\mathbf{X}} = 2a(\pm \cos \theta \sin \phi - \sin \theta \cos \phi)\mathbf{i}$$

$$+ 2a(\mp \cos \theta \cos \phi - \sin \theta \sin \phi)\mathbf{j}$$

$$- a \cos \theta \, \mathbf{k}.$$

Apply the formula of Section 2.5 Exercise 8.

## Exercises 3.4b (p. 289)

1. The mapping is conformal everywhere except at $u = v = 0$ because the Cauchy-Riemann equations are satisfied. At the origin all first derivatives vanish. In polar coordinates $u = r \cos \theta$, $v = r \sin \theta$ the mapping becomes $x = r^2 \cos 2\theta$, $y = r^2 \sin 2\theta$; thus, at the origin, all angles are doubled.

2. Whenever it is defined; that is, everywhere except on the line $u = 0$.

3. Verify the Cauchy-Riemann equations with $p = x\xi - y\eta$, $q = x\eta + y\xi$,

$$\frac{\partial p}{\partial u} = x\frac{\partial \xi}{\partial u} + \xi\frac{\partial x}{\partial u} - y\frac{\partial \eta}{\partial u} - \eta\frac{\partial y}{\partial u}$$

$$= x\frac{\partial \eta}{\partial v} + \xi\frac{\partial y}{\partial v} + y\frac{\partial \xi}{\partial v} + \eta\frac{\partial x}{\partial v} = \frac{\partial q}{\partial v}.$$

4. (a) From (40f) it follows that $\mathbf{X}_u \cdot \mathbf{X}_u = \mathbf{X}_v \cdot \mathbf{X}_v = 4r^4/(u^2 + v^2 + r^2)^2$ and $\mathbf{X}_u \cdot \mathbf{X}_v = 0$. Set $E = G$ and $F = 0$ in (48) to obtain the desired result.

   (b) A circle on the sphere is the intersection of the sphere with a plane, say $P$. If the plane $P$ passes through the north pole, stereographic projection maps the circle onto the intersection line of $P$ with the $x, y$-plane. More generally, if $P$ has the equation $ax + by + cz = d$, then, from (40f),

$$(c - d)(u^2 + v^2) + 2ar^2u + 2br^2v = r^2(cr + d),$$

   which is the equation of a line if $c = d$ and a circle if $c \neq d$.

   (c) From (40f)

$$u = x\left(1 - \frac{z}{r}\right) ; \quad v = y\left(1 - \frac{z}{r}\right)$$

   Reflection in the equatorial plane yields the transformation $(u, v) \rightarrow (\xi, \eta)$, where

$$\xi = \frac{x}{1 + z/r} ; \quad \eta = \frac{y}{1 + z/r}.$$

   Substituting for $x$ and $z$ from (40f), we find

$$\xi = \frac{r^2u}{u^2 + v^2} ; \quad \eta = \frac{r^2v}{u^2 + v^2}.$$

   which are the equations of inversion in a circle of radius $r$.

   (d) From the result of part (a),

$$ds^2 = \frac{4r^4}{(u^2 + v^2 + r^2)^2}(du^2 + dv^2).$$

5. The angle given by (48) must satisfy

$$\cos \omega = \frac{du/dt \ du/d\tau + dv/dt \ dv/d\tau}{\sqrt{[(du/dt)^2 + (dv/dt)^2][(du/d\tau)^2 + (dv/d\tau)^2]}}$$

Taking orthogonal pairs of vectors $(du/dt, dv/dt) = (0, 1)$ and $(du/d\tau, dv/d\tau) = (1, 0)$ yields $F = 0$. Similarly, the pair $(1, 1)$, $(1, -1)$ yields $E = G$. If $E$ and $G$ are not 0, the conditions

$$E = G, \quad F = 0$$

are sufficient.

6. From the solution of Exercise 5, we require

$$E = \sin^2\phi = \phi'^2 = G.$$

Solving the equation $\phi' = \sin \phi$, we obtain

$$v = \log \tan \frac{\phi}{2} \quad \text{or} \quad \phi = 2 \text{ arc tan } e^v.$$

## Exercises 3.5a (p. 292)

1. (a) A family of similar ellipses centered at the origin with axes aligned with the coordinate axes.
   (b) The family of circles tangent to the $x$-axis with centers on the $y$-axis.
   (c) Not a family. Each value of $c$ yields the same curve, the unit circle $x^2 + y^2 = 1$.
2. The spheres of radius 1 with centers on the line

$$x = y - 1 = \frac{1}{2}(z + \sqrt{2}).$$

## Exercises 3.5b (p. 295)

1. No. For example, consider the normals to a straight line or circle.
2. An envelope satisfies the parametric equations

$$x = -\psi'(c), \quad y = -c\psi'(c) + \psi(c).$$

If $\psi'$ has an inverse $\phi$, we may set $\phi(-x) = (\psi')^{-1}(-x)$ and use $c = \phi(-x)$ to obtain the nonparametric equation

$$y = x\phi(-x) + \psi(\phi(-x)),$$

from which

$$y' = \phi(-x) - x'\phi'(-x) - \psi'(\phi(-x))\,\phi'(-x)$$
$$= \phi(-x).$$

Entering $c = \phi(-x) = y'$ in the expression for $y$, we obtain the desired result.

## Exercises 3.5c (p. 302)

1. (a) Eliminate $t$ to obtain

$$y = x \tan \alpha - \frac{g}{2v^2} x^2(1 + \tan^2\alpha).$$

Let $c = \tan \alpha$ be the parameter of the family:

(a)
$$y = cx - \frac{(1 + c^2)}{2v^2} gx^2.$$

The envelope has the equation

$$y = \frac{v^2}{2g} - \frac{gx^2}{2v^2}$$

(b) For a fixed $x$, $dy/dc = x - cgx^2/v^2$ and $d^2y/dc^2 = -gx^2/v^2 < 0$. Since $dy/dc = 0$ on the envelope we conclude that for a given $x$ the point on the envelope is the highest reachable target.

(c) For $(x, y)$ with $y$ below the maximum, the quadratic equation ($\alpha$) has two solutions for $c$.

2. (a) The parabola $y^2 = 4x$.
   (b) The straight lines $x = \pm 2y$.
   (c) The hyperbolas $xy = \pm\frac{1}{2}$.
   (d) The straight lines $y = \pm ax$.

3. Let the equation of the curve be given parametrically by $x = \phi(t)$, $y = \psi(t)$. The envelope of the family of circles satisfies

$$[x - \phi(t)]^2 + [y - \psi(t)]^2 = p^2$$

and

$$[x - \phi(t)]\phi'(t) + [y - \psi(t)]\,\psi'(t) = 0.$$

These are precisely the conditions that $(x, y)$ lie at the distance $p$ from the point $(\phi(t), \psi(t))$ in a normal direction.

4. We may introduce $t$ as parameter on the curve, so that the latter is given by $x = x(t)$, $y = y(t)$, $z = z(t)$ and the tangent at the point with parameter $t$ lies in the two planes corresponding to $t$; this gives the relations

$$ax' + by' + cz' = 0, \quad dx' + ey' + fz' = 0.$$

By differentiating the equations of the straight lines with respect to $t$, we thus obtain

$$a'x + b'y + c'z = 0, \quad d'x + e'y + f'z = 0.$$

With the relation

$$ax + by + cz = dx + ey + fz$$

we then have three homogeneous equations in $x, y, z$, and the determinant must vanish.

5. (a) The parametric equations for $C'$ with $t$ as parameter are defined by the equations

$$\xi x + \eta y = 1, \quad \xi x' + \eta y' = 0.$$

Taking the ordinary derivative in the first equation with respect to $t$, we find, in view of the second equation,

$$\xi'x + \eta'y = 0.$$

This, coupled with the first equation, defines the polar reciprocal of $C'$ which is clearly the curve $C$.

(b) $\xi^2(1 - a^2) + \eta^2(1 - b^2) - 2ab\xi\eta + 2a\xi + 2b\eta = 1$:
(c) $a^2\xi^2 + b^2\eta^2 = 1$.

6. The equation of the generating tangent is

$$x \sin \theta + y \cos \theta = a(\theta \sin \theta + \cos \theta - 1).$$

7. If $(x^2/a^2) \pm (y^2/b^2) = 1$ is the equation of the conic, then $(x^2 + y^2)^2 = 4(a^2x^2 + b^2y^2)$ is the equation of the envelope. Note that if the conic is a rectangular hyperbola, this envelope is an ordinary lemniscate $(x^2 + y^2)^2 = 4a^2(x^2 - y^2)$.

8. (a) If $\Gamma$ is given parametrically by the vector equation $\mathbf{X} = \Phi(t)$, the points $\mathbf{Y}$ of the pedal curve are defined by the conditions

$$(\mathbf{Y} - \mathbf{X}) \cdot \mathbf{Y} = 0, \quad \mathbf{Y} \cdot \mathbf{X}' = 0,$$

A point $\mathbf{Z}$ on the circle must satisfy $(\mathbf{Z} - \frac{1}{2}\mathbf{X})^2 = \frac{1}{4}\mathbf{X}^2$ or $\mathbf{Z}^2 - \mathbf{Z} \cdot \mathbf{X} = 0$. To be on the envelope, then, $\mathbf{Z}$ must satisfy $\mathbf{Z} \cdot \mathbf{X}' = 0$. These are the conditions that $\mathbf{Z}$ be on the pedal curve.

   (b) From the original definition of pedal curve, a cardioid $r = a(1 + \cos \theta)$, where $a$ is the radius of the circle and $\theta$ is the azimuth with respect to the direction of the center from 0.

9. If the ellipse has equation $(x^2/a^2) + (y^2/b^2) = 1$, the envelope is the ellipse with equation

$$\frac{u^2}{b^2(a^2 + b^2)} + \frac{v^2}{b^2} = 1.$$

## Exercises 3.5d (p. 306)

1. These are ellipsoids $(x^2/a^2) + (y^2/b^2) + (z^2/c^2) = 1$, with $abc = k$, where $k$ is fixed. The envelope is $xyz = k^2/3\sqrt{27}$.

2. These are planes with unit distance from 0. Envelope, the unit sphere $x^2 + y^2 + z^2 = 1$.

3. (a) $\sqrt{x} + \sqrt{y} + \sqrt{z} = 1$.

   (b) $x^{2/3} + y^{2/3} + z^{2/3} = 1$.

4. For the envelope we have the two equations

$$x \cos t + y \sin t + z = t$$

$$-x \sin t + y \cos t = 1.$$

These two equations give a family of straight lines with parameter $t$; if a curve having these lines as tangents exists, it must also satisfy the equations obtained by differentiating once again.

   (a) $r \sin [z + \sqrt{r^2 - 1} - \theta] + 1 = 0$.

   (b) The curve is given by $z = \theta - \pi/2$, $r = 1$.

5. Let $P(x, y, z)$ be a point on the tube-surface $\Sigma$, and let $S$ be the sphere of the family that has the point $P$ in common with $\Sigma$. Then $S$ and $\Sigma$ have the same tangent plane at $P$, that is, the same values of $x$, $y$, $z$, $z_x$, $z_y$ at that point. It is therefore sufficient to prove that the relation is true for any sphere of unit radius that has its center in the $x$, $y$-plane, that is, for $u(x, y) = \sqrt{1 - (x - a)^2 - (y - b)^2}$.

6. Use inversion. Since $S_1$, $S_2$, $S_3$ pass through the origin, they are transformed into planes; we have then merely to find the envelope of the spheres touching three planes (i.e., a certain circular cone), which we reinvert:

$$(x^2 + y^2 + z^2)^2 - 2(x^2 + y^2 + z^2)(x + y + z)$$
$$- 3(x^2 + y^2 + z^2 - 2xy - 2xz - 2yz) = 0.$$

7. (a) If $P$ describes the pedal curve $\Gamma'$ of $\Gamma$, construct on $OP$ as diameter a circle in the plane perpendicular to the plane of $\Gamma$; the envelope is the surface generated by this variable circle.

   (b) See the solutions of part (a) and Exercise 8(b) of section 3.5c.

8. This is the family $(x/a) + (y/b) + (z/c) = 1$, with $abc = k$. The envelope is defined by these equations together with

$$-\frac{x}{a^2} + \frac{zk}{c^2a^2b} = 0; \quad -\frac{y}{b^2} + \frac{zk}{c^2ab^2} = 0$$

which yield, with the first equation $x/a = y/b = z/c = \frac{1}{3}$, whence, $xyz = k/27$.

9. Such a plane must contain the tangent vectors $T_1 = (a, 1, 0)$ at the point $(a^2, 2a, 0)$ of the first parabola and $T_2 = (b, 0, 1)$ at the point $(b^2, 0, 2b)$ of the second. The condition that the tangents intersect yields $b = + a$, with the intersection point $(-a^2, 0, 0)$. Using $T_1 \times T_2 = (1, -a, -b)$ as a normal to the plane, we then obtain its equation in the form $x - a(y + z) + a^2 = 0$, with $a$ as parameter and, as an envelope, the parabolic cylinders $4x = (y + z)^2$.

## Exercises 3.6a (p. 310)

1. (a) $-\sin v.$

   (b) $(a^3 + b^3 + c^3)(u - v) + 3abcv.$

   (c) $4uv.$

## Exercises 3.6b (p. 312)

1. (a) $-2xy\,dx\,dy.$

   (b) $(x^4 - 4x^2y^2 + y^4)\,dx\,dy.$

   (c) $(a^2 + b^2)\,dx\,dy\,dz.$

2. For $\omega = A\,dx + B\,dy + C\,dz$,

$$\omega^2 = A^2\,dx\,dx + B^2\,dy\,dy + C^2\,dz\,dz$$
$$+ AB(dx\,dy + dy\,dx)$$
$$+ BC(dydz + dz\,dy)$$
$$+ CA(dz\,dx + dx\,dz)$$

and each term in $\omega^2$ clearly vanishes.

Alternatively, since we know for any two such forms that $\omega_1\omega_2 = -\omega_2\omega_1$, it follows that $\omega^2 = -\omega^2$; hence, $\omega^2 = 0$

3. Use the result of Exercise 2.

4. Rewrite the left side in the form

$$[(\omega_1 + \omega_3) + (\omega_2 + \omega_4)] \, [(\omega_1 + \omega_3) - (\omega_2 + \omega_4)]$$

and apply the result of Exercise 3.

5. $L_1(L_2L_3) = (A_1 \, dx + B_1 \, dy + C_1 \, dz) \left\{ \left| \begin{matrix} B_2 & B_3 \\ C_2 & C_3 \end{matrix} \right| dy \, dz \right.$

$$\left. + \left| \begin{matrix} C_2 & C_3 \\ A_2 & A_3 \end{matrix} \right| dz \, dx + \left| \begin{matrix} A_2 & A_3 \\ B_2 & B_3 \end{matrix} \right| dx \, dy \right\}$$

$$= \left\{ A_1 \left| \begin{matrix} B_2 & B_3 \\ C_2 & C_3 \end{matrix} \right| + B_1 \left| \begin{matrix} C_2 & C_3 \\ A_2 & A_3 \end{matrix} \right| + C_1 \left| \begin{matrix} A_2 & A_3 \\ B_2 & B_3 \end{matrix} \right| \right\} dx \, dy \, dz,$$

where the coefficient of $dx \, dy \, dz$ is the expansion in minors of the first row for the determinant $\left| \begin{matrix} A_1 & B_1 & C_1 \\ A_2 & B_2 & C_2 \\ A_3 & B_3 & C_3 \end{matrix} \right|$.

## Exercises 3.6c (p. 316)

1. (a) $-\dfrac{y}{x^2 + y^2} dx + \dfrac{x}{x^2 + y^2} dy$

   (b) $2 \, dx \, dy$

   (c) $0$

   (d) $x \, (\cos y - 1) \sin z$

   (e) $0$.

2. For $\omega_i = A_i \, dx + B_i \, dy + C_i \, dz$, $(i = 1, 2)$,

$$d(\omega_1\omega_2) = \left\{ \left( \frac{\partial B_1}{\partial x} C_2 + B_1 \frac{\partial C_2}{\partial x} - \frac{\partial C_1}{\partial x} B_2 - C_1 \frac{\partial B_2}{\partial x} \right) \right.$$

$$+ \left( \frac{\partial C_1}{\partial y} A_2 + C_1 \frac{\partial A_2}{\partial y} - \frac{\partial A_1}{\partial y} C_2 - A_1 \frac{\partial C_2}{\partial y} \right)$$

$$+ \left. \left( \frac{\partial A_1}{\partial z} B_1 + A_1 \frac{\partial B_2}{\partial z} - \frac{\partial B_2}{\partial z} A_2 - B_1 \frac{\partial A_2}{\partial z} \right) \right\} dx \, dy \, dz$$

$$= \left\{ \left( \frac{\partial C_1}{\partial y} - \frac{\partial B_1}{\partial z} \right) A_2 + \left( \frac{\partial A_1}{\partial z} - \frac{\partial C_1}{\partial x} \right) B_2 \right.$$

$$+ \left. \left( \frac{\partial B_1}{\partial x} - \frac{\partial A_1}{\partial y} \right) C_2 \right\} dx \, dy \, dz$$

$$+ \left\{ A_1 \left( \frac{\partial B_2}{\partial z} - \frac{\partial C_2}{\partial y} \right) + B_1 \left( \frac{\partial C_2}{\partial x} - \frac{\partial A_1}{\partial z} \right) \right.$$

$$+ \left. C_1 \left( \frac{\partial A_2}{\partial y} - \frac{\partial B_2}{\partial x} \right) \right\} dx \, dy \, dz$$

$$= (d\omega_1)\omega_2 + \omega_1(d\omega_2).$$

3. From Exercise 2, if $d\omega_1 = d\omega_2 = 0$, then $d(\omega_1\omega_2) = 0$.

## Exercises 3.6d (p. 325)

1. Considering $F(\mathbf{X}) = f(\rho, \phi, \theta) = g(x, y, z)$ as a function of a point in space, we know from the invariance of the differential form that

$$dF = dg = \frac{\partial g}{\partial x}\,dx + \frac{\partial g}{\partial y}\,dy + \frac{\partial g}{\partial z}\,dz$$

$$= \nabla F \cdot d\mathbf{X}$$

$$= \frac{\partial f}{\partial \rho}\,d\rho + \frac{\partial f}{\partial \phi}\,d\phi + \frac{df}{d\theta}\,d\theta.$$

Consequently,

$$\nabla F \cdot d\mathbf{X} = \left(\frac{\partial f}{\partial \rho}\mathbf{u} + \frac{1}{\rho}\frac{\partial f}{\partial \phi}\mathbf{v} + \frac{1}{\rho \sin \phi}\frac{\partial f}{\partial \theta}\mathbf{w}\right) \cdot d\mathbf{X},$$

whence

$$\nabla f = \frac{\partial f}{\partial \rho}\mathbf{u} + \frac{1}{\rho}\frac{\partial f}{\partial \phi}\mathbf{v} + \frac{1}{\rho \sin \phi}\frac{\partial f}{\partial \theta}\mathbf{w}.$$

## Exercises 3.7b (p. 329)

1. (a) Saddles at $y = 0$, $x = \pi/3 + 2n\pi$; minima at $y = 0$, $x = -\pi/3 + 2n\pi$.
   (b) Maxima at $x = \pi/4 + 2n\pi$, $y = \pi/4 + 2n\pi$, and $x = 3\pi/4 + 2n\pi$, $y = 3\pi/4 + 2n\pi$; minima at $x = \pi/4 + 2n\pi$, $y = 3\pi/4 + 2n\pi$, and $x = 3\pi/4 + 2n\pi$, $y = \pi/4 + 2n\pi$.
   (c) Saddle at $x = 0$, $y = 1$.
   (d) No stationary points.
   (e) Saddle at $x = 0$, $y = 0$.
2. Maxima for $x = 0$, $y = \pm 1$; minimum for $x = y = 0$.
3. Minimum for $x = 1$, $y = 4$, saddle point for $x = -1$, $y = 2$.
4. $a/20$,  $a/10$,  $a/10$.
5. Improper minima on the planes $x = 0$, $y = 1$, $z = -\frac{1}{2}$.
6. Maximize $V = xy[100 - 2(x + y)]$. Maximum volume for $x = y = 50/3$, $z = 100/3$; $V_{\max} = (25/27) \times 10^4$ in$^3 \approx 5.4$ ft$^3$.
7. Set $\mathbf{X} = (x, y, z)$ and let the $n$ points be $(a_i, b_i, c_i)$, where $i = 1, 2, \ldots, n$. To minimize $\Sigma[(x - a_i)^2 + (y - b_i)^2 + (z - c_i)^2]$, set

$$2\Sigma(x - a_i) = 2\Sigma(y - b_i) = 2\Sigma(z - c_i) = 0$$

Hence, $x = (1/n)\,\Sigma a_i$, $y = (1/n)\,\Sigma b_i$, $z = (1/n)\,\Sigma c_i$. The sum is minimized at the center of gravity of the $n$ points.

## Exercises 3.7c (p. 334)

1. Take

$$F(x, y, z) = xyz + \lambda[2(x + y) + z - 100].$$

From

$$F_x = yz + 2\lambda, \; F_y = zx + 2\lambda, \; F_z = xy + \lambda,$$

the extremum occurs when

$$V = xyz = -2\lambda x = -2\lambda y = -\lambda z.$$

Thus, $z = 2x = 2y$. Entering this in the subsidiary condition, we obtain $z = 100/3$, $x = y = 50/3$, as before.

2. $x = y = \frac{1}{2}$, $z = \frac{1}{16}$.

3. $x = -y = 1/\sqrt{2}$, $z = 1$.

4. Take the center of gravity of the $n$ points as the origin and let their coordinates be $(a_i, b_i)$. Set $\mathbf{X} = (x, y)$ and let the line be given by $Ax + By = C$. Applying the method of Lagrange multipliers to

$$\Sigma[(x - a_i)^2 + (y - b_i)^2] + (C - Ax - By),$$

we obtain

$$2nx - \lambda A = 2ny - \lambda B = 0;$$

whence,

$$\lambda = \frac{2nC}{A^2 + B^2}.$$

Thus,

$$x = \frac{AC}{A^2 + B^2}, \; y = \frac{BC}{A^2 + B^2};$$

that is, $\mathbf{X}$ is the nearest point on the line to the center of gravity.

5. Let $S$ denote the curve $f(x, y) = C$ and $S'$ the curve $\phi(x, y) = C'$. $S$ and $S'$ have a point of contact in $(a, b)$. In general, $f(x, y) - C$ is positive on one side of $S$ and negative on the other side in some neighborhood; similarly, with $\phi(x, y) - C'$ and $S'$. If, for example, $f(a, b)$ is a maximum of $f$, then $f(x, y) - C \leqq 0$ on $S'$ i.e., $S'$ is wholly on one side of $S$, then $S$ is also on one side of $S'$. That is, $\phi(x, y) - C'$ has a constant sign on $S$, and as it is equal to 0 at $(a, b)$, it has either a maximum or a minimum there.

## Exercises 3.7e (p. 340)

1. For smooth $f$ and $\phi$, the minimum $c$ characterizes a level surface $f(x, y, z) = c$ tangent to the surface $\phi(x, y, z) = 0$.

2. Find a point on the intersection of the two cylinders $\phi(x, y) = 0$ and $\psi(y, z) = 0$ where $f(x, y, z)$ is an extremum. Assuming $f$ is smooth and the intersection is a smooth curve, this occurs where a level surface of $f$ touches the curve.

**Exercises 3.7f** (p. 344)

1. Extremize
$$(x - a)^2 + (y - b)^2 + (z - c)^2 + \lambda(D - Ax - By - Cz)$$
to obtain the conditions
$$2(x - a) - \lambda A = 2(y - b) - \lambda B = 2(z - c) - \lambda C = 0,$$
whence
$$\lambda = \frac{2(D - aA - bB - cC)}{A^2 + B^2 + C^2}.$$

This yields
$$x = a + \frac{A(D - aA - bB - cC)}{A^2 + B^2 + C^2}, \ldots$$
and the minimum distance $p$ is given by
$$p = \frac{|D - aA - bB - cC|}{\sqrt{A^2 + B^2 + C^2}}.$$

2. $(4 + \sqrt{5})/\sqrt{2}$, $(4 - \sqrt{5})/\sqrt{2}$.

3. The maximum value is the same as for the expression $ax^2 + 2bxy + cy^2$ subject to the subsidiary condition $ex^2 + 2fxy + gy^2 = 1$.

4. Cf. Exercise 3.

   (a) $14/3 + 2\sqrt{67}/3$.

   (b) The function has a non-strict maximum (p. 325) equal to 1.95, when $y/x = 0.64$.

5. The ellipse obviously touches the circle; that is, the two equations must give a double root in $x$. Hence, the condition for contact is $a^2(b^2 - 1) = b^4$: $a = 3/\sqrt{2}$, $b = \sqrt{3/2}$.

6. $(-1/\sqrt{14}, -2/\sqrt{14}, -3/\sqrt{14})$. This is on the line joining the given point to the center.

7. $A = a^2/x$, $B = b^2/y$, $C = c^2/z$, together with the subsidiary condition $(x^2/a^2) + (y^2/b^2) + (z^2/c^2) = 1$:

   (a) $x = \dfrac{a^{4/3}}{\sqrt{a^{2/3} + b^{2/3} + c^{2/3}}}, \ldots$

   (b) $x = \dfrac{a^{3/2}}{\sqrt{a + b + c}}, \ldots$

8. The vertices are given by $x = \pm a/\sqrt{3}$, $y = \pm b/\sqrt{3}$, $z = c/\sqrt{3}$.

9. The vertices are given by $x = a^2/\sqrt{a^2 + b^2}$, $y = b^2/\sqrt{a^2 + b^2}$.

10. $x = 1$, $y = 1$.

11. The greatest axis is given by the maximum of $\sqrt{x^2 + y^2 + z^2}$, with the subsidiary condition that $(x, y, z)$ lies on the ellipsoid. Hence, we have the three equations

$$\frac{x}{\sqrt{x^2 + y^2 + z^2}} = \frac{x}{l} = \lambda(ax + dy + ez), \ldots .$$

Multiplying these by $(x, y, z)$, respectively, and adding, we have $\lambda = \sqrt{x^2 + y^2 + z^2} = l$. On the other hand, we may regard the equations as three linear homogeneous equations in $x, y, z$ whose determinant must vanish.

12. (a) Equivalently, maximize

$$a \log x + b \log y + c \log z + \lambda(1 - x^k - y^k - z^k).$$

This yields

$$\lambda x^k = \frac{a}{k}, \ \lambda y^k = \frac{b}{k}, \ \lambda z^k = \frac{c}{k} \ ;$$

whence,

$$\lambda = \frac{1}{k}(a + b + c).$$

The maximum is attained when

$$x^k = \frac{a}{a + b + c}, \ y^k = \frac{b}{a + b + c}, \ z^k = \frac{c}{a + b + c}$$

and is equal to $\sqrt[k]{\dfrac{a^a \, b^b \, c^c}{(a + b + c)^{a+b+c}}}.$

(b) Set $x^k = u/(u + v + w)$, $y^k = v/(u + v + w)$, $z^k = w/(u + v + w)$ in

$$(x^a y^b z^c)^k \leq \frac{a^a \, b^b \, c^c}{(a + b + c)^{a+b+c}}.$$

13. Compare the similar proof for triangles on p. 328. A minimum point 0 does exist. First show that if 0 is not one of the vertices, then it can only be the point of intersection of the diagonals. Use the fact that the final points of four unit vectors whose vector sum is **0** form a rectangle. Then prove that the sum of the distances from the vertices is less for the point of intersection of the diagonals than for any of the vertices.

14. Suppose the pairs $a, b$ and $c, d$ are adjacent. Let $\phi$ be the angle between $a$ and $b$, $\psi$ that between $c$ and $d$. The problem is to maximize

$$A(\phi, \psi) = \frac{1}{2}(ab \sin \phi + cd \sin \psi)$$

subject to

$$f(\phi, \psi) = (a^2 + b^2 - 2ab \cos \phi) - (c^2 + d^2 - 2cd \cos \psi) = 0.$$

Setting the respective derivatives $(\partial/\partial\phi)(A + \lambda f)$ and $(\partial/\partial\psi)(A + \lambda f)$ equal to 0 we obtain

$$\lambda = -\frac{1}{4 \tan \phi} = \frac{1}{4 \tan \psi},$$

whence $\phi + \psi = \pi$. Consequently,

$$A = \frac{1}{2}(ab + cd) \sin \phi,$$

where $\cos \phi = \frac{1}{2}(a^2 + b^2 - c^2 - d^2)/(ab + cd)$. Eliminating $\phi$, we obtain the maximum area

$$A = \frac{1}{4} \sqrt{4\,(ab + cd)^2 - (a^2 + b^2 - c^2 - b^2)^2}$$

$$= \frac{1}{4} \sqrt{8abcd - (a^2 + b^2 + c^2 + d^2)^2},$$

which is clearly independent of our assumption concerning the order of the sides.

The conclusion that the maximum is independent of the order of the sides is geometrically obvious since any pair of adjacent sides may be interchanged without affecting the area of a convex polygon.

## Exercises A.1 (p. 350)

1. (a) Minimum at the origin.

   (b) For simplicity, introduce new variables $u = x + y$, $v = x - y$. We seek extreme values of

   $$f(u, v) = \cos u + \sin v + \frac{1}{4}(u + v)^2.$$

   The conditions $f_u = f_v = 0$ yield (i) $\cos v = -\sin u = -\frac{1}{2}(u + v)$. We must entertain two possibilities:

   1. $\sin v = -\cos u$. In this case

   $$f_{uv}^2 - f_{uu}f_{vv} = \cos^2 u$$

   and only saddles are found.

   2. $\sin v = \cos u$. In this case, (i) yields $u + v = -\pi/2$, we may have either $u = -\alpha$ or $u = \pi + \alpha$. In the former case, $f_{uv}^2 - f_{uu}f_{vv} = \cos u\,(1 - \cos u)$ is positive and we obtain a saddle; in the latter case, it is negative and we obtain a minimum from $f_{uu} = f_{vv} = \cos \alpha + \frac{1}{2}$.

   (c) No extreme, since $f_x > 0$ everywhere.

2. $f(x) + f(y) + f(z)$

   $$= 3f(a) + \{(x - a) + (y - a) + (z - a)\}\,f'(a) + \frac{1}{2}\rho^2\{f''(a) + \varepsilon\},$$

   where $\rho^2 = (x - a)^2 + (y - a)^2 + (z - a)^2$. On the other hand, the subsidiary condition gives

   $$(x - a) + (y - a) + (z - a)$$

   $$= \rho^2 \left(-\frac{\phi''(a)}{2\phi'(a)} + \varepsilon\right) - \frac{\phi'(a)}{\phi(a)}\,\{(x - a)\,(y - a)$$

   $$+ (x - a)\,(z - a) + (y - a)\,(z - a)\}$$

   $$= \left(-\frac{\phi''(a)}{2\phi'(a)} + \frac{\phi'(a)}{2\phi(a)} + \varepsilon\right)\rho^2,$$

   where $\lim_{x,y,z \to a} \varepsilon = 0$.

3. If $P_i = (x_i, y_i)$, $r_i = PP_i$, we have

$$d^2f = \sum_1^3 d^2r_i = \sum_{i=1}^3 r_i^{-3}[(y - y_i)dx - (x - x_i)dy]^2$$

which is positive definite.

4. At the point $P_1$. Note that the function $f = r_1 + r_2 + r_3$ is continuous in the whole plane but not differentiable at the points $P_1, P_2, P_3$, where it has conical points (like the function $z = \sqrt{(x - x_1)^2 + (y - y_1)^2}$, which geometrically represents a circular cone). Investigate the derivative of $f$ at $P_1$ in all directions around this point.

5. (a) If we put $f = lx + my + nz$, $\phi = x^p + y^p + z^p - c^p$, $F = f - \lambda\phi$, then the conditions for stationary values are

   (1)          $l = \lambda p x^{p-1}$, $m = \lambda p y^{p-1}$, $n = \lambda p z^{p-1}$

   Multiplying these equations by $x, y, z$, respectively, and adding, we have

   (2)                    $lx + my + nz = \lambda p c^p$.

   Calculating $x, y, z$ from (1) and substituting in $\phi = 0$, we get

   $$\lambda p = (l^q + m^q + n^q)^{1/q} c^{1-p}.$$

   Substitution of this expression for $\lambda p$ in (2) gives the stationary value.

   (b) Cf. Exercise 6. Here we have

   $$d^2F = -\lambda p(p - 1)(x^{p-2} dx^2 + y^{p-2} dy^2 + z^{p-2} dz^2);$$

   as $p > 0$, this quadratic form is positive or negative definite according to whether $p \gtrless 1$.

6. The proof resembles that for $n = 2$ (p. 347). A positive definite quadratic form $\sum a_{ik}x_ix_k$ can be brought by a suitable transformation

$$x_i = \sum_{k=1}^n c_{ik}y_k \qquad\qquad (i = 1, \ldots, n)$$

with a nonvanishing determinant into the form $\sum a_{ik}x_ix_k = y_1^2 + y_2^2 + \cdots + y_n^2 > m(x_1^2 + \cdots + x_n^2)$, where $m$ is a suitable positive constant. For the applications, it is important to remember that a necessary and sufficient condition that a form $\Phi = \sum a_{ik}x_ix_k$ shall be positive definite is that its principal first minors of order 1, 2, . . . , $n$, as indicated below,

$$
\begin{vmatrix}
a_{11} & a_{12} & a_{13} & a_{1n} \\
a_{21} & a_{22} & a_{23} & \\
a_{31} & a_{32} & a_{33} & \\
a_{n1} & & & a_{nn}
\end{vmatrix}
$$

shall all be positive. $\Phi$ is negative definite if $-\Phi$ is positive definite.

7. According to the first rule, we have to compute $d^2f$ from (3), with $dx_1$, ..., $dx_m$, $d^2x_1$, ..., $d^2x_m$ substituted from (1). Note that (1) implies that

$$d^2\phi_\mu = \Sigma\phi_{\mu x_i x_k}\, dx_i\, dx_k + \phi_{\mu x_1} d^2x_1 + \cdots + \phi_{\mu x_m} d^2x_m$$

$$= 0 \qquad\qquad (\mu = 1, \ldots, m);$$

if this is multiplied by $\lambda_\mu$ and added to (3) for all values of $\mu$, we have $d^2f = d^2F = \Sigma F_{x_1 x_k}\, dx_i\, dx_k$, because $d^2x_1, \ldots, d^2x_m$ drop out on account of the relations (2).

8. For $F = f + \lambda\phi$ (disregarding a positive factor), we get

$$d^2F = \sum_{i,k=1,\ldots,n} dx_i\, dx_k \qquad (d\phi = dx_1 + \cdots + dx_n = 0).$$

Eliminating $dx_n$, we have to show that the quadratic form

$$-d^2F = (dx_1 + \cdots + dx_{n-1})^2 - \sum_{i,k=1,\ldots,n-1} dx_i\, dx_k$$

$$= \sum_{i=1,\ldots,n} dx_i{}^2 + \sum_{i,k}^{1,n-1} dx_i\, dx_k$$

is positive definite.

9. From $dx = -dy - dz$,

$$d^2F = -2s[(s-z)dy^2 + (s-x)dy\, dz + (s-y)dz^2].$$

When $x = y = z$ the discriminant of $d^2F$ is positive and $d^2F$ is negative definite.

## Exercises A.2 (p. 359)

1. (c) Using polar coordinates $x = r\cos\theta$, $y = r\sin\theta$, take

$$f(x, y) = r^{n+1}\sin(n+1)\theta,$$

for which

$$\nabla f = (n+1)r^n(\sin n\theta, \cos n\theta).$$

2. (b) Extend the solution of Exercise 1:

$$f(x, y) = r^{-n+1}\sin(-n+1)\theta$$

and

$$\nabla f = (n-1)r^{-n}(\sin n\theta, -\cos n\theta).$$

3. If there is no fixed point, we have $u^2 + v^2 \neq 0$ everywhere in $R$. Since the convex region $R$ is simply connected, it follows as on p. 358 that the index $I_C$ of the curve $C$ with respect to the vector field is zero. On the other hand, since $R$ is mapped into itself, the vector $(u, v)$ for every point on $C$ points into $R$ or is tangential. This implies that $I_C = 1/2\pi \int_C d\theta = 1$ if $C$ has the usual orientation determined by the $x$, $y$-coordinate system.

## Exercises A.3 (p. 362)

1. (a) A node at $(0, 0)$, with tangents $x = \pm y$.
   (b) The equations

$$f_x = 2x - 6x^2 + 4xy^2 = 0,$$
$$f_y = 2y - 6y^2 + 4x^2y = 0$$

have the common solutions $(0, 0)$, $(\sqrt{\frac{1}{2}}, 0)$, $(0, \sqrt{\frac{1}{2}})$, $(\frac{1}{2}, \frac{1}{2})$, and $(1, 1)$, of which only the first and last are points of the curve. At $(0, 0)$ the singularity is an isolated point. At $(1, 1)$, $f_{xx} = f_{yy} = 0$ and $f_{xy} = 8$; the singularity is a node with tangents $x = 1$ and $y = 1$.
   (c) A double tangent $y = x$ at $(0, 0)$. The curve has two branches; to second order $y = x \pm x^2$
   (d) A double tangent $y = 0$ at $(0, 0)$. The curve has a cusp. This is the same curve as that of Section 3.2b, Exercise 3.

## Exercises A.4 (p. 363)

1. If the quadratic form is nondegenerate and definite, the singularity is an isolated point; if nondegenerate and indefinite, the tangent lines at the singularity form a cone. If the form is degenerate and semidefinite, the tangent lines may lie in a plane where two branches are tangent to each other, like the plane $z = 0$ for the surfaces

$$z^{2/3} + (x^2 + y^2)^{2/3} = a^{2/3}$$

at $(a, 0, 0)$ (a line cusp),

$$z^4 = (x^2 + y^2)^3$$

at $(0, 0, 0)$ (two tangent branches). Or there may be a point cusp where only one tangent line exists, like the line $x = y = 0$ for the former surface at $(0, 0, a)$. If the form is degenerate and indefinite, the tangent lines lie in two planes, like the planes $x = \pm y$ at $(0, 0, 0)$ for the surface $x^2 - y^2 + z^3 = 0$.

## Exercises A.5 (p. 364)

1. The flow is stationary; that is, the fluid velocity is constant in time at each point of space.
2. If $\mathbf{U} = (u, v, w)$ is the velocity of the particle passing through the point $\mathbf{X} = (x, y, z)$ at time $t$, its acceleration is

$$\frac{d^2\mathbf{X}}{dt^2} = \frac{d\mathbf{U}}{dt} = \frac{d\mathbf{X}}{dt} \cdot \nabla\mathbf{U} + \frac{\partial\mathbf{U}}{\partial t}$$

$$= \mathbf{U} \cdot \nabla\mathbf{U} + \frac{\partial\mathbf{U}}{\partial t}.$$

## Exercises A.6 (p. 366)

1. (a) $x = -2 - 2\cos\alpha$, $y = -2\sin\alpha$ or $(x+2)^2 + y^2 = 4$; $L = 4\pi$; $A = 4\pi$.

   (b) $x = -\sin^3\alpha$, $y = -\cos^3\alpha$ or $x^{2/3} + y^{2/3} = 1$,

   $$L = \frac{3}{2}\int_0^{2\pi}|\sin 2\alpha\,|d\alpha| = 6\int_0^{\pi/2}\sin 2\alpha\,d\alpha = 6.$$

   $A = -(3/8)\pi$, where the sign comes from the clockwise orientation of the curve.

2. Yes. Consider the right triangle with vertices $(0,0)$, $(0,c)$, $(c^{-2}, 0)$ for large $c$.

3. For the curve to be expressible as the envelope of its tangents, it must be piecewise smooth.

## Exercises 4.1 (p. 374)

1. In the $n$th subdivision, any square that contains points of $S$ contains points of $T$, $A_n^+(S) \leqq A_n^+(T)$. On passing to the limit as $n \to \infty$, we obtain the result.

2. In the $n$th subdivision, any square that contains points of $T - S$ may not be one that consists entirely of points of $S$, and both kinds of squares contain points of $T$; therefore,

   $$A_n^+(T) \geqq A_n^+(T - S) + A_n^-(S).$$

   Similarly,

   $$A_n^+(T) \leqq A_n^-(T - S) + A_n^+(S).$$

   Combining these results with $A_n^-(T - S) \leqq A_n^+(T - S)$, we find

   $$A_n^+(T) - A_n^+(S) \leqq A_n^-(T - S) \leqq A_n^+(T - S)$$
   $$\leqq A_n^+(T) - A_n^-(S),$$

   from which the result follows on passing to the limit as $n \to \infty$.

3. For the proof of (a), observe that any square of the $n$th subdivision that enters in $A_n^+(S)$ or $A_n^+(T)$ may enter in only one or in both of these; if a square enters into only one, it enters in $A_n^+(S \cup T)$; if it enters in both, it enters in $A_n^+(S \cup T)$; but need not enter in $A_n^+(S \cap T)$, because the square may contain points of both $S$ and $T$ without containing points common to the two, Consequently,

   $$A_n^+(S \cup T) + A_n^+(S \cap T) \leqq A_n^+(S) + A_n^+(T),$$

   from which (a) follows.

   For (b) we observe that any square that enters in one sum but not the other, say, $A_n^-(S)$ but not $A_n^-(T)$, will enter in $A_n^-(S \cup T)$ but not $A_n^-(S \cap T)$ and any square that enters in both $A_n^-(S)$ and $A_n^-(T)$ also enters in both $A_n^-(S \cap T)$ and $A_n^-(S \cup T)$. Thus,

$$A_n^- (S) + A_n^- (T) \leqq A_n^- (S \cap T) + A_n^- (S \cup T),$$

from which (b) follows.

Note that a square consisting of points of $S \cup T$ need not consist wholly of points of $S$ or wholly of those of $T$; consequently, the inequality sign can not be removed.

4. In the $n$th subdivision, consider any square that consists entirely of points of $S \cup T$. If it contains any point of $S$, the square enters in $A_n^+(S)$, but it cannot enter in $A_n^- (T)$, because it cannot consist wholly of points of $T$. If the square contains no points of $S$, it must consist wholly of points of $T$ and, thus, enters in $A_n^- (T)$. Finally, we observe that any square that enters in $A_n^+(S)$ but does not lie wholly in $S \cup T$ must contain a boundary point of $S \cup T$ and therefore enter $A_n^+ (\partial[S \cup T])$. Combining these results, we find

$$A_n^-(S \cup T) \leqq A_n^+(S) + A_n^-(T) \leqq A_n^-(S \cup T) + A_n^+(\partial[S \cup T]).$$

Since $\lim\limits_{n \to \infty} A_n^- (S \cup T) = A (S \cup T)$ and $\lim\limits_{n \to \infty} A_n^+ (\partial[S \cup T]) = 0$, the desired result follows.

5. (a) Let Jordan content in the original system be denoted by $A$, and in the transformed system, by $B$. Since $A(\partial S) = 0$, $\lim\limits_{n \to \infty} A_n^+(\partial S) = 0$.

Let $P$ be any point of $\partial S$. Note that in the $n$th subdivision, the maximum distance from $P$ of any point of a square that contains $P$ is $2^{-n} \sqrt{2}$. Now, in the $n$th subdivision with respect to the new coordinate system, let $R_B$ be any square containing $P$. Form a larger square $R_B^*$ with $R_B$ at its center and five subdivision squares on a side. The smallest distance from any point of $R_B$ to the boundary of $R_B^*$ is $2 \cdot 2^{-n}$. Thus, $R_B^*$ contains each square $R_A$ that contains $P$ in the subdivision with respect to the original system. We conclude that for each square that enters into $A_n^* (\partial S)$ no more than 25 squares enter $B_n^+(\partial S)$. Since $0 \leqq B_n^+(\partial S) \leqq A_n^+(\partial S)$, it follows that $\lim\limits_{n \to \infty} B_n^+(\partial S) = 0$.

(b) Observe that in the $n$th subdivision with respect to the two systems, any square that enters in $A_n^-(S)$ is covered by squares that enter into $B_n^+(S)$. It follows that $A_n^-(S) \leqq B_n^+(S)$ and, passing to the limit as $n \to \infty$, $A(S) \leqq B(S)$. By a parallel argument, $B(S) \leqq A(S)$. Consequently, $A(S) = B(S)$.

The foregoing argument makes tacit use of the assumption that if two sets $U$ and $V$ are made up of nonoverlapping congruent squares from respective grids and $U \subset V$, then the number of squares in $U$ is less than, or equal to, the number of squares in $V$. We prove this inductively as follows: Let $u$ and $v$ be two finite collections of nonoverlapping squares of side length $a$ from respective grids such that the union $U$ of squares of $u$ is contained in the union $V$ of squares of $v$. If $p$ is the number of squares of $u$, and $q$, the number of squares of $v$, then $p \leqq q$ and equality holds if and only if $u = v$. For the proof, we use induction on $p$.

If $p = 1$, we cannot have $q < p$; for, then, $q = 0$ and $V$ does not contain $U$. Moreover, if $q = p = 1$, we note that opposite vertices of the square of $u$ must be opposite vertices of the square of $v$, since the maximum distance $a\sqrt{2}$ between any two points of either square is attained only at opposite vertices. Consequently, the squares are the same and $u = v$.

Now we prove that the truth of the hypothesis for a fixed $p$ implies its truth for $p + 1$: Let $u$ be a collection of $p + 1$ squares and let $u^*$ be any subcollection of $p$ squares. Suppose $q < p + 1$. Since $V \supset U \supset U^*$, $q \geq p$ by the induction hypothesis. However, $p \leq q < p + 1$ implies $q = p$, and hence, by the induction hypothesis, $v = u^*$. But, then $V$ cannot contain the one square of $u$ that does not belong to $u^*$, contradicting that $V \supset U$. We conclude that $q \geq p + 1$. If equality holds, $q = p + 1$, we now show that $v = u$. We shall show that the set $U(=V)$ must have a corner on the boundary; that is, at least one of the squares $R$ of $u$ must have a vertex with its adjacent edges on the boundary of $U$. The square $R$ must also belong to $v$, as we shall prove. By the induction hypothesis, the collections $u^*$ and $v^*$, obtained from $u$ and $v$ by deleting $R$, must be the same. Consequently, $u = v$.

To prove that $U$ has a corner, let $P$ be any point of $U$ most distant from an arbitrary given point $Q$. The point $P$ must lie on the boundary of $U$, otherwise it would be an interior point and its neighborhood within $U$ would contain points more distant from $Q$. Furthermore, $P$ must be a vertex of one of the squares of $u$, because if it were an inner point of an edge, at least one of the two vertices on the edge would be farther from $Q$ than $P$, since it would be farther than $P$ from the perpendicular from $Q$ to the line of the edge. No two edges meeting at $P$ can be aligned, for the same argument shows that one of the end points of the segment made up of the two edges must be more remote from $Q$ than $P$. It follows that $P$ and its adjacent edges can belong to only one square $R$ of $u$. (The figure shows all possible configurations in the neighborhood of a boundary vertex.) Exactly the same argument applies to $v$, but then, $R$ must belong to $v$, as claimed.

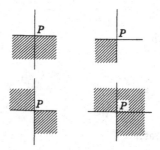

6. If $P$ is a boundary point of $S$, it is either a point of $S$ and covered or a limit point of $S$ such that every deleted neighborhood of $P$ contains infinitely many points of $S$. Thus, $P$ is the limit of a convergent sequence

of distinct points of $S$. Since the collection of covering sets is finite, at least one of these sets must contain a subsequence, and because this set is closed, it must contain the limit of the subsequence, $P$.

7. The area of the set is zero. Let $S_n$ be the set of points for which both $p$ and $q$ are greater than $n$ and $T_k$ the set for which either $p$ or $q$ is equal to $k$.

$$S = S_n \cup T_1 \cup T_2 \cup \cdots \cup T_n.$$

Note that $S_n$ is contained in the square

$$\left\{ (x, y) \,|\, 0 \leq x < \frac{1}{n}, \ 0 \leq y < \frac{1}{n} \right\}.$$

Consequently,

$$A_n^+ (S^n) \leq \left( \frac{1}{n} + \frac{1}{2^n} \right)^2.$$

Observe also that $T_k$ contains $2k - 1$ points, each of which may lie in no more than four squares of the $n$th subdivision. Consequently,

$$A_n^+ (T_k) \leq \frac{4(2k - 1)}{2^{2n}}.$$

Summing, we see that

$$A_n^+ (S) \leq A_n^+ (S_n) + \sum_{k=1}^{n} A_n^+ (T_k)$$

$$\leq \left( \frac{1}{n} + \frac{1}{2^n} \right)^2 + \frac{4n^2}{2^{2n}};$$

whence, $\lim\limits_{n \to \infty} A_n^+ (S) = 0$.

## Exercises 4.6 (p. 405)

1. (a) $a^2 b^2 (a^2 - b^2)/8$.
   (b) $-4$.
   (c) $\log 2$.
   (d) $-a + (e^{ab} - 1)/b$.
   (e) $\pi/16$.
   (f) $4/3$.
2. $\pi/2$
3. $0$.
4. $2\pi$.
5. Use polar coordinates:

   (a) $\displaystyle\int_{-4/\pi}^{\pi/4} \int_0^{\sqrt{\cos 2\theta}} \frac{r}{(1 + r^2)^2} \, dr \, d\theta = \frac{\pi}{4} - \frac{1}{2}$

   (b) $\displaystyle\int_0^{\pi/3} \int_0^{\sqrt{3}/\cos(\theta - \pi/6)} \frac{r}{(1 + r^2)^2} \, dr \, d\theta = \frac{\sqrt{3}}{2} \arctan \frac{1}{2}$.

6. Use the substitution $x = a\xi$, $y = b\eta$, $z = c\zeta$; then use polar coordinates and symmetry to obtain

$$8a^2b^2c^2 \int_0^{\pi/2} \int_0^{\pi/2} \int_0^1 \rho^5 \cos\phi \sin\phi \sin^3\theta \cos\theta \, d\rho \, d\phi \, d\theta$$

$$= \frac{a^2b^2c^2}{6}.$$

7. Use the fact that the figure is symmetrical; 1/16 of the volume lies above the triangle with vertices $(0, 0)$, $(1, 0)$, $(1, 1)$ and below the surface $x^2 + z^2 = 1$; 16/3.

8. $\pi\,(2r^3 - 3r^2\,h + h^3)$.

9. 0.

10. 0. With the additional restriction $z \geqq 0$; $\pi/8$.

11. 1/50,400.

12. Use cylindrical coordinates and integrate with respect to $\theta$, $r$, and $z$ in that order; $\pi[2 - (3/2)\log 3]$.

13. Use spherical coordinates with origin at $(0, 0, \frac{1}{2})$. With $\alpha = \cos^{-1}[\rho - (3/4\rho)]$ for $\frac{1}{2} \leqq \rho \leqq 3/2$,

$$\int_{1/2}^{3/2} \int_0^{\alpha} \int_0^{2\pi} + \int_0^{1/2} \int_0^{\pi} \int_0^{2\pi} \sin\theta \, d\phi \, d\theta \, d\rho$$

$$= \pi\left\{2 + \frac{3}{2}\log 3\right\}.$$

14. Use polar coordinates: $4\log(1 + \sqrt{2})$.

15. Let $(a, b)$ be any point of the domain and choose a $\delta$-neighborhood $R_\delta$ of $(a, b)$ within $D$ so small that $|f(x, y) - f(a, b)| < \varepsilon$ in the neighborhood. By the mean value theorem,

$$\int_{R_\delta} f(x, y) \, dx \, dy = \mu\delta^2,$$

where $|\mu - f(a, b)| < \varepsilon$. Since the integral vanishes, $\mu = 0$. Consequently, $|f(a, b)| < \varepsilon$ for arbitrary positive $\varepsilon$, and hence, $f(a, b) = 0$.

16. Using $d(x, y)/d(u, v) = u/(1 + v^2)$, we obtain

$$\iint_R e^{-(x^2+y^2)} \, dx \, dy = \int_0^\infty \int_{-u/a}^{u/a} \frac{e^{-(u^2+a^2)}u}{1 + v^2} \, dv \, du$$

$$= 2e^{-a^2} \int_0^\infty ue^{-u^2} \arctan\frac{u}{a}\, du.$$

Integration by parts yields the result.

17. Set $\rho^2 = \xi^2 + \eta^2$. From $\xi_x = \eta^2 - \xi^2$, $\xi_y = -2\xi\eta$, $\eta_x = -2\xi\eta$, $\eta_y = \xi^2 - \eta^2$, it follows that $|d(x, y)/d(\xi, \eta)| = 1/\rho^4$ and also that $u_x^2 + u_y^2 = \rho^4(u_\xi^2 + u_\eta^2)$.

18. For new Cartesian coordinates to the same scale, the Jacobian of the transformation is 1. With $r = (x^2 + y^2 + z^2)^{1/2}$, choose Cartesian

coordinates $u$, $v$, $w$ for which $u = (x\xi + y\eta + z\zeta)/r$. The integral becomes

$$I = \iiint \cos ru \, du \, dv \, dw$$

over the sphere $u^2 + v^2 + w^2 \leqq 1$. In cylindrical coordinates $u$, $v = \rho \cos \theta$, $w = \rho \sin \theta$, we find.

$$I = \int_{-1}^{1} \int_{0}^{2\pi} \int_{0}^{\sqrt{1-u^2}} \rho \cos ru \, d\rho \, d\theta \, du$$

$$= 4\pi \left( \frac{\sin r}{r^3} - \frac{\cos r}{r^2} \right).$$

19. $-\int_{1}^{2} (4 - y) \int_{4/y}^{(20-8y)/(4-y)} dx \, dy = 16 \log 2 - 12.$

## Exercises 4.7 (p. 416)

1. (a) $K = \lim\limits_{\varepsilon \to 0} \int_{0}^{\beta} \int_{\varepsilon}^{a} r \log r^2 \, dr \, d\theta.$

   (b) $K = \left( \int_{0}^{a \cos \beta} \int_{0}^{x \tan \beta} + \int_{a \cos \beta}^{a} \int_{0}^{\sqrt{a^2-x^2}} \right) \log (x^2 + y^2) \, dy \, dx.$

2. (a) $\pi$.    (b) $\pi^2$.

3. Symmetry shows that reversal of the order of integration reverses the sign. Since $I$ is not zero, $I = \frac{1}{2}$, the result is established. Alternately, for $0 < a, b \leqq 1$, set

$$J = \int_{b}^{1} \int_{a}^{1} \frac{y - x}{(x + y)^3} \, dx \, dy = \frac{(1 - a)(1 - b)(b - a)}{2(1 + a)(1 + b)(a + b)}.$$

Integrating first with respect to $x$, then $y$, is equivalent to taking

$$I = \lim_{b \to 0} \lim_{a \to 0} J = \frac{1}{2};$$

integrating first with respect to y, then x, to taking

$$\lim_{a \to 0} \lim_{b \to 0} J = -\frac{1}{2}.$$

## Exercises 4.8 (p. 430)

1. Apply Guldin's rule; $2\pi^2 ab$.

2. $\frac{1}{2}\pi abh^2$.

3. Set $x = a\xi$, $y = b\eta$, $z = c\zeta$. With $d = p/\sqrt{a^2 l^2 + b^2 m^2 + c^2 n^2}$, the volume is $\pi abc(2 - 3d + d^3)/3$.

4. (a) With $\theta$ and $\phi$ as parameters for both surfaces, $\sqrt{EG - F^2} = a^2 \sin \theta$.

   (b) $a^2 \int_{0}^{2\pi} \int_{0}^{f(\phi)} a^2 \sin \theta \, d\phi \, d\theta = a^2 \int_{0}^{2\pi} \{1 - \cos f(\phi)\} \, d\phi.$

(c) Take $f(\phi) = \pi/4$; $\pi a^2(2 - \sqrt{2})$.

5. Let $a$, $b$, $c$ be the lengths of the sides opposite $A$, $B$, $C$ respectively, and $p$ the altitude from $C$. Apply Guldin's rule.

  (a) $\frac{1}{3}\pi c p^2$,

  (b) $\pi p(a + b)$.

6. $\frac{1}{3}\pi (n - m) (4n^2 + 4mn + 4m^2 - 6n - 6m + 3)$.

7. Take polar coordinates in the $x$, $y$-plane as surface parameter for the cylinder $x^2 + z^2 = a^2$. Thus, $x = r \cos\theta$, $y = r \sin\theta$, $z = \sqrt{a^2 - r^2}$ and $E = a^2/(a^2 - r^2)$, $F = 0$, $G = r^2$. The surface area is then

$$S = 8\int_0^{\pi/4} \int_0^{b \sec\theta} \frac{ar}{\sqrt{a^2 - r^2}}\, dr\, d\theta$$

$$= -8a \int_0^{\pi/4} \sqrt{a^2 - r^2}\,\Big|_0^{b \sec\theta}\, d\theta$$

$$= 2a^2\pi - 8aI,$$

where

$$I = \int_0^{\pi/4} \sqrt{a^2 - b^2 \sec^2\theta}\, d\theta.$$

Set $\theta = \arctan(\sqrt{(a^2 - b^2)/b^2}\, \sin\omega)$ to obtain

$$I = \int_0^\lambda \frac{(a^2 - b^2) \cos^2\omega}{a^2 \sin^2\omega + b^2 \cos^2\omega}\, d\omega,$$

where $\tan\lambda = b/\sqrt{a^2 - 2b^2}$. The explicit integral is

$$I = a \arctan\left(\frac{a}{b}\tan\omega\right) - b\omega\ \Big|_0^\lambda.$$

Hence,

$$S = 8a^2\left[\frac{\pi}{4} - \arctan\frac{a}{\sqrt{a^2 - 2b^2}}\right] - 8ab \arctan\frac{b}{\sqrt{a^2 - 2b^2}}.$$

8.
$$\Sigma = \iint \sqrt{EG - F^2}\, dr\, d\theta$$

$$= \int_{\theta_1}^{\theta_2} d\theta \int_0^{f'(\theta)} \sqrt{r^2 + f'^2}\, dr$$

$$= [\sqrt{2} + \log(1 + \sqrt{2})] \int_{\theta_1}^{\theta_2} \frac{1}{2}f'^2\, d\theta,$$

(cf. Volume I, p. 215), which is $[\sqrt{2} + \log(1 + \sqrt{2})]$ times the area of the projection

$$\theta_1 \leqq \theta \leqq \theta_2,\ 0 \leqq r \leqq f'(\theta).$$

## Exercises 4.9 (p. 442)

1. (a) Use cylindrical coordinates. On the axis of the cone, three-fourths of the way from the vertex to the base.

(b) On the axis of the cone, two-thirds of the way from the vertex to the base.

2. $x = 2x_0/3$, where $y = z = 0$.

3. Let $(\xi, \eta, \zeta)$ be the centroid:

$$\xi = \frac{1}{V} \int_0^a \int_0^{b\left(1-\frac{x}{a}\right)} \int_0^{c\left(1-\frac{x}{a}-\frac{y}{b}\right)} x \; dz \; dy \; dx,$$

where $V$, the volume of the tetrahedron is obtained by replacing the integrand $x$ by unity in the above triple integral. Integrate to obtain $\xi = a^2bc/24V$, where $V = abc/6$. Hence, by algebraic symmetry, $\xi = a/4$, $\eta = b/4$, $\zeta = c/4$.

4. (a) Use spherical coordinates, $z = 3(b^4 - a^4)/8(b^3 - a^3)$, $x = y = 0$.

(b) Factor $b - a$ out of the numerator and denominator in the solution of part (a) and take the limit.

5. $m (b^2 + c^2)/3$.

6. If $\mu$ is the density,

(a) $\pi\mu h(R^2 - R'^2)$,

(b) $2\pi\mu h(R - R') \left[\frac{1}{4} (R + R') + \frac{1}{3} h^2\right]$.

7. Use spherical coordinates. Mass, $\frac{1}{3}\pi a^3[\mu_0 + 3\mu_1]$. Moment of inertia, $4\pi a^5 [\mu_0 + 5\mu_1]/45$.

8. Substitute $x = a\xi$, $y = b\eta$, $z = c\zeta$; use the expressions for the moments of inertia given in the text and the properties of symmetry of the ellipsoid:

(a) $\dfrac{4}{15} \pi abc (a^2 + b^2)$,

(b) $\dfrac{4}{15} \pi abc \{(1 - \alpha^2)a^2 + (1 - \beta^2)b^2 + (1 - \gamma^2)c^2\}$.

9. For example, with $A = \int_R (y^2 + z^2) \, dV$, $B = \int_R (z^2 + x^2) \, dV$, and $C = \int_R (x^2 + y^2) \, dV$,

$$A + B = \int_R (x^2 + y^2 + 2z^2) \, dV$$

$$= C + \int_R 2z^2 \, dV > C.$$

10. Let $(\xi, \eta, \zeta)$ be the point on the ray at distance $1/\sqrt{I}$ from $O$. The squared distance of a point $(x, y, z)$ from the line is

$$x^2 + y^2 + z^2 - (\xi x + \eta y + \zeta z)^2/(\xi^2 + \eta^2 + \zeta^2).$$

Consequently,

$$I = \iiint_R \left[x^2 + y^2 + z^2 - \frac{(\xi x + \eta y + \zeta z)^2}{\xi^2 + \eta^2 + \zeta^2}\right] dx \; dy \; dz$$

$$= \frac{1}{\xi^2 + \eta^2 + \zeta^2} \, .$$

Multiplying both sides of this equation by $\xi^2 + \eta^2 + \zeta^2$, we obtain a positive definite quadratic expression in $\xi$, $\eta$, $\zeta$ set equal to unity; hence, the equation is that of an ellipsoid.

11. $a^2(x - \xi)^2 + b^2(y - \eta)^2 + c^2(z - \zeta)^2$
$= \{a^2 + b^2 + c^2 + 5(\xi^2 + \eta^2 + \zeta^2)\} \{(x - \xi)^2 + (y - \eta)^2 + (z - \zeta)^2\} \, .$

12. $(\frac{1}{8}, 0, 0)$

13. $x = \dfrac{5a}{16} \dfrac{2a^2 + b^2 + c^2}{a^2 + b^2 + c^2} \, .$

14. $I = (I_1 + m_1 r_1{}^2) + (I_2 + m_2 r_2{}^2)$, where $r_1$ and $r_2$ are the distances from the axes through the centers of mass of the respective parts from the axis through the center of the system. Use $m_1 r_1 = m_2 r_2$ and $r_1 + r_2 = d$.

15. The distance of the point $(x, y, z)$ from the plane $ux + vy + wz = -1$ is given by

$$\frac{ux + vy + wz + 1}{\sqrt{u^2 + v^2 + w^2}} \, .$$

The moment of inertia of the ellipsoid with respect to this plane is therefore given by

$$\frac{Au^2 + Bv^2 + Cw^2 + V}{u^2 + v^2 + w^2} \, ,$$

where $A$, $B$, $C$ denote the moments of inertia with respect to the co-ordinate planes and $V$ is the volume of the ellipsoid, that is, $B = 4ab^3c/15$, $C = 4abc^3/15$, and $V = 4abc/3$. We have now to find the envelope of the planes for which this expression is equal to $h$. The envelope is given by the equations

$$(A - h)u = \lambda x, \quad (B - h)v = \lambda y, \quad (C - h)w = \lambda z.$$

where $\lambda$ denotes a common multiplier, which from the expression for the moment of inertia and the equation of the plane is found to be $V$. By squaring the three equations we obtain the equation of the envelope, namely,

$$\frac{x^2}{h - A} + \frac{y^2}{h - B} + \frac{z^2}{h - C} = \frac{1}{V} \, ,$$

16.
$$\frac{2\pi a^2 b\mu}{\sqrt{b^2 - a^2}} \log \left\{ \frac{1}{a} (b + \sqrt{b^2 - a^2}) \right\},$$

where $\mu$ is the constant density.

17. $2\pi\mu \int_a^b \sqrt{z^2 + \{f(z)\}^2} \, dz - \pi\mu \, |b^2 \pm a^2|$, where the lower or upper sign is to be taken according as the origin is inside the body or not.

18. Let $\mathbf{X}$ be a variable point of the solid, $\mathbf{O}$ its center of mass and $\mathbf{Y}$ a variable point of the space where the potential is calculated. The potential at $\mathbf{Y}$ is

$$U(\mathbf{Y}) = \iiint_S \frac{\mu \, dV}{|\mathbf{Y} - \mathbf{X}|}.$$

Let $a$ be the maximum value of $|\mathbf{X}|$ in $S$, $|\mathbf{X}| \leqq a$, and suppose $|\mathbf{Y}| > a$. Then, if $M$ is the mass of the solid,

$$\left| U(\mathbf{Y}) - \frac{M}{|\mathbf{Y}|} \right| = \left| \iiint_S \mu \left( \frac{1}{|\mathbf{Y} - \mathbf{X}|} - \frac{1}{|\mathbf{Y}|} \right) dV \right|$$

$$\leqq \iiint_S \mu \left| \frac{1}{|\mathbf{Y} - \mathbf{X}|} - \frac{1}{|\mathbf{Y}|} \right| dV$$

$$\leqq \iiint \mu \, \frac{|\mathbf{X}|}{|\mathbf{Y}|(|\mathbf{Y}| - |\mathbf{X}|)} dV$$

(since $\big| |\mathbf{Y}| - |\mathbf{Y} - \mathbf{X}| \big| \leqq |\mathbf{X}|$ by the triangle inequality)

$$\leqq \iiint \mu \, \frac{a}{|\mathbf{Y}|(|\mathbf{Y}| - a)} dV$$

$$\leqq \frac{2a}{|\mathbf{Y}|^2} \iiint \mu \, dV$$

(where we suppose $|\mathbf{Y}| \geqq 2a$)

$$\leqq \frac{2aM}{|\mathbf{Y}|^2}.$$

19. As $A - BR^2 = \dfrac{5}{2}$, $A - \dfrac{3}{5} BR^2 = \dfrac{11}{2}$, we have $A = 10$, $B = \dfrac{15/2}{R^2}$. The attraction at an internal point is equal to the attraction of the total mass of the points inside of the sphere of radius $r$ concentrated at the center of the sphere.

20. Use cylindrical or spherical coordinates.

21. By translation we can ensure that the triangle lies in the upper half-plane. Then its moment of inertia is equal to

$$\phi(x_1 y_1, x_2 y_2) + \phi(x_2 y_2, x_3 y_3) + \phi(x_3 y_3, x_1 y_1),$$

where $\phi(x_1 y_1, x_2 y_2)$ denotes the moment of inertia of the quadrilateral with vertices $(x_1, 0)$, $(x_1, y_1)$, $(x_2, 0)$ multiplied by the sign of $(x_1 - x_2)$. Then show that

$$\phi(x_1 y_1, x_2 y_2) = \frac{1}{12} (x_1 - x_2)(y_1^3 + y_1^2 y_2 + y_1 y_2^2 + y_2^3).$$

22. $I = \displaystyle\int_1^2 (y - 4) \, dy \int_{(8y-20)/(y-4)}^{4/y} dx = 12 - 16 \log 2$.

23. Let $f(\rho)$ be the potential associated with a unit point charge. The potential at a point $(0, 0, z)$ in the interior of a spherical lamina centered at the origin and carrying unit-charge density is

$$U(z) = \int_0^{2\pi} \int_0^{\pi} f(\rho) a^2 \sin \theta \, d\theta \, d\phi$$

where, in the integrand, if $a$ is the radius of the sphere, $\rho$ is given by

$$\rho = \sqrt{a^2 + z^2 - 2az\cos\theta}.$$

If $g$ is a function such that $g'(\rho) = \rho f(\rho)/z$, where $z$ is kept constant, then

$$U(z) = 2\pi a g(\rho)\Big|_{\theta=0}^{\pi}$$

$$= 2\pi a[g(a+z) - g(a-z)].$$

Since the force vanishes for $|z| < a$, we obtain

$$U'(z) = 2\pi a[g'(a+z) + g'(a-z)] = 0;$$

consequently,

$$(a+z)f(a+z) = (a-z)f(a-z).$$

This is a relation holding for all positive $a$ and all $z$ with $|z| < a$. Introducing new independent variables $\xi$ and $\eta$ with $\xi = a + z$ and $\eta = a - z$, we obtain

$$\xi f(\xi) = \eta f(\eta)$$

for all positive $\xi$ and $\eta$. Consequently, $\rho f(\rho) = c$, where $c$ is constant. Thus, we conclude that

$$f(\rho) = \frac{c}{\rho} \qquad\qquad (c = \text{constant}),$$

which is the potential for the inverse square force law.

## Exercises 4.11 (p. 462)

1. Substitute $x_1 = a_1\xi_1, \ldots, x_n = a_n\xi_n$: $\dfrac{\sqrt{\pi}^n}{\Gamma\left(\dfrac{n+2}{2}\right)} a_1 a_2 \cdots a_n.$

2. $$I = \int \cdots \int \frac{f(x_1) + f(-x_1)}{\sqrt{1 - x_2^2 - \cdots - x_n^2}} dx_2 \cdots dx_n$$

taken throughout the interior of the $(n-1)$-dimensional unit sphere in $x_2 \cdots x_n$ space. Introducing polar coordinates, we obtain

$$I = \int_0^1 dr \int_{S(r)} \frac{f(\sqrt{1-r^2}) + f(-\sqrt{1-r^2})}{\sqrt{1-r^2}} d\sigma,$$

where $S(r)$ denotes the sphere of radius $r$ and center $0$ in $x_2 \cdots x_n$-space. As the integrand depends on $r$ only,

$$I = \omega_{n-1} \int_0^1 \frac{f(\sqrt{1-r^2}) + f(-\sqrt{1-r^2})}{\sqrt{1-r^2}} r^{n-2}\, dr.$$

Putting $y = \sqrt{1-r^2}$, we have

$$I = \omega_{n-1} \int_{-1}^{+1} f(y)\,(1-y^2)^{(n-3)/2}\, dy.$$

3. $a_1a_2 \cdots a_n/n!$

## Exercises 4.12 (p. 474)

1. Put $I_n(a) = \int_0^\infty x^n e^{-ax^2}\, dx$; then $I_n(a) = -I''_{n-2}(a)$, where primes denote differentiation with respect to $a$. Alternatively, integrate by parts.

$$\frac{1}{2}\left(\frac{n-1}{2}\right)! \text{ when } n \text{ is odd, } \sqrt{\pi}\,\frac{1.3.\cdots(n-1)}{2^{(n+2)/2}} \text{ when } n \text{ is even.}$$

2. Integrate by parts. Diverges for $y \leq 0$; for $y > 0$, $F(y) = 0$.

3. Use the relation

$$\frac{1}{z}(f_x \cos \phi + f_y \sin \phi) = f_{xx} \sin^2 \phi - 2f_{xy} \sin \phi \cos \phi + f_{yy} \cos^2 \phi$$

$$+ \frac{1}{z}\frac{d}{d\phi}(f_x \sin \phi - f_y \cos \phi).$$

4. Integrate $u_{xx}$ by parts twice (special precautions necessary in the case where $p < 5/2$).

5. Substitute $\xi = \alpha x + \beta y$, $\eta = \gamma x + \delta y$, where $\alpha$, $\beta$, $\gamma$, $\delta$ are chosen so that

$$\xi^2 + \eta^2 = ax^2 + 2bxy + cy^2.$$

Then $(\alpha\delta - \beta\gamma)^2 = ac - b^2$, and the integral is transformed into

$$\frac{1}{\sqrt{ac - b^2}} \int_{-\infty}^{\infty} \int_{-\infty}^{\infty} e^{-(\xi^2+\eta^2)}\, d\xi\, d\eta.$$

$ac - b^2 = \pi^2$, $a > 0$.

6. Make the same substitution as in Exercise 5 and evaluate the resulting integrals, (a) using the result of Exercise 1, (b) introducing polar coordinates.

(a) $\dfrac{\pi(aC + cA + 2bB)}{(ac - b^2)^{3/2}}$.

(b) $\dfrac{2\pi}{(ac - b^2)^{1/2}}$.

7. Differentiate with respect to $x$ and integrate by parts to obtain

$$J_0' = -\frac{1}{\pi}\int_{-1}^{1} \sin xt\, \frac{t\, dt}{\sqrt{1 - t^2}}$$

$$= -\frac{x}{\pi}\int_{-1}^{1} \sqrt{1 - t^2}\, \cos xt\, dt.$$

Differentiate the first of these expressions with respect to $x$ to obtain

$$J_0'' = -\frac{1}{\pi}\int_{-1}^{1} \frac{t^2}{\sqrt{1 - t^2}}\, \cos xt\, dt.$$

Now combine the integral representations with the cosine factor in the integrand.

8. Compare the answer to Exercise 7.

9. (a) Forming $K'(a)$, where the dash denotes differentiation with respect to $a$, and integrating by parts twice (taking $xe^{-ax^2}$ as one factor), we have $K'(a) = -K(a)/2a + K(a)/4a^2$ ; that is,

$$K(a) = Ca^{-1/2} e^{-1/4a},$$

where $C$ is given by $C = \lim_{a \to \infty} \sqrt{a}\, K(a) = \lim_{a \to \infty} \int_0^\infty e^{-t^2} \cos \frac{t}{\sqrt{a}}\, dt = \frac{1}{2}\sqrt{\pi}.$

$$K(a) = \frac{1}{2}\sqrt{\frac{\pi}{a}}\, e^{-1/4a},$$

(b) Integrate the formula $t/(1+t^2) = \int_0^\infty e^{-tx} \cos x\, dx$ with respect to $t$ from $a$ to $b$.

$$\frac{1}{2} \log \frac{1+a^2}{1+b^2}.$$

(c) Substituting $x = 1/t$ in the expression for $I'(a)$, prove that $I' = -2I$, that is,

$$I = Ce^{-2a},$$

where $C = \lim_{a \to 0} I = \int_0^\infty e^{-x^2}\, dx.$

$$\frac{1}{2}\sqrt{\pi}e^{-2a}.$$

(d) Substitute the integral expression for $J_0$ and change the order of integration. Use the formula $2 \sin ax \cos bxt = \sin (a + bt)x + \sin(a-bt) x$; cf. the expression for $\int_0^\infty \frac{\sin xy}{y}\, dy$ on pp. 463.

$\pi/2$ when $a > b$; arc sin $a/b$ when $a < b$.

10. Set $\sin^2 ax = (1 - \cos 2ax)/2$. Compare Volume I, Section 3.15, p. 322; Exercise 8 and 9b.

11. There exists an $\varepsilon > 0$ such that for every $A$ there is an $A' > A$ such that

$$\left| \int_{A'}^\infty f(x, y)\, dy \right| \geq \varepsilon$$

for some value of $x$.

## Exercises 4.13 (p. 497)

1. (a) $ic\, (e^{-ia\tau} - 1)/\sqrt{2\pi}\, \tau.$

   (b) $1/\sqrt{2\pi}\, (a + i\tau).$

(c) From 4.12, Exercise 8, $J_n(x)/x^n$ is the Fourier transform of the function

$$f(x) = \begin{cases} \dfrac{n!\ 2^n}{\sqrt{2\pi}\ (2n!)}\ (1-t^2)^{n-1}, & |x|<1 \\[2mm] 0, & |x|>1. \end{cases}$$

Consequently, by Fourier's integral theorem $f(-t) = f(t)$ is the Fourier transform of $J_n(x)$.

## Exercises 4.14 (p. 513)

1. From (97b),

$$\Gamma\left(n+\frac{1}{2}\right) = \frac{2n\ (2n-1)\ (2n-2)\cdots 3\cdot 2\cdot 1\ \sqrt{\pi}}{2^n(2n)\ (2n-2)\cdots 2},$$

which immediately yields the desired result.

2. Form (97a),

$$\Gamma\left(n+\frac{1}{2}\right)\Gamma\left(\frac{1}{2}-n\right) = \frac{\pi}{\sin\pi\left(n+\frac{1}{2}\right)} = (-1)^n\pi.$$

Insert the result of (97b) to obtain

$$\Gamma\left(\frac{1}{2}-n\right) = \frac{(-2)^n\ \sqrt{\pi}}{1\cdot 3\cdot 5\cdots(2n-1)}.$$

3. From (98d)

$$\begin{aligned} B(x,\ x) &= 2\int_0^{\pi/2} \frac{(\sin 2t)^{2x-1}}{2^{2x-1}}\ dt \\[2mm] &= \int_0^{\pi} \frac{(\sin s)^{2x-1}}{2^{2x-1}}\ ds \qquad\qquad (s=2t) \\[2mm] &= 2\int_0^{\pi/2} \frac{(\sin s)^{2x-1}}{2^{2x-1}}\ ds \\[2mm] &= 2^{1-2x}\ B\left(x, \frac{1}{2}\right). \end{aligned}$$

4. Set $s = t^x$ in the integral to obtain

$$\begin{aligned} I &= \frac{1}{x}\int_0^1 s^{(1/x)-1}\ (1-s)^{-1/2}\ ds \\[2mm] &= \frac{1}{x}B\left(\frac{1}{x},\frac{1}{2}\right) = \frac{1}{x}\frac{\Gamma(1/x)\ \Gamma(1/2)}{\Gamma(1/x+1/2)}. \end{aligned}$$

5. Set $t = x^2$ in the integral

$$I = \int_0^1 \frac{x^2}{\sqrt{1-x^2}}\ dx$$

to obtain

$$I = \frac{1}{2}\int t^{(a-1)/2}(1-t)^{-1/2}\,dt = \frac{1}{2}B\left(\frac{\alpha+1}{2},\frac{1}{2}\right)$$

$$= 2^{a-1}B\left(\frac{\alpha+1}{2},\frac{\alpha+1}{2}\right),$$

where the result of Exercise 3 is employed at the end.

(a) For $\alpha = 2n+1$, this yields

$$I = 2^{2n}\frac{\Gamma(n+1)\,\Gamma(n+1)}{\Gamma(2n+2)} = \frac{2^{2n}(n!)^2}{(2n+1)!}.$$

(b) For $\alpha = 2n$, with the result of Exercise 1, we obtain

$$I = 2^{2n-1}\frac{\Gamma(n+1/2)\,\Gamma(n+1/2)}{\Gamma(2n+1)}$$

$$= 2^{2n-1}\left[\frac{(2n)!\,\sqrt{\pi}}{n!\,4^n}\right]^2/(2n)!,$$

which immediately yields the desired result.

6. Set $x^m = a^m h\xi/c$, $y^m = b^m h\eta/c$, and $z = h\zeta$ to obtain the volume integral

$$V = \frac{abh}{m^2}\left(\frac{h}{c}\right)^{2/m}\int_0^1\int_0^{1-\xi}\int_{\xi+\eta}^1 \xi^{(1/m)-1}\,\eta^{(1/m)-1}\,d\zeta\,d\eta\,d\xi.$$

Then, on integrating with respect to $\zeta$ and $\eta$,

$$V = \frac{abh}{m}\left(\frac{h}{c}\right)^{2/m}\left[B\left(\frac{1}{m},\frac{1}{m}+1\right) - B\left(\frac{1}{m}+1,\frac{1}{m}+1\right)\right.$$

$$\left. - \frac{1}{m+1}B\left(\frac{1}{m},\frac{1}{m}+2\right)\right]$$

$$= abh\left(\frac{h}{c}\right)^{2/m}B\left(\frac{1}{m}+1,\frac{1}{m}+1\right).$$

7. Set $x^2 = a^2\xi$, $y^2 = b^2\eta$, $z^2 = c^2\zeta$ to reduce the integral to

$$I = \frac{a^p b^q c^r}{8}\iiint f(\xi+\eta+\zeta)\,\xi^{(p/2)-1}\,\eta^{(q/2)-1}\,\zeta^{(r/2)-1}\,d\xi\,d\eta\,d\zeta$$

over the tetrahedron bounded by the coordinate planes and the plane $\xi+\eta+\zeta = 1$. Now replace $\zeta$ by the new variable $t$ with $\zeta = t-\xi-\eta$ to obtain

$$I = \frac{a^p b^q c^r}{8}\int_0^1\int_0^t\int_0^{t-\eta} f(t)\,\xi^{(p/2)-1}\,\eta^{(q/2)-1}\,(t-\xi-\eta)^{(r/2)-1}\,d\xi\,d\eta\,dt$$

$$= \frac{a^p b^q c^r}{8}\int_0^1\int_0^t f(t)\eta^{(q/2)-1}\,(t-\eta)^{(p/2)+(r/2)-1}\int_0^1 u^{(p/2)-1}(1-u)^{(r/2)-1}$$

$$du\,d\eta\,dt$$

where we have put $\xi = (t-\eta)u$. Thus,

$$I = \frac{a^p b^q c^r}{8} B\left(\frac{p}{2}, \frac{r}{2}\right) \int_0^1 \int_0^t f(t) \eta^{(q/2)-1} (t - \eta)^{(p/2)+(r/2)-1} \, d\eta \, dt.$$

Now, setting $\eta = tv$ in this, we obtain

$$I = \frac{a^p b^q c^r}{8} B\left(\frac{p}{2}, \frac{r}{2}\right) B\left(\frac{q}{2}, \frac{p+r}{2} - 1\right) \int_0^1 f(t) t^{(p+q+r)/2-1} \, dt,$$

which immediately gives the desired result. Note the general result implied by the foregoing:

$$J = \iiint f(\xi + \eta + \zeta) \, \xi^{\alpha-1} \eta^{\beta-1} \zeta^{\gamma-1} \, d\xi \, d\eta \, d\zeta$$

$$= \frac{\Gamma(\alpha) \, \Gamma(\beta) \, \Gamma(\gamma)}{\Gamma(\alpha + \beta + \gamma)} \int_0^1 f(t) t^{\alpha+\beta+\gamma-1} \, dt,$$

where the triple integral is taken in the positive octant bounded by the plane $\xi + \eta + \zeta = 1$. Many integrals can be reduced to this form, as seen in the following exercises.

8. Set $x = a\xi^n$, $y = b\eta^n$, $z = c\zeta^n$ to obtain

$$\bar{x} = \frac{a \iiint \xi^{2n-1} \eta^{n-1} \zeta^{n-1} \, d\xi \, d\eta \, d\zeta}{\iiint \xi^{n-1} \eta^{n-1} \zeta^{n-1} \, d\xi \, d\eta \, d\zeta}$$

where the integrals are taken over the positive octant bounded by the plane $\xi + \eta + \zeta \leq 1$ and have the form of the integral $J$ in the solution of Exercise 7. Consequently,

$$\bar{x} = \frac{3a}{4} \frac{\Gamma(2n) \, \Gamma(3n)}{\Gamma(n) \, \Gamma(4n)} .$$

9. Set $x = R\xi^{2/3}$, $y = R\eta^{3/2}$ to obtain

$$I = 4 \iiint x^2 \, dx \, dy = 9R^4 \iint \xi^{7/2} \eta^{1/2} \, d\xi \, d\eta,$$

where the latter double integral is taken over the positive quadrant in the $\xi$, $\eta$-plane bounded by the line $\xi + \eta = 1$. As in Exercise 8, this yields

$$I = 2R^4 B\left(\frac{11}{2}, \frac{3}{2}\right) = \frac{21}{2^9} \pi R^4.$$

10. As in Exercise 7, replace $x_0$ through $x_0 = t - x_1 - \cdots - x_n$. Then,

$$I = \int_0^1 \int_0^{1-x_0} \cdots \int_0^{1-x_0 \cdots -x_{k-1}} \cdots \int_0^{1-x_0 \cdots x_{n-1}} f(x_0 + \cdots + x_n)$$

$$x_0{}^{a_0-1} \cdots x_n{}^{a_n-1} \, dx_n \cdots dx_k \cdots dx_1 \, dx_0$$

$$= \int_0^1 \int_0^t \cdots \int_0^{t-x_1 \cdots x_{k-1}} \int_0^{t-x_1 \cdots x_{n-2}} x_1{}^{a_1-1} \cdots x_{n-1}{}^{a_{n-1}-1} f(t)$$

$$\int_0^{t-x_1 \cdots -x_{n-1}} x_n{}^{a_n-1} (t - x_1 \cdots -x_n)^{a_0-1} \, dx_n \, dx_{n-1} \cdots dx_k$$

$$\cdots dx_1 dt.$$

In the integral with respect to $x_n$, set $x_n = (t - x_1 \cdots x_{n-1})u_n$, which yields

$$\int_0^{t-x_1 \cdots \; x_{n-1}} x_n^{a_n-1}(t - x_1 \cdots -x_n)^{a_0-1} \, dx_n$$

$$= (t - x_1 \cdots -x_{n-1})^{a_0+a_n-1} \int_0^1 u_n^{a_n-1}(1 - u_n)^{a_0-1} \, du_n$$

$$= (t - x_1 \cdots -x_{n-1})^{a_0+a_n-1} B(a_n, a_0).$$

Iterating this procedure with $x_k = (t - x_1 \cdots - x_{k-1})u_k$ for $k = 2, \ldots$ $n$ and $x_1 = tu_1$, we finally obtain

$$I = B(a_n, a_0) \; B(a_{n-1}, a_n + a_0) \cdots B(a_1, a_2 + \cdots + a_n + a_0)$$

$$\int_0^1 f(t)t^{a_0+a_1+\cdots a_{n-1}} \, dt,$$

which immediately yields the desired result.

11. Show that for $G_n(x)$ defined by the expression following the limit sign in the right hand side of formula (86e), p. 506,

$$G_{2n}(2x) = \frac{1}{2} 2^{2x} G_n(x) G_n\left(x + \frac{1}{2}\right) \frac{(2n)! \sqrt{n}}{2^{2n}(n!)^2};$$

then let $n \to \infty$ and apply Wallis's formula (Volume I, p. 282).

12. (a) Set $u = \alpha - p$, $v = \beta - q$. Integrating $D^{-u} f(x)$ repeatedly by parts, we obtain

(i) $\quad D^{-u} f(x) = \dfrac{f(0)x^u}{\Gamma(u+1)} + \cdots + \dfrac{f^{(p-1)}(0)x^{u+p-1}}{\Gamma(u+p)}$

$$+ \frac{1}{\Gamma(u+p)} \int_0^x (x-t)^{u+p-1} f^{(p)}(t) \, dt.$$

Noting that the derivatives at 0 vanish and differentiating $p$ times with respect to $x$, we then find

(ii) $\quad g(x) = D^a f(x) = \dfrac{d^p}{dx^p} [D^{-u} f(x)] = D^{-u} f^{(p)}(x).$

$$= \frac{1}{\Gamma(u)} \int_0^x (x-t)^{u-1} f^{(p)}(t) \, dt.$$

Further integrations by parts yield

$$g(x) = \frac{f^{(p)}(0)x^u}{\Gamma(u+1)} + \cdots + \frac{f^{(p+q-1)}(0)x^{u+q-1}}{\Gamma(u+q)}$$

$$+ \frac{1}{\Gamma(u+1)} \int_0^x (x-t)^{u+q-1} f^{(p+q)}(t) \, dt.$$

Since the derivatives of $f$ at the origin vanish, we then find

$D^{-v}D^a f(x) = D^{-v} g(x)$

$$= \int_0^x \frac{(x-t)^{v-1}}{\Gamma(v)} \int_0^t \frac{(t-s)^{u+q-1} f^{(p+q)}(s)}{\Gamma(u+q)} \, ds \, dt$$

$$= \frac{1}{\Gamma(v)\,\Gamma(u+q)} \int_0^x f^{(p+q)}\,(s) \int_0^x (x-t)^{v-1}\,(t-s)^{u+q-1}\,dt\,ds.$$

We evaluate the inner integral by introducing a new variable of integration, $z = (t-s)\,/\,(x-s)$ to obtain

$$D^{-v}D^a f(x) = \frac{B(u+q,v)}{\Gamma(v)\,\Gamma(u+q)} \int_0^x (x-s)^{u+v+q-1}\,f^{(p+q)}(s)\,ds$$

$$= \frac{1}{\Gamma(u+v+q)} \int_0^x (x-s)^{u+v+q-1}\,f^{(p+q)}(s)\,ds.$$

Now differentiating $q$ times, we find

(iii)    $D^\beta D^a f(x) = D^q D^{-v}\,g(x)$

$$= \frac{1}{\Gamma(u+v)} \int_0^x (x-s)^{u+v-1}\,f^{(p+q)}(s)\,ds.$$

The final result is symmetric in $u$ and $v$ and, hence, independent of the order in which the operators $D^a$ and $D^\beta$ are applied; hence, $D^a D^\beta\,f(x) = D^\beta D^a\,f(x)$.

(b) Let $r$ be the smallest integer greater than $\alpha + \beta$, $w = r - \alpha - \beta$. Then (ii) yields

$$D^{\alpha+\beta}\,f(x) = \frac{1}{\Gamma(w)} \int_0^x (x-t)^{w-1}\,f^{(r)}(t)\,dt.$$

If $u + v \le 1$, then $r = p + q$, $w = u + v$, and this integral is the same as that for $D^\beta D^a f(x)$ obtained in (iii). However, if $1 < u + v \le 2$, then $w = u + v - 1$ and $r = p + q + 1$. Now we only carry the expansion (i) out to the $(r-1)$-th derivative, namely,

$$D^{-w} f(x) = \frac{1}{\Gamma(w+r-1)} \int_0^x (x-t)^{w+r-2}\,f^{(r-1)}(t)\,dt$$

and differentiate $r - 2$ times with respect to $x$ to obtain

$$D^{r-2}D^{-w}\,f(x) = D^{\alpha+\beta-2}\,f(x)$$

$$= \frac{1}{\Gamma(w+1)} \int_0^x (x-t)^w\,f^{(r-1)}(t)\,dt$$

$$= \frac{1}{\Gamma(u+v)} \int_0^x (x-t)^{u+v-1}\,f^{(p+q)}(t)\,dt.$$

Thus, in this case, $D^a D^\beta f(x) \ne D^{\alpha+\beta} f(x)$.

## Exercises 5.2 (p. 555)

1. (a) $-b/2\alpha^2\beta^2$.

   (b) 0.

   (c) 0.

4. Write $d(u,v)/d(x,y) = (uv_y)_x - (uv_x)_y = \text{curl}\,(u\,\text{grad}\,v)$.

**Exercises 5.7** (p. 588)

1. Observe that $\xi = \mathbf{X}_u + \mathbf{X}_v$, $\eta = \mathbf{X}_u - \mathbf{X}_v$.
2. Compare the direction $\mathbf{X}_r$ of the exterior normal with the normal direction represented by $\mathbf{X}_\theta \times \mathbf{X}_\phi$.
3. (a) The line $v = a/2$ divides $S$ into a portion $S'$ given by $a/2 < v < a$ (or, equivalently, by $-a < v < -a/2$) and oriented by $\xi = \mathbf{X}_u$, $\eta = \mathbf{X}_v$, and a portion $S''$ given by $-a/2 < v < a/2$, which is just another Möbius band.
   (b) $S_1$ is representable in the form (40a) with $v$ restricted to the interval $0 < v < a$. Obviously, any two points on $S_1$ can be joined by the curve on $S_1$ that is the image of the line segment joining the corresponding points $(u, v)$ in the parameter plane.
   (c) $S_1$ is oriented by $\xi = \mathbf{X}_u$, $\eta = \mathbf{X}_v$.
4. One easily verifies that $\mathbf{R}(t)$ has length $|\xi|$ and is linearly dependent on $\xi$, $\eta$ and, hence, lies in $\pi$. Moreover, $\mathbf{R}(t) \cdot \xi/|\xi|^2 = \cos t$. The vector $\mathbf{R}(t)$ coincides with $\xi$ for $t = 0$ and has the direction of $\eta$ for a certain $t$ between 0 and 180°, namely, for that $t$ determined by the relations

$$\cos t = b/\sqrt{ac}, \quad \sin t = \sqrt{1 - b^2/ac}.$$

**Exercises 5.9a** (p. 602)

1.
$$\iint \frac{z}{p}\, dS = \left(\frac{1}{a^2} + \frac{1}{b^2} + \frac{2}{c^2}\right) \iiint z\, dx\, dy\, dz,$$

where the volume integral is to be extended throughout the upper half of the ellipsoid. (The base of this half-ellipsoid contributes nothing to the surface integral): $\dfrac{\pi}{4}\left(\dfrac{1}{a^2} + \dfrac{1}{b^2} + \dfrac{2}{c^2}\right)abc^2$.

2. Since $H$ is a homogeneous function of the fourth degree, we have

$$4\iint H\, dS = \iint (xH_x + yH_y + zH_z)\, dS$$

$$= \iint \frac{\partial H}{\partial n}\, dS = \iiint \Delta H\, dx\, dy\, dz$$

$$= 6 \iiint [x^2(2a_1 + a_4 + a_6) + y^2(2a_2 + a_4 + a_5)$$

$$+ z^2(2a_3 + a_5 + a_6)]\, dx\, dy\, dz.$$

$$\frac{4\pi}{5}(a_1 + a_2 + a_3 + a_4 + a_5 + a_6).$$

**Exercises 5.9e** (p. 610)

1. (a) Compare Exercise 8, Section 2.4, p. 203.
   (c) Let $R$ be an arbitrary region and $v$ an arbitrary function vanishing on the boundary of $R$. Then, by Green's first formula,

$$\iiint_R (u_{x_1} v_{x_1} + u_{x_2} v_{x_2} + u_{x_3} v_{x_3}) \; dx_1 \; dx_2 \; dx_3$$

$$= - \iiint_R v \; \Delta u \; dx_1 \; dx_2 \; dx_3$$

$$= - \iiint_R v \; \Delta u \; \sqrt{e_1 e_2 e_3} \; dp_1 \; dp_2 \; dp_3.$$

Now

$$u_{x_i} = u_{p_1} \frac{\partial p_1}{\partial x_i} + u_{p_2} \frac{\partial p_2}{\partial x_i} + u_{p_3} \frac{\partial p_3}{\partial x_i}$$

$$= u_{p_1} \frac{a_{i1}}{e_1} + u_{p_2} \frac{a_{i2}}{e_2} + u_{p_3} \frac{a_{i3}}{e_3}$$

and

$$v_{x_i} = v_{p_1} \frac{a_{i1}}{e_1} + v_{p_2} \frac{a_{i2}}{e_2} + v_{p_3} \frac{a_{i3}}{e_3} \,.$$

Hence,

$$\iiint_R (u_{x_1} v_{x_1} + u_{x_2} v_{x_2} + u_{x_3} v_{x_3}) \; dx_1 \; dx_2 \; dx_3$$

$$= \iiint \left( \frac{1}{e_1} u_{p_1} v_{p_1} + \frac{1}{e_2} u_{p_2} v_{p_2} + \frac{1}{e_3} u_{p_3} v_{p_3} \right) dx_1 \; dx_2 \; dx_3$$

$$= \iiint \left( \sqrt{\frac{e_2 e_3}{e_1}} \, u_{p_1} v_{p_1} + \sqrt{\frac{e_3 e_1}{e_2}} \, u_{p_2} v_{p_2} + \sqrt{\frac{e_1 e_2}{e_3}} \, u_{p_3} v_{p_3} \right) dp_1 \; dp_2 \; dp_3$$

$$= \iiint (U_1 v_{p_1} + U_2 v_{p_2} + U_3 v_{p_3}) \; dp_1 \; dp_2 \; dp_3,$$

where we write $U_i = \dfrac{\sqrt{e_1 e_2 e_3}}{e_i} u_{p_i}$.

Applying Gauss's theorem to the vector $(U_1 v, U_2 v, U_3 v)$, we obtain

$$- \iiint \left( \frac{\partial U_1}{\partial p_1} + \frac{\partial U_2}{\partial p_2} + \frac{\partial U_3}{\partial p_3} \right) v \; dp_1 \; dp_2 \; dp_3.$$

Thus, for an arbitrary $v$ vanishing on the boundary of $R$ we have

$$\iiint v \; \Delta u \; \sqrt{e_1 e_2 e_3} \; dp_1 \; dp_2 \; dp_3$$

$$= \iiint v \left( \frac{\partial U_1}{\partial p_1} + \frac{\partial U_2}{\partial p_2} + \frac{\partial U_3}{\partial p_3} \right) dp_1 \; dp_2 \; dp_3$$

and, hence (cf. Lemma I, p. 744),

$$\Delta u = \left( \frac{\partial U_1}{\partial p_1} + \frac{\partial U_2}{\partial p_2} + \frac{\partial U_3}{\partial p_3} \right) \frac{1}{\sqrt{e_1 e_2 e_3}}$$

$$= \frac{1}{\sqrt{e_1 e_2 e_3}} \left[ \frac{\partial}{\partial p_1} \left( \sqrt{\frac{e_2 e_3}{e_1}} \frac{\partial u}{\partial p_1} \right) + \frac{\partial}{\partial p_2} \left( \sqrt{\frac{e_3 e_1}{e_2}} \frac{\partial u}{\partial p_2} \right) + \frac{\partial}{\partial p_3} \left( \sqrt{\frac{e_1 e_2}{e_3}} \frac{\partial u}{\partial p_3} \right) \right].$$

(d) Use Exercise 9c, Section 3. 3d, p. 257:

$$\frac{1}{4}(t_2 - t_1)(t_3 - t_1)(t_3 - t_2)\,\Delta u = (t_3 - t_2)\sqrt{\phi(t_1)}\,\frac{\partial}{\partial t_1}\left(\sqrt{\phi(t_1)}\,\frac{\partial u}{\partial t_1}\right)$$

$$+ (t_3 - t_1)\sqrt{-\phi(t_2)}\,\frac{\partial}{\partial t_2}\left(\sqrt{-\phi(t_2)}\,\frac{\partial u}{\partial t_2}\right)$$

$$+ (t_2 - t_1)\sqrt{\phi(t_3)}\,\frac{\partial}{\partial t_3}\left(\sqrt{\phi(t_3)}\,\frac{\partial u}{\partial t_3}\right),$$

where $\phi(x) = (a - x)(b - x)(c - x)$.

## Exercises 5.10a (p. 615)

1. (a) $I = -\iint_{y^2+z^2<1/4}(zx_z + x)\,dy\,dz$, where $x = \sqrt{1 - y^2 - z^2}$.

   (b) $I = \int_{\partial S^*} L = -x\int_{\partial S^*} y\,dz = -\frac{1}{2}\int_0^{2\pi}\frac{3}{4}\cos^2\theta\,d\theta = -\frac{3}{8}\pi$.

## Exercises 5.10b (p. 617)

2. If $(\xi, \eta)$ and $(x, y)$ are rectangular coordinates in $\Pi$ and $P$, respectively, then the motion of the point $M(x, y)$ can be described by the equations

   $$\xi = x\cos\phi - y\sin\phi + a, \qquad \eta = x\sin\phi + y\cos\phi + b$$

   (i.e., by a rotation and a translation). Then

   $$S(M) = A(x^2 + y^2) + Bx + Cy + D.$$

   ($\alpha$)  If $A = n\pi \neq 0$, we have $S(M) = n\pi[(x - x_0)^2 + (y - y_0)^2] + S(C)$, where $C$ is the point $x = x_0 = -B/2n\pi$, $y = y_0 = -C/2n\pi$, hence $A, B, C, D$ have the values in Exercise 1.

   ($\beta_1$) If $A = n\pi = 0$ but $B^2 + C^2 > 0$, then

   $$S_M = \sqrt{B^2 + C^2}\,\frac{Bx + Cy + D}{\sqrt{B^2 + C^2}} = \lambda\,d(M),$$

   where $\lambda = \sqrt{B^2 + C^2}$ and $\Delta$ is the line $Bx + Cy + D = 0$.

   ($\beta_2$) If $A = B = C = 0$, we have $S(M) = D = $ constant.

3. For the motion of the plane $P$ rigidly attached to the connecting-rod $AB$, we have $n = 0$, $S(A) = 0$, $S(B) = \pi\overline{CB}^2 = \pi\gamma^2$. Hence, $\Delta$ passes through $A$, and by symmetry, $\Delta$ is perpendicular to $AB$ at $A$. Hence, $S(M) = \pi\gamma^2 l^{-1}\,d(M)$, where $l = \overline{AB}$.

4. For the motion of the plane $P$ rigidly attached to the chord $AB$, we have $n = 1$, $S(A) = S(B) = S = $ area of $\Gamma$. The point $C$ of Steiner's theorem is therefore equidistant from $A$ and $B$ and $S(A) = \pi\overline{CA}^2 + S(C)$, $S(M) = \pi\overline{CM}^2 + S(C)$; hence, $S(A) - S(M) = $ area of $\Gamma - $ area of $\Gamma'$ $= \pi(\overline{CA}^2 - \overline{CM}^2) = \pi ab$.

5. If $l$ is the length of $\Gamma$, the Frenet formulae (Exercise 16, Section 2.5, p. 216) give

$$\int \frac{\mathbf{n}}{\rho} \, ds = \int \frac{\xi_2}{\rho} \, ds = \int \xi_1 \, ds = \int \frac{d^2\mathbf{x}}{ds^2} \, ds = 0;$$

$$\int \frac{\mathbf{x} \times \mathbf{n}}{\rho} \, ds = \int \mathbf{x} \times \xi_1 \, ds = \mathbf{x} \times \xi_1 \Big|_0^l - \int \mathbf{x} \times \xi_1 \, ds$$

$$= - \int \xi_1 \times \xi_1 \, ds = 0$$

6. Let $\mathbf{n}' = (\alpha, \beta, \gamma)$, $\mathbf{x} = (x, y, z)$. If in Gauss's formula

$$\iint (a\alpha + b\beta + c\gamma) \, d\sigma = - \iiint \left( \frac{\partial a}{\partial x} + \frac{\partial b}{\partial y} + \frac{\partial c}{\partial z} \right) dx \, dy \, dz,$$

we substitute $a = 1$, $b = c = 0$, and $a = 0$, $b = -z$, $c = y$, we get

$$\iint \alpha \, d\sigma = 0 \quad \text{and} \quad \iint (y\gamma - z\beta) \, d\sigma = 0,$$

respectively.

7. Take rectangular coordinates $(x, y, z)$ such that $z = 0$ is the free horizontal surface of the fluid and $Oz$ points downward. The pressure on $d\sigma$ is $\mathbf{n}z \, d\sigma$, where $z$ is the depth of $d\sigma$. By repeated applications of Gauss's formula in three dimensions, with obvious choices of the functions $a$, $b$, $c$ we find for the components of the resultant of the fluid pressure

$$\iint \alpha z \, d\sigma = 0, \ \iint \beta z \, d\sigma = 0, \ \iint \gamma z \, d\sigma = -\iint dx \, dy \, dz = -V.$$

For the components of the resultant moment with respect to the origin 0 we find, again by Gauss's formula,

$$\iint (yz\gamma - z^2\beta) d\sigma = \iiint y \, dx \, dy \, dz = V y_0,$$

$$\iint (z^2\alpha - xz\gamma) d\sigma = -\iiint x \, dx \, dy \, dz = -V x_0,$$

$$\iint (xz\beta - yz\alpha) d\sigma = 0,$$

$(x_0, y_0, z_0$ are the coordinates of the centroid $C$). Now we note that the components of the force $\mathbf{f}$ are $0, 0, -V$, and the components of its moment with respect to 0 are $V y_0, -V x_0, 0$.

8. From the parametric equations

$$x = a \cos u \cos v, \ y = b \sin u \cos v, \ z = c \sin v$$

$$\left( 0 \leq u < 2\pi, \ -\frac{\pi}{2} \leq v < \frac{\pi}{2} \right)$$

of the ellipsoid we readily obtain the formulae

$$p \, dS = abc \cos v \, du \, dv, \ \frac{dS}{p} = \frac{D^2 \, du \, dv}{abc \cos v},$$

where

$$D^2 = b^2c^2 \, \cos{}^2u \, \cos^2v + a^2c^2 \, \sin^2u \, \cos^2v + a^2b^2 \, \sin{}^2v \, \cos^2v.$$

10. The integral represents the flat solid angle which the plane $z = 0$ subtends at the point $M = (0, 0, 1)$. For a direct analytical proof, use plane polar coordinates.

12. Verify the identity

$$\frac{\partial}{\partial x}\left(\frac{a-x}{\gamma^3}\right) + \frac{\partial}{\partial y}\left(\frac{b-y}{\gamma^3}\right) + \frac{\partial}{\partial z}\left(\frac{c-z}{\gamma^3}\right) = 0,$$

$$\gamma^2 = (x-a)^2 + (y-b)^2 + (z-c)^2,$$

for all points $(x, y, z)$ different from $(a, b, c)$. From Gauss's formula in three dimensions we conclude (i) that $\Omega = 0$ if $\Sigma$ is a closed surface such that $A = (a, b, c)$ is outside the volume bounded by $\Sigma$; (ii) that if $A$ is within $\Sigma$, the value of the integral is independent of the shape of $\Sigma$. Taking for $\Sigma$ a sphere with center $A$, we easily see that $\Omega = 4\pi$.

13. The integral, writing $\gamma$ for $r$,

$$\frac{\partial\Omega}{\partial a} = \iint_{\Sigma}\frac{\partial}{\partial a}\left(\frac{a-x}{\gamma^3}\right)dy\,dz + \frac{\partial}{\partial a}\left(\frac{b-x}{\gamma^3}\right)dz\,dx + \frac{\partial}{\partial a}\left(\frac{c-z}{\gamma^3}\right)dx\,dy$$

is independent of $\Sigma$ and depends only on the boundary $\Gamma$ of $\Sigma$, for the identity given in the answer to Exercise 12 implies that

$$\frac{\partial}{\partial x}\left[\frac{\partial}{\partial a}\left(\frac{a-x}{\gamma^3}\right)\right] + \frac{\partial}{\partial y}\left[\frac{\partial}{\partial a}\left(\frac{b-y}{\gamma^3}\right)\right] + \frac{\partial}{\partial z}\left[\frac{\partial}{\partial a}\left(\frac{c-z}{\gamma^3}\right)\right] = 0.$$

By Stokes's theorem (p. 611) and the discussion of Chapter 5, pp. 613–614, the surface integral expression for $\partial\Omega/\partial a$ may be expressed as a line integral $\int u\,dx + v\,dy + w\,dz$ along $\Gamma$. Verify that the functions

$$u = 0, \quad v = \frac{z-c}{\gamma^3}, \quad w = -\frac{y-b}{\gamma^3}$$

satisfy the identities

$$\frac{\partial w}{\partial y} - \frac{\partial v}{\partial z} = \frac{\partial}{\partial a}\left(\frac{a-x}{\gamma^3}\right), \quad \frac{\partial u}{\partial z} - \frac{\partial w}{\partial x} = \frac{\partial}{\partial a}\left(\frac{b-y}{\gamma^3}\right), \quad \frac{\partial v}{\partial x} - \frac{\partial u}{\partial y} = \frac{\partial}{\partial a}\left(\frac{c-z}{\gamma^3}\right).$$

14. Note the following facts: (1) the value of the line integral $\theta$ remains unchanged if $\Gamma$ is deformed in such a way that $\Gamma$ never sweeps over any of the points $(-1, 0)$ or $(1, 0)$ during its deformation; (2) $\theta = 2\pi$ if $\Gamma$ is a small circle around $(1, 0)$ oriented counterclockwise; (3) $\theta = 2\pi$ if $\Gamma$ is a small circle around $(-1, 0)$ oriented clockwise.

15. Think of $C$ as being a rigid circle made of wire and of $\Gamma$ as being a string. Now deform the string $\Gamma$ to a new position $\Gamma'$ lying entirely within the plane $y = 0$. The numbers $p$ and $n$ are not changed during this deformation, and the first formula now follows directly if Exercise 14 is applied to the curve $\Gamma'$ within the plane $y = 0$ and the line segment $-1 < x < 1$, $y = 0$, $z = 0$ of this plane. The factor $4\pi$ (instead of $2\pi$, as in the previous example) results from the solid angle $\Omega$ increasing by $4\pi$ along a closed path for which $p = 1$, $n = 0$. One way of carrying out the above deformation of $\Gamma$ into $\Gamma'$ analytically is as follows. Assume that $\Gamma$ does not meet the $z$-axis and let

$$x = \gamma(t) \cos \phi(t), \qquad y = \gamma(t) \sin \phi(t), \qquad z = z(t) \qquad (0 \leqq t \leqq 2\pi)$$

be the parametric equations of $\Gamma$. Consider now the family of curves

$$\Gamma(\tau): x = \gamma(t) \cos [\tau\phi(t)], \qquad y = \gamma(t) \sin [\tau\phi(t)], \qquad z = z(t),$$

depending on the parameter $\tau$, which decreases from $\tau = 1$ to $\tau = 0$. Note that $\Gamma(1) = \Gamma$ and that $\Gamma' = \Gamma(0)$ is a closed curve that lies in the plane $y = 0$. Note also that (for a fixed value of $z$) each point $P$ of $\Gamma(\tau)$ rotates about the $z$-axis as $\tau$ varies; hence, the solid angle $\Omega$ that $C$ subtends at $P$ does not vary with $\tau$. This implies that $\Omega_1 - \Omega_0$ will have the same value for $\Gamma(0)$ as for $\Gamma(1) = \Gamma$. To prove the second formula, note that

$$\Omega_1 - \Omega_0 = \int_\Gamma d\Omega = \int_\Gamma \operatorname{grad} \Omega \cdot dP = -\int_\Gamma dP \cdot \int_C \frac{\overline{PP'} \times dP'}{|\overline{PP'}|^3}$$

$$= -\int_\Gamma \int_C \frac{dP \cdot (\overline{PP'} \times dP')}{|\overline{PP'}|^3} = \int_\Gamma \int_C \frac{\overline{PP'} \cdot (dP \times dP')}{|\overline{PP'}|^3}.$$

16. Take a coordinate system $Ox_1, Ox_2, Ox_3$, and denote the position vector of a variable point on $\Gamma$ by $\mathbf{x}$. Then

$$\mathbf{a} = \frac{1}{2} \int_\Gamma \mathbf{x} \times d\mathbf{x}$$

has the required properties, for

$$\mathbf{a} \cdot \mathbf{x}_3 = \frac{1}{2} \int_\Gamma (x_1 \, dx_2 - x_2 \, dx_1)$$

is the area of the projection of $\Gamma$ on the plane $Ox_1 x_2$.

17. The two equations $u = f_x$, $v = f_y$ can be solved for $x$ and $y$, since $\partial(u, v)/\partial(x, y) \neq 0$. Let $x = \sigma(u, v)$, $y = \tau(u, v)$; since $u_y = v_x$, we have (cf. p. 261) $x_v = y_u$, $\sigma_v = \tau_u$. Hence, a function $g$ exists such that $x = g_u(u, v)$, $y = g_v(u, v)$.

18. $u = \dfrac{yz}{(x^2 + y^2)\sqrt{x^2 + y^2 + z^2}}$,

$v = \dfrac{-xz}{(x^2 + y^2)\sqrt{x^2 + y^2 + z^2}}$, $\qquad w = 0$.

## Exercises 6.1e (p. 671)

1. With $\theta = 0$, equation (17c) takes the form

   (i) $$\dot{r}^2 = c + \frac{b}{r},$$

   where $c = 2C/m$ and $b = 2\gamma\mu$. Writing this in the form

   $$\sqrt{\frac{r}{cr + b}} \frac{dr}{dt} = 1$$

   and integrating, we obtain if $c \neq 0$,

(iia)
$$t = k + \frac{\sqrt{cr^2 + br}}{c} - \frac{b}{2c} f(r),$$

where

(iib)
$$f(r) = \begin{cases} \dfrac{1}{\sqrt{c}} \text{ ar sinh } (1 + 2cr/b) & \text{for} \quad c > 0 \\[2ex] \dfrac{-1}{\sqrt{-c}} \text{ arc sin } (1-2cr/b) & \text{for} \quad c < 0, \end{cases}$$

and if $c = 0$,

(iic)
$$r = \left( \frac{3\sqrt{b}}{2} t + k \right)^{2/3}.$$

Returning to the differential equation (i), we determine the integration constant $c$ by

$$c = \dot{r}_0{}^2 - \frac{b}{r_0}.$$

If $c < 0$, we see that $r$ is bounded, $r \leqq -b/c$. If $\dot{r}_0 > 0$, $r$ increases to this value and then decreases as the orbiting body falls toward the sun. If $\dot{r}_0 < 0$, the body moves directly toward the sun until collision.

If $c = 0$, we observe that the constant of integration $k$ in (iic) is $k = \pm r_0{}^{3/2} = b^{3/2}/\dot{r}_0{}^3$, where the plus or minus sign is taken according to whether $\dot{r}_0$ is positive or negative. If $\dot{r}_0$ is negative, we again get a solution in which the body accelerates into the sun. If $\dot{r}_0$ is positive, the body escapes to infinity but with limiting velocity zero.

If $k > 0$ and $\dot{r}_0 < 0$, the body accelerates into collision with the sun as before. But if $\dot{r}_0 > 0$, the body escapes and it can be seen from (i) and (iii) that it has a positive limiting velocity, namely,

$$\dot{r}_\infty = c = \dot{r}_0{}^2 - \frac{b}{r_0}.$$

2. For both the parabola and the hyperbola, the orbit is nonperiodic and $\theta$ is bounded. Consequently, from $\int_{\theta_0}^{\theta} r^2 \, d\theta = h(t - t_0)$, for $t$ to approach $\infty$, $r$ also must approach $\infty$. From (17d) we conclude that $\dot{\theta} = 0$ as $t \to \infty$; hence in (17c), from

$$\lim_{t \to \infty} r^2 \dot{\theta}^2 = \left( \lim_{t \to \infty} r^2 \dot{\theta} \right) \left( \lim_{t \to \infty} \dot{\theta} \right) = h \lim_{t \to \infty} \dot{\theta} = 0,$$

we conclude that $\lim\limits_{t \to \infty} \dot{r}^2 = 2C/m$. However, from the definition of $\varepsilon$, for the parabola ($\varepsilon = 1$) $C$ has the value 0 and for the hyperbola ($\varepsilon > 1$), a positive value.

3. The force is $-m/2 \text{ grad } \dot{r}^2$. Hence, by conservation of energy,

$$\frac{1}{2} m(\dot{r}^2 + r^2 \dot{\theta}^2) + \frac{1}{2} mr^2 = C$$

and the moment equations, as for any centrally directed force, yield

$$r^2\dot\theta = h.$$

We eliminate $t$ from these equations, as we did from the equations (17c) and (17d) for planetary motion, to obtain

$$\frac{dr}{d\theta} = \frac{r}{h}\sqrt{\frac{2Cr^2}{m} - h^2 - r^4}.$$

This is easily integrated to give

$$r^2 = \frac{a}{b + \sin 2\theta},$$

where $a = 2h^2$ and $b = \sqrt{1 - h^2m^2/C^2}$. In Cartesian coordinates this becomes

$$b(x^2 + y^2) + 2xy = a,$$

which is the equation of a conic section.

4. The force is $-\operatorname{grad} U$, where $U = -\int f(r)\, dr$. As for planetary motion we may apply conservation of energy and the moment equation (17d), namely,

$$\frac{1}{2}m(\dot r^2 + r^2\dot\theta^2) - \int f(r)\, dr = C$$

$$r^2\dot\theta = h.$$

We may now proceed in the same way to the desired result.

5. Apply the result of Exercise 4.

6. If $(\xi, \eta)$ are the coordinates with respect to the axes of the ellipse, then

$$\xi = a \cos \omega = x + \varepsilon a$$

$$\eta = b \sin \omega = y$$

give the equation of the ellipse and by the law of areas

$$h(t - t_s) = \int_0^\omega \left( x\frac{\partial y}{\partial \omega} - y\frac{\partial x}{\partial \omega} \right) d\omega$$

$$= ab \int_0^\omega (1 - \varepsilon \cos \omega)\, d\omega.$$

7. The motion takes place in a plane, since $p$ is a central force (proved for the case $p = 1/r^2$ on pp. 666). Hence,

$$\ddot x = -\frac{x}{r}p,$$

$$\ddot y = -\frac{y}{r}p.$$

It follows that

$$x\dot y - \dot x y = \text{constant} = h,$$

$$\ddot{x}\dot{x} + \ddot{y}\dot{y} = \frac{-x\dot{x} - y\dot{y}}{r} p = -\dot{r}p.$$

Hence,

$$\frac{1}{2}\frac{d}{dt}(\dot{x}^2 + \dot{y}^2) = -\dot{r}p.$$

The distance of the tangent from the origin is

$$q = \frac{|x\dot{y} - \dot{x}y|}{\sqrt{\dot{x}^2 + \dot{y}^2}} = \frac{h}{\sqrt{\dot{x}^2 + \dot{y}^2}};$$

therefore,

$$\frac{1}{2}\frac{d}{dt}\frac{h^2}{q^2} = -p\frac{dr}{dt}$$

or

$$\frac{1}{2}\frac{d}{dr}\frac{h^2}{q^2} = -p,$$

which proves the first statement. For the cardioid we have $q = r^2/\sqrt{2ar}$.

8. By definition

(A)
$$\ddot{x} = -\lambda^2 x - 2\mu\dot{y}$$
$$\ddot{y} = -\lambda^2 y + 2\mu\dot{x}.$$

On differentiating the two equations twice and combining them, we get an equation involving $x$ only,

$$\overset{....}{x} + (2\lambda^2 + 4\mu^2)\ddot{x} + \lambda^4 x = 0$$

and a corresponding equation involving $y$ only,

$$\overset{....}{y} + (2\lambda^2 + 4\mu^2)\ddot{y} + \lambda^4 y = 0.$$

Thus, $x$ and $y$ are linear combinations of $\exp[\pm i(\mu \pm \sqrt{\lambda^2 + \mu^2})t]$ (cf. Exercise 2, p. 696) or of $\cos(\mu + \sqrt{\lambda^2 + \mu^2})t$, $\cos(\mu - \sqrt{\lambda^2 + \mu^2})t$, $\sin(\mu + \sqrt{\lambda^2 + \mu^2})t$, $\sin(\mu - \sqrt{\lambda^2 + \mu^2})t$, with constant coefficients $a, b, c, d$, and $a', b', c', d'$. From (A) it follows that $a' = -c$, $b' = -d$, $c' = a$, $d' = b$. Using the initial conditions $x(0) = y(0) = \dot{y}(0) = 0$, $\dot{x}(0) = u$, we obtain the result given.

9. Let $(x_1, y_1), \ldots, (x_n, y_n)$ be the attracting particles. Then the resultant force at a point $(x, y)$ has the components

$$X = \sum_\nu \frac{x - x_\nu}{\sqrt{(x - x_\nu)^2 + (y - y_\nu)^2}}, \quad Y = \sum_\nu \frac{y - y_\nu}{\sqrt{(x - x_\nu)^2 + (y - y_\nu)^2}}.$$

If we introduce the complex quantities $z_1 = x_1 + iy_1, \ldots, z_n = x_n + iy_n$, $z = x + iy$, $Z = X + iY$, we have

$$Z = \sum_\nu \frac{1}{z - \bar{z}_\nu} = \frac{\overline{f'(z)}}{\overline{f(z)}},$$

where $f(z)$ denotes the polynomial $(z - z_1) \cdots (z - z_n)$ and $\bar{z}$ the complex quantity conjugate to $z$. The positions of equilibrium correspond to $Z = 0$, that is, to the zeros of the polynomial $f'(z)$ of which there are $n - 1$ at most.

Positions of equilibrium in the particular case: $(0, 0)$, $(\sqrt{a^2 - b^2}, 0)$, $(-\sqrt{a^2 - b^2}, 0)$.

## Exercises 6.2 (p. 682)

1. (a) $y = \tan \log (c/\sqrt{1 + x^2})$.

   (b) $y = c\sqrt{1 + e^{2x}}$.

2. (a) $y = ce^{y/x}$.

   (b) $y^2(2x^2 + y^2) = c^2$.

   (c) $x^2 - 2cx + y^2 = 0$ (circles).

   (d) arc $\tan (y/x) + c = \log \sqrt{x^2 + y^2}$ or, in polar coordinates $r = e^{\phi+c}$ (logarithmic spirals).

   (e) $c + \log |x| = $ arc $\sin(y/x) - \dfrac{1}{x}\sqrt{x^2 - y^2}$.

3. If $ab_1 - a_1b \neq 0$, we have

$$\frac{d\eta}{d\xi} = \frac{a + by'}{a_1 + b_1y'} = \frac{a + b\phi(\eta/\xi)}{a_1 + b_1\phi(\eta/\xi)},$$

which is a homogeneous equation.

$$\text{If } ab_1 - a_1b = 0 \quad \text{or} \quad a_1/a = b_1/b = k, \quad \text{then}$$

$$\frac{d\eta}{dx} = a + b\frac{dy}{dx} = a + b\phi\left(\frac{\eta + c}{k\eta + c_1}\right),$$

and the variables are separated.

4. (a) $4x + 8y + 5 = ce^{4x-8y}$.

   (b) $x = c - \dfrac{1}{4}(3y - 7x) - \dfrac{3}{4}\log (3y - 7x)$.

5. (a) $y = ce^{-\sin x} + \sin x - 1$.

   (b) $y = (x + 1)^n(e^x + c)$.

   (c) $y = cx(x - 1) + x$.

   (d) $y = \dfrac{1}{3}x^5 + cx^2$.

   (e) $y = \dfrac{c}{\sqrt{1 + x^2}} - \dfrac{1}{(1 + x^2)(x + \sqrt{1 + x^2})}$.

6. Introduce $1/y$ as a new unknown function; the equation then becomes homogeneous:

$$\frac{1}{x} \frac{1 - cx^{\sqrt{5}}}{cx^{\sqrt{5}} \left(\frac{1}{2} - \frac{1}{2}\sqrt{5}\right) - \frac{1}{2} - \frac{1}{2}\sqrt{5}}.$$

7. With this substitution, the equation becomes

$$v' = v^n g(x) F(x)^{n-1}.$$

8. See Exercise 7. Eliminate $y$ through $v = xy$, $y' = v'/x - v/x^2$ to obtain a separable equation;

$$y = \frac{1}{x(c - \log x)}.$$

9. Following the idea of the substitution in Exercise 7, seek a function $f(x)$ such that $v = yf(x)$ and $v' = (y' + y \sin x) f(x)$. From $f' = y'f(x) + yf'(x)$, we have

$$f'(x) = f(x) \sin x;$$

whence,

$$f(x) = ae^{-\cos x}.$$

The constant $a$ is irrelevant for our purpose, and we set $a = 1$. We then obtain the separable equation

$$v' = -e^{(n-1)\cos x} \sin 2x,$$

which is easily integrated by separation of variables. The final result is

$$y = \begin{cases} \sqrt[n-1]{2\left[\dfrac{1}{(n-1)} - \cos x\right] + ke^{-(n-1)\cos x}} & (n \neq 1) \\ ke^{\cos x + (\cos 2x)/2} & (n = 1). \end{cases}$$

### Exercises 6.3b (p. 690)

1. If any linear combination of these were to vanish, say

$$c_1 \sin n_1 x + c_2 \sin n_2 x + \cdots + c_k \sin n_k x = 0,$$

then, on multiplication by $\sin n_j(x)$, where $j = 1, \ldots, k$, and integration over $[0, \pi]$, we would obtain

$$c_j \int_0^\pi \sin^2 n_j x \, dx = 0;$$

whence $c_j = 0$ for all $j$.

2. Use induction. Suppose that a linear relation $c_1\phi_1 + \cdots + c_k\phi_k = 0$ holds. Divide by $e^{a_k x}$ and differentiate $(n_k + 1)$ times if $P_k(x)$ is of degree $n_k$. The degree of the coefficients of the other $e^{a_i x}$ is unchanged, so that they remain different from zero.

3. Multiply both sides of the equation by $(1 - n)y^{-n}$.

(a) $y^{-1} = cx + \log x + 1.$

(b) $y^3 = cx^{-3} + \dfrac{3a^2}{2x}$.

(c) $(y^{-1} + a)^2 = c(x^2 - 1)$.

4. If we put $y = y_1 + u^{-1}$, the equation reduces to the linear equation $u' - (2Py_1 + Q)\,u = P$.

$$y = x - \frac{\exp\,[(1/2)x^4]}{c + \int_0^x x^2\,\exp\,[(1/2)x^4]\,dx}$$

5. Equate the right sides of the two equations to obtain $y = x^2$ and verify directly that this is an integral of both equations.

6. Note that this is equation (a) of Exercise 5 and is therefore a Riccati equation with one solution known. Then apply the result of Exercise 4.

$$y = x^2 - \frac{\exp\,[(2/3)x^3]}{c + \int_{-\infty}^x \exp\,[(2/3)x^3]\,dx} \qquad\qquad [= f(x, c)].$$

To draw the graphs of the corresponding family of curves, first plot the two branches of the curve

$$y^2 + 2x - x^4 = 0 \qquad , \qquad y = \pm\sqrt{(x^3 - 2)x},$$

which divides the plane into two regions where $y' < 0$ and one region where $y' > 0$. The two infinite branches of this curve are asymptotic to the two parabolas $y = \pm x^2$. Show that all the integral curves are asymptotic to these parabolas by proving the two relations

$$f(x, c) = -x^2 + o(1) \qquad \text{as } x \to +\infty \qquad (-\infty < c < \infty)$$

and

$$f(x, c) = x^2 + o(1) \qquad \text{as } x \to -\infty \qquad\qquad (c \neq 0),$$

where $o\,(1)$ denotes a function that tends to zero.

7. Put

$$y_1 - y_3 = a, \qquad y_1 - y_4 = b, \qquad y_2 - y_3 = c, \qquad y_2 - y_4 = d.$$

Then

$$a' + Pa(y_1 + y_3) + Qa = 0,$$

so that

$$P(y_1 + y_3) = -Q - \frac{a'}{a},$$

$$P(y_1 - y_3) = aP$$

or

$$2Py_1 = aP - Q - \frac{a'}{a}.$$

Similarly,

$$2Py_1 = bP - Q - \frac{b'}{b}.$$

Hence,

$$\frac{d \log (a/b)}{dx} = P(a - b) = -P(y_3 - y_4),$$

and similarly,

$$\frac{d \log (c/d)}{dx} = -P(y_3 - y_4);$$

by subtraction,

$$\log \frac{a/b}{c/d} = \text{constant}.$$

8. Compare the relation

$$\frac{d \log (a/b)}{dx} = P(y_4 - y_3),$$

in the proof of the preceding example.

Particular solutions of the special equation are $y_1 = 1/\cos x$ and $y_2 = -1/\cos x$;

$$y = \frac{1 + ce^{2x}}{(1 - ce^{2x})\cos x}.$$

9. The common solution $e^x$ of (a) and (b) is obtained by eliminating $y''$ from the two equations.

(a) $c_1 e^x + c_2 x.$

(b) $c_1 e^x + c_2 \sqrt{x}.$

10. The curve satisfies the differential equation

$$n \left( x \frac{dx}{dy} - y \right) = r$$

or in polar coordinates, $r. \theta$, with $\theta$ as independent variable,

$$\frac{nr^2}{\cos \theta \dfrac{dr}{d\theta} - r \sin \theta} = r;$$

that is,

$$\frac{d \log r}{d\theta} = \frac{n}{\cos \theta} + \tan \theta,$$

whence,

$$r = a \frac{[\tan(\theta/2 + \pi/4)]^n}{\cos \theta} = a \frac{(1 + \sin \theta)^n}{\cos^{n+1} \theta}.$$

(cf. Volume I, pp. 271–272.)

## Exercises 6.3c (p. 695)

1. (a) $y = c_1 e^x + c_2 e^{-(1/2)x} \cos \dfrac{\sqrt{3}\, x}{2} + c_3 e^{-(1/2)x} \sin \dfrac{\sqrt{3}\, x}{2}$.

   (b) $y = c_1 e^x + c_2 x e^x + c_3 e^{2x}$.

   (c) $y = c_1 e^x + c_2 x e^x + c_3 x^2 e^x$.

   (d) $y = c_1 e^x + c_2 e^{-x} + c_3 e^{\sqrt{2}x} + c_4 e^{-\sqrt{2}x}$.

   (e) Substitute $x = e^t$:

$$y = c_1 x + c_2/x.$$

2. From the fundamental theorem of algebra, it follows that $f(z)$ may be written

$$f(z) = (z - a_1)^{\mu_1}(z - a_2)^{\mu_2} \cdots (z - a^k)^{\mu_k}$$

(cf. Volume I, p. 286; Volume II, p. 806), where the $\mu_v$'s are positive integers such tha $\mu_1 + \cdots + \mu_k = n$ and

$$f(a_v) = f'(a_v) = \cdots = f^{(\mu_v - 1)}(a_v) = 0.$$

Now

$$L(e^{\lambda x}) = f(\lambda)e^{\lambda x}.$$

On differentiating this relation $(\mu_v - 1)$ times and putting $\lambda = a_v$ in the result, we get (cf. Leibnitz's rule, Volume I, p. 203)

$$L(e^{a_v x}) = f(a_v)\, e^{a_v x} = 0$$
$$L(xe^{a_v x}) = [f'(a_v) + xf(a_v)]e^{a_v x} = 0$$
$$L(x^2 a^{a_v x}) = [f''(a_v) + 2xf'(a_v) + x^2 f(a_v)]e^{a_v x} = 0$$

$$\cdot \quad \cdot \quad \cdot \quad \cdot \quad \cdot \quad \cdot \quad \cdot \quad \cdot \quad \cdot \quad \cdot \quad \cdot \quad \cdot \quad \cdot$$

$$L(x^{\mu_v-1}e^{a_v x}) = \left[ \binom{\mu_v - 1}{0} f^{(\mu_v-1)}(a_v) + \binom{\mu_v - 1}{1} f^{(\mu_v-2)}(a_v)x \right.$$
$$\left. + \cdots + \binom{\mu_v - 1}{\mu_v - 1} f(a_v)x^{\mu_v-1} \right] e^{a_v x} = 0.$$

So we have $n$ particular solutions

$$e^{a_1 x},\ xe^{a_1 x},\ \ldots,\ x^{\mu_1-1}e^{a_1 x}$$
$$e^{a_2 x},\ xe^{a_2 x},\ \ldots,\ x^{\mu_2-1}e^{a_2 x}$$

$$\cdot \quad \cdot \quad \cdot \quad \cdot \quad \cdot \quad \cdot \quad \cdot \quad \cdot \quad \cdot$$

$$e^{a_k x},\ xe^{a_k x},\ \ldots,\ x^{\mu_k-1}e^{a_k x},$$

which are linearly independent by Exercise 2, p. 690.

3. On substituting in the differential equation, we get

$$(a_0 b_0 - 1)P(x) + (a_0 b_1 + a_1 b_0)P'(x)$$
$$+ (a_0 b_2 + a_1 b_1 + a_2 b_0)P''(x) + \cdots = 0,$$

and this is an identity if $a_0b_0 = 1$, $a_0b_1 + a_1b_0 = 0$, . . . , from the expansion. The second case reduces to the first if we substitute $y'$ for $y$.

4. (a) $1/(1 + t^2) = 1 - t^2 + t^4 - \cdots$ ; hence,

$$y = P(x) - P''(x) = 3x^2 - 5x - 6.$$

(b) $1/(t + t^2) = (1/t) - 1 + t - t^2 + \cdots$ ; hence,

$$y = \int P(x)\, dx - P(x) + P'(x) - P''(x) = -\frac{2}{3} + x + \frac{1}{3}x^3.$$

5. (a) $y = \frac{3}{8} e^x.$  (b) $y = \frac{1}{6} x^3 e^x.$

6. $y = e^x\left(\frac{x^2}{2} + \frac{3}{2}x + \frac{7}{4}\right) + c_1 e^{3x} + c_2 e^{2x}.$

7. (b) The equation becomes of the form treated in (a) if we multiply it by $x^3$. It has the particular solutions $u = x^3$ and $y = x^5$; hence, by (a), a third solution is given by $w = 1 + x^2$; the general solution is then

$$A(1 + x^2) + Bx^3 + Cx^5.$$

**Exercises 6.4** (p. 706)

1. (a) $x^2 + y^2 + cx + 1 = 0$ $(-\infty < c < \infty)$ and the line $x = 0$.
   (b) $x^2 + 2y^2 = c^2$.
   (c) The differential equation of the family of confocal conics (cf. p. 256) is found to be

$$y'^2 + \frac{x^2 - y^2 - a^2 + b^2}{xy}y' - 1 = 0,$$

   which is unaltered if $y'$ is replaced by $-1/y'$; the family of ellipses $(-b^2 < c < \infty)$ is orthogonal to the family of hyperbolas $(-a^2 < c < -b^2)$.
   (d) $y = \log|\tan(x/2)| + c$ and the vertical lines $x = k\pi$ ($k$ an integer).
   (e) The family of curves (tractrix)

$$x - c = \pm[\sqrt{a^2 - y^2} - a \operatorname{ar cosh}(a/y)]$$

   and the same family reflected in the $x$-axis.

2. (a) The family of parabolas $y = cx^2$.
   (b) The family of hyperbolas $xy = c$.

3. (a) $y = x^2$. (b) $y = -x + x \log(-x)$, $(0 > x > -\infty)$.

4. $y = xp + a\sqrt{1 + p^2} - ap \operatorname{ar sinh} p.$

5. $x = ce^{-p/a} + \frac{1}{2}p$

$$y = c(p + a)e^{-p/a} + \frac{1}{2}p(p + a) - \frac{1}{4}(p + a)^2.$$

Note that for $c = 0$ this gives the parabola $y = x^2 - (a^2/4)$. What is the geometrical meaning of this result?

6. (a) $y = \sin(x + c)$, singular solutions $y = \pm 1$.

   (b) $x = \pm \dfrac{1}{2}(\text{arc} \sin y + y\sqrt{1 - y^2}) + c.$

   (c) $x = \pm \left( \sqrt{(2a - y)y} - 2a \text{ arc} \tan \sqrt{\dfrac{y}{2a - y}} \right) + c,$

   which is a family of cycloids and can be expressed in the parametric form $x = c + a(\phi - \sin \phi)$, $y = a(1 - \cos \phi)$. Singular solution $y = 2a$.

   (d)
   $$x = \pm \int_0^y \sqrt{\dfrac{1 + y^2}{1 - y^2}}\, dy + c \qquad\qquad (-1 \leqq y \leqq 1);$$

   singular solutions $y = \pm 1$. (The reader should prove that these curves are not sine curves. The expression for $x$ can be expressed in terms of elliptic integrals of the second kind; see Volume I, pp. 436 ff. Section 4.1g, Problem 1.)

7. $y = x \sin ax$; singular solutions $y = x$ and $y = -x$.

8. In each case, let the equation of the tangent line be given in the form $x/a + y/b = 1$.

   (a) Clairaut equation, $y = xp + kp/(p - 1)$, where $k = a + b$. The singular integral is the parabola $x^2 - 2xy + y^2 - 2kx - 2ky + k^2 = 0$ symmetric about the line $x = y$ and tangent to the x- and y-axes at the points $(k, 0)$ and $(0, k)$, respectively.

   (b) Set $a = k \cos \theta$ and $b = k \sin \theta$, where $k$ is the intercepted length on the tangent, and use $\theta$ as the parameter along the curve. The Clairaut equation is $y = xp \pm kp/\sqrt{1 + p^2}$. The parametric equations of the curve are $x = k \cos^3 \theta$, $y = k \sin^3 \theta$. This is the astroid of Volume I, p. 436, Section 4.1e, Problem 7.

   (c) Set $|ab| = k$. The Clairaut equation is $y = xp + \sqrt{k|p|}$. The curve is the union of two rectangular hyperbolas $4xy = \pm k$.

## Exercises 6.5 (p. 710)

1. (a) Rewrite as $(\tfrac{1}{2}y'^2)' = x$;
   $$y = \dfrac{1}{2}x\sqrt{x^2 + a} + \dfrac{1}{2}a \log(x + \sqrt{x^2 + a}).$$

   (b) Rewrite as $(y''^2)' = 1$;
   $$y = \dfrac{4}{15}(x + a)^{5/2} + bx + c.$$

   (c) Rewrite as $(xy')' = 2$;
   $$y = 2x + a \log x + b.$$

(d) Rewrite as $x (y''^2)' = y''^2 - 2$ and introduce $y''^2$ as a new independent variable. $y = x^2 + \dfrac{1}{6} ax^3 + bx + c$.

2. (a) $y = (ax + b)^{2/3}$.

   (b) $y = \sqrt{a + (x + b)^2}$.

   (c) $y = \sqrt{a(x + b)^2 + a^{-1}}$.

   (d) The equation can be expressed in the form $p(d/dy) (p/y) = 1$. $y = a/(1 - be^{ax})$. Note solutions $p = 0$, $y = $ constant.

   (e) Introduce new variables $z$ and $q$, where $z = y''$, $q = y'''$ and $q(dq/dz) = y^{iv}$.

$$y = ax^2 + bx + c + \frac{2}{15} \left( \frac{x}{2} + b \right)^5$$

   (f) Proceed as in part (e):

$$y = ax + b + c \sin (x + d).$$

3. $MN = y\sqrt{1 + y'^2}$, $MC = - [(1 + y'^2)^{3/2}/y'']$, and the differential equation is

$$(1 + y'^2)^2 y + ky'' = 0.$$

By the general method this is easily reduced to

$$\left( \frac{dy}{dx} \right)^2 = \frac{k + c - y^2}{y^2 - c} \qquad (c \text{ an arbitrary constant}).$$

The various cases, all of importance in the differential geometry of surfaces, [1] are as follows:

(1) $k = x^2 (> 0)$, $c = -\gamma^2 (< 0, \ \gamma^2 < x^2)$. The curve is everywhere smooth and oscillates, alternately touching the lines $y = \pm\sqrt{x^2 - \gamma^2}$. It looks like a sine curve, but is not one.

(2) $k = x^2$, $c = 0$. The curve is a circle of radius $x$ with center on the $x$-axis.

(3) $k = x^2$, $c = \gamma^2 (> 0)$. The curve consists of a sequence of identical arcs, joined by cusps lying on the line $y = \gamma$, and all touched by $y = \sqrt{x^2 + \gamma^2}$. It looks like a cycloid but is not one.

(4) $k = -x^2 (< 0)$, $c = \gamma^2 > x^2$. The curve consists of a sequence of identical arcs upside-down, with their cusps on $y = \gamma$ and touched by $y = \sqrt{\gamma^2 - x^2}$.

(5) $k = -x^2$, $c = \gamma^2 = x^2$. The curve is a tractrix.

(6) $k = -x^2$, $c = \gamma^2 < x^2$. The curve has an infinity of cusps perpendicular to the lines $y = \gamma$ and $y = -\gamma$ alternately.

4. Eliminate $a$, $b$, $c$ by using the equations obtained by differentiating the equation of the circle three times successively.

---

[1] See L. P. Eisenhart, *A Treatise on the Differential Geometry of Curves and Surfaces,* reprinted by Dover (N.Y., 1960), pp. 270–274.

$$(1 + y^2) \, y''' - 3y'y''^2 = 0.$$

## Exercises 6.6 (p. 713)

1. (a) $c_0 = a$, $c_1 = a$, $c_v = \dfrac{a+1}{v!}$    ($v \geq 2$).

   (b) $c_0 = \dfrac{\pi}{2}$, $c_1 = 1$, $c_{2v} = 0$, $c_{2v+1} = \dfrac{2(-1)^v}{2v+1}$    ($v \geq 1$).

   (c) $c_0 = 0$, $c_1 = 1$, $c_2 = 0$, $c_3 = \dfrac{1}{3}$.

   (d) $1 + x + \dfrac{x^2}{2} + \dfrac{x^3}{4} + \cdots$.

2. If $y(x) = \Sigma c_v x^v$, then

$$c_{v+2} = -\frac{c_v}{(v+2)^2} \qquad \text{and} \qquad c_0 = 1, \ c_1 = 0;$$

$$y(x) = \sum_{v=0}^{\infty} \frac{(-1)^v}{2^{2v} \, v!^2} x^{2v}.$$

If we substitute the power series for cos $xt$ in the expression for $J_0(x)$ in Exercise 7, p. 475, and interchange summation and integration (Why is this permissible?), we get

$$J_0(x) = \frac{1}{\pi} \sum_{v=0}^{\infty} \frac{x^{2v}}{(2v)!} (-1)^v \int_{-1}^{+1} \frac{t^{2v}}{\sqrt{1 - t^2}} \, dt;$$

the value of

$$\int_{-1}^{+1} \frac{t^{2v}}{\sqrt{1 - t^2}} \, dt \qquad \text{is} \qquad \frac{(2v)! \, \pi}{v!^2 \, 2^{2v}},$$

as is found by putting $t = \sin \tau$ and referring to Volume I, p. 280. The power series for $y(x)$ and $J_0(x)$ are therefore identical.

## Exercises 6.7 (p. 726)

1. Poisson's formula gives a potential function $u(r, \theta)$ inside the unit circle, with boundary values $f(\theta)$. Now $u(1/r, \theta)$ is also a potential function (cf. p. 58, Exercise 4) with the same boundary values, and it is bounded in the region outside the unit circle; thus, the expression

$$\frac{r^2 - 1}{2\pi} \int_0^{2\pi} f(\alpha) \frac{d\alpha}{1 - 2r\cos(\theta - \alpha) + r^2}$$

is a solution of the problem.

2. The potential is

$$\mu \log \frac{z + l + \sqrt{(z + l)^2 + x^2 + y^2}}{z - l + \sqrt{(z - l)^2 + x^2 + y^2}}.$$

Since on the ellipsoid $z = l\alpha \cos \phi$, $\sqrt{x^2 + y^2} = l\sqrt{\alpha^2 - 1} \sin \phi$, the potential is

$$\mu \log \frac{\alpha + 1}{\alpha - 1},$$

the confocal ellipsoids

$$\frac{z^2}{l^2\alpha^2} + \frac{x^2 + y^2}{l^2(\alpha^2 - 1)} = 1 \qquad (1 \leq \alpha \leq \infty)$$

are equipotential surfaces. The lines of force are the orthogonal trajectories and hence (cf. Exercise 1.c. p. 707) are the confocal hyperbolas given by the same equation when $0 \leq \alpha \leq 1$ and the ratio of $x$ to $y$ is constant.

3. Let $\Sigma$ be a sphere of radius $\rho$ and center $(x, y, z)$, lying inside $S$. Since $\Delta(1/r) = 0$ and $\Delta u = 0$ in the region bounded by $\Sigma$ and $S$, by Green's theorem (cf. p. 608) we have

$$0 = \iint_S \left( \frac{1}{r} \frac{\partial u}{\partial n} - u \frac{\partial(1/r)}{\partial n} \right) d\sigma - \iint_\Sigma \left( \frac{1}{r} \frac{\partial u}{\partial n} - u \frac{\partial(1/r)}{\partial n} \right) d\sigma,$$

where in the first integral $n$ is the outward normal to $S$ and in the second the outward normal to $\Sigma$. Now on the sphere $\Sigma$ we have $\dfrac{\partial(1/r)}{\partial n} = \dfrac{\partial(1/r)}{\partial r} = -\dfrac{1}{\rho^2}$, $r = \text{constant} = \rho$; therefore,

$$\iint_\Sigma \frac{1}{r} \frac{\partial u}{\partial n} d\sigma = \frac{1}{\rho} \iint_\Sigma \frac{\partial u}{\partial n} d\sigma = 0,$$

since $u$ is a harmonic function (cf. p. 720); in addition,

$$-\frac{1}{4\pi} \iint_\Sigma u \frac{\partial(1/r)}{\partial n} d\sigma = \frac{1}{4\pi\rho^2} \iint_\Sigma u \, d\sigma,$$

and as $\rho \to 0$, this expression obviously tends to $u(x, y, z)$, for it is the mean value of $u$ on $\Sigma$.

## Exercises 6.8 (p. 734)

1. (a) $u = f(x) + g(y)$; $f$ and $g$ are arbitrary functions.
   (b) $u = f(x, y) + g(x, z) + h(y, z)$; $f, g, h$ are arbitrary functions.
   (c) The most general solution is obtained from a particular solution by adding the general solution of the homogeneous equation $u_{xy} = 0$.

$$u = \int_0^x d\xi \int_0^y a(\xi, \eta) \, d\eta + f(x) + g(y),$$

   where $f$ and $g$ are arbitrary.
2. If $u(x, y) = \sum \alpha_{\nu\mu} x^\nu y^\mu$, then

$$\alpha_{\nu+1,\,\mu+1} = \frac{\alpha_{\nu\mu}}{(\nu+1)(\mu+1)};$$

in addition,

$$\alpha_{\nu 0} = \alpha_{0\nu} = 0$$

for $\nu \geq 1$ and $\alpha_{00} = 1$.  Hence,

$$u(x,y) = \sum_{\nu=0}^{\infty} \frac{x^\nu y^\nu}{\nu!^2} = J_0(2i\sqrt{xy}),$$

where $J_0$ is the Bessel function of Exercise 2, p. 713.

3. $z^2(z_x^2 + z_y^2 + 1) = 1.$

4. A one-parameter family is obtained from the two-parameter family of solutions $z = u(x, y, a, b)$ by making $a$ and $b$ depend in some way on a parameter $t$:

$$a = f(t)$$
$$b = g(t),$$
$$z = u(x, y, f(t), g(t)).$$

The envelope of this one-parameter family is obtained by finding $t$ from the equation

$$0 = z_t = u_a f' + u_b g',$$

and substituting this expression for $t$ in $z = u(x, y, f(t), g(t))$. The result is again a solution of $F(x, y, z, z_x, z_y) = 0$, as

$$z = u(x, y, a, b)$$
$$z_x = u_x + u_t t_x = = u_x(x, y, a, b)$$
$$z_y = u_y + u_t t_y = u_y(x, y, a, b)$$

and $z = u(x, y, a, b)$ satisfies the equation $F(x, y, z, z_x, z_y) = 0.$

5. (a) From the differential equation we get

$$[f'(x)]^2 + [g'(y)]^2 = 1$$

or

$$[f'(x)]^2 = 1 - [g'(y)^2].$$

As the left-hand side does not depend on $y$, nor the right-hand side on $x$, both sides are equal to a constant (which has to be positive or zero), say $c^2$; that is,

$$[f'(x)]^2 = c^2,\ 1 - [g'(y)]^2 = c^2.$$

Hence,

$$u = cx + \sqrt{1 - c^2}\, y + b$$

is a solution, where $c$ and $b$ are arbitrary and $c^2 \leq 1$.

(b) $u = f(x) + g(y)$ gives

$$f(x) = \frac{1}{g'(y)} = \text{constant} = a,$$

so that

$$u = ax + \frac{1}{a}y + b$$

(where $a$ and $b$ are constants).

If $u = f(x) g(y)$, then

$$\frac{d}{dx}[f(x)]^2 = 4\Big/\frac{d}{dy}[g(y)]^2 = \text{constant} = 2c;$$

so, in this case,

$$u = \sqrt{(2cx + a)\left(\frac{2}{c}y + b\right)},$$

where $a$, $b$, $c$ are arbitrary constants.

(c) $u = x\sqrt{\dfrac{y}{x+k}} + y\sqrt{\dfrac{x+k}{y}} + k\sqrt{\dfrac{y}{x+k}}.$

6. Apply the linear transformation

$$x = \xi + \eta,$$
$$y = 3\xi + 2\eta,$$
$$u = f(y - 2x) + g(3x - y) + \frac{1}{12}e^{x+y}.$$

7. Put $u = (x^2 + y^2 + z^2)^{n/2}$ and let $K$ be of degree $h$. Then,

$$\Delta u = u_{xx} + u_{yy} + u_{zz} = n(n+1)(x^2 + y^2 + z^2)^{(n-2)/2},$$

$$x\frac{\partial K}{\partial y} + y\frac{\partial K}{\partial y} + z\frac{\partial K}{\partial z} = hK$$

(cf. p. 120). Hence, $u = (x^2 + y^2 + z^2)^{-(1+h)/2}$ is a solution.

8. According to p. 728, a solution of the first equation is of the form

$$z = f(x + at) + g(x - at).$$

On substituting this expression in the second equation, we have

$$f'g' = 0;$$

that is, either $f = \text{constant}$ or $g = \text{constant}$. Hence, $z = f(x + at)$ or $z = f(x - at)$ is the most general solution of both equations.

9. (a) From the differential equation

$$\frac{\phi_{xx}}{\phi} = \frac{1}{c^2}\frac{\psi_{tt}}{\psi} = \lambda,$$

a constant. The boundary conditions can be satisfied only if $\lambda = -n^2$, where $n$ is an integer and

$$\phi(x) = \alpha \sin nx,$$

whence,

$$\psi(t) = a \sin nct + b \cos nct.$$

Thus, the most general particular solution of the specified type is

$$u(x, t) = \sin nx \, (a \sin nct + b \cos nct).$$

(b) Using $\sin A \sin B = \frac{1}{2} [\cos (A - B) - \cos (A + B)]$ and $\sin A \cos B = \frac{1}{2} [\sin(A + B) + \sin(A - B)]$, we obtain

$$u(x, t) = \frac{1}{2} [a \cos n(x - ct) + b \sin n(x - ct)]$$

$$- \frac{1}{2} [a \cos n(x + ct) - b \sin n(x + ct)].$$

(c) Assume a solution in the form of a sum of solutions of the type obtained in part (a), that is,

$$u(x, t) = \sum_{n=1}^{\infty} \sin nx(a_n \sin nct + b_n \cos nct).$$

In order to satisfy the initial conditions in (ii), we must have $b_n = \alpha_n$, $a_n = 0$.

For the solution of (i), observe from Volume I, p. 587, (17), that

$$\alpha_n = \frac{1}{\pi} \left[ \int_{-\pi}^{0} -f(-x) \sin nx \, dx + \int_{0}^{\pi} f(x) \sin nx \, dx \right]$$

$$= \frac{2}{\pi} \int_{0}^{\pi} f(x) \sin nx \, dx.$$

For the particular function in (i), we find $\alpha_{2\nu} = 0$, $\alpha_{2\nu+1} = (-1)^\nu / \pi(2\nu + 1)^2$, where $\nu = 0, 1, 2, \ldots$;

whence

$$u(x, t) = \frac{1}{\pi} \left[ \frac{\sin x \cos ct}{1^2} - \frac{\sin 3x \cos 3ct}{3^2} \right.$$

$$\left. + \frac{\sin 5x \cos 5ct}{5^2} - \cdots \right].$$

10. $u(x, t) = f(x - at) + g(x + at)$; then, for $x \geq 0$,

$$0 = u(x, 0) = f(x) + g(x)$$

$$0 = u_t(x, 0) = -af'(x) + ag'(x);$$

by differentiating the first equation and comparing with the second, we have

$$f'(x) = 0, \qquad g'(x) = 0,$$

or

$$f(x) = \text{constant} = c, \qquad g(x) = -c \qquad \text{for} \qquad x \geq 0.$$

For $t \geq 0$, moreover,

$$\phi(t) = u(0, t) = f(-at) + g(at) = f(-at) - c;$$

that is, $f(\xi) = c + \phi(\xi/-a)$ if $\xi < 0$. As $x + at \geq 0$ always, and, hence, $g(x + at) = -c$, it follows that

$$u(x, t) = \begin{cases} 0 & \text{for } x - at \geq 0 \\ \phi\left(\dfrac{x - at}{-a}\right) & \text{for } x - at \leq 0 \end{cases}$$

if both $x$ and $t$ are nonnegative.

## Exercises 7.2a (p. 743)

1. $\dfrac{2}{\sqrt{2g}} \sqrt{\dfrac{(x_1 - x_0)^2 + (y_1 - y_0)^2}{y_1 - y_0}}$.

2. $T = \displaystyle\int_{\sigma_0}^{\sigma_1} f(r) \sqrt{\dot{r}^2 + r^2\dot{\theta}^2 + r^2 \sin^2\theta\dot{\phi}^2} \, d\sigma$.

## Exercises 7.2d (p. 751)

1. (a) Parabolas $y = c^2 + \dfrac{x^2}{4c^2}$.

   (b) Circle with center on $x$-axis.

   (c) $y = c \sin \dfrac{x - a}{c}$.

2. $y = \dfrac{a}{x^{n-1}} + b$ for $n > 1$, and $y = a \log x + b$ for $n = 1$.

3. $y = a(x - b)^{n(n+m)}$ if $n + m \neq 0$; $y = ae^{bx}$ if $n = -m$.

4. $ay'' + a'y' + (b' - c) y = 0$; for $b = $ constant,

$$\int_{x_2}^{x_1} byy' \, dx = \frac{b}{2}(y_2{}^2 - y_1{}^2)$$

only depends on the end points of the curve $y = y(x)$.

5. $y_1 - y_0 < \dfrac{\pi}{2}$.

6. Consider $F(x, y)$ for fixed $x$ as a function of $y$; let this function of $y$ have a minimum for $y = \bar{y}$. Then, $F(x, y) \geq F(x, \bar{y})$ for a certain neighborhood of $\bar{y}$ and $F_y(x, \bar{y}) = 0$. $\bar{y}$ will depend on the parameter $x$; [i.e., $\bar{y} = \bar{y}(x)$]. Then, for any neighboring function $y$, we have

$$\int_{x_0}^{x_1} F(x, y(x)) \, dx \geq \int_{x_0}^{x_1} F(x, \bar{y}(x)) \, dx,$$

where $\bar{y}(x)$ satisfies the equation $F_y(x, \bar{y}(x)) = 0$.

7. (a) $y = 0$.

(b) Use Cauchy's inequality. For any admissible $x$,

$$1 = y(1) - y(0) = \int_0^1 y'\ dx \leq \sqrt{\int_0^1 1^2\ dx} \sqrt{\int_0^1 y'^2\ dx} = \sqrt{I},$$

and the equality sign holds for $y = x$.

8. Introduce $1/r$ as new dependent variable in Euler's equation. The general solution is the line $1/r = a\cos\theta + b\sin\theta$.

## Exercises 7.3b (p. 757)

1. If $v = 1/f(r)$, then $T$ is given by Exercise 2, p. 743:

$$F = f(r)\sqrt{\dot{r}^2 + r^2\dot{\theta}^2 + r^2\sin^2\theta\ \dot{\phi}^2}.$$

Euler's equation for the variable $\phi$ gives

$$F_{\dot\phi} = \frac{\dot\phi f^2 r^2 \sin^2\phi}{F} = \text{constant} = C$$

along a ray. Now let the polar coordinates be chosen in such a way that the plane $\phi = 0$ passes through the initial point and the end point; since $\phi = 0$ at both these points, we have $\dot\phi = 0$ for some intermediate point, by the mean value theorem, that is, $C = 0$; but then $\dot\phi = 0$ for the whole ray, that is, $\phi \equiv 0$. Hence the whole ray must lie in the plane $\phi = 0$.

2. See Exercise 1. Using $\phi$ as parameter, we have to minimize $r\int\sqrt{\dot\theta^2 + \sin^2\theta}\,d\phi$, where $r = $ constant. Introducing $\cot\theta$ as new dependent variable in Euler's equation leads to the general solution $\cot\theta = a\cos\phi + b\sin\phi$, corresponding to a curve of intersection of the sphere with a plane through the center.

3. See Exercise 1 above. Here in spherical coordinates we have $\theta = $ constant. Introducing $r$ as dependent and $\phi\sin\theta$ as independent variable yields the same integral to be minimized as in Exercise 8, p. 752. (The mapping of the point of the cone with spherical coordinates $r$, $\theta$, $\phi$ onto the point in the plane with polar coordinates $r$, $\phi\sin\theta$ preserves arc length).

$$1/r = a\cos(\phi\sin\theta) + b\sin(\phi\sin\theta).$$

4. The path has to be straight, since it has to have minimum length for given end points. We only have to find the minimum distance between two points constrained to move on two given curves, which is a minimum problem for a function of several variables with subsidiary conditions (cf. Chapter 3, p. 337).

5. See solution to next problem.

6. Let the end points be constrained to lie on the curves $y = f(x)$ and $y = g(x)$, respectively. Let the minimizing curve have end points $(a_0, f(a_0))$, $(b_0, g(b_0))$, and an equation $y = u(x)$, where $u(a_0) = f(a_0)$, $u(b_0)$ $g(b_0)$. Since $u$ also is an extremal for fixed end points, it satisfies Euler's equation. Consider a family of curves $y = u(x) + \varepsilon\eta(x)$ with parameter $\varepsilon$ and end points $(a, f(a))$, $(b, g(b))$, where $a = a(\varepsilon)$, $b = b(\varepsilon)$ are solu-

tions of $f(a) = u(a) + \varepsilon\eta(a)$, $g(b) = u(b) + \varepsilon\eta(b)$. The corresponding integral is

$$G(\varepsilon) = \int_{a(\varepsilon)}^{b(\varepsilon)} F(x, u(x) + \varepsilon\eta(x)) \sqrt{1 + [u'(x) + \varepsilon\eta'(x)]^2}\, dx.$$

For the extremal $u$ we have the condition $0 = G'(0)$. We evaluate $G'(0)$ as on pp. 743–744, using integration by parts to eliminate $\eta'(x)$. Because $u$ satisfies Euler's equation the only contributions arise from differentiating the limits in the integral for $G$ and from the boundary terms in the integration by parts. Noticing that, for $\varepsilon = 0$,

$$[f'(a) - u'(a)]\frac{da}{d\varepsilon} = \eta(a),\quad [g'(b) - u'(b)]\frac{db}{d\varepsilon} = \eta(b)$$

and that $\eta(a)$, $\eta(b)$ are arbitrary, we find the relations

$$0 = 1 + u'(a_0)\, f'(a_0) = 1 + u'(b_0)\, g'(b_0)$$

expressing orthogonality at the end points.

## Exercises 7.4a (p. 765)

1. The law of conservation of energy gives

$$T + U = T = \frac{1}{2}\left(\frac{ds}{dt}\right)^2 = \text{constant} = \frac{1}{2}C^2;$$

hence, $ds/dt = \text{constant} = C = $ initial velocity.

Then Hamilton's principle asserts the stationary character of

$$\int_{t_0}^{t_1} (T - U)\, dt = \int_{t_0}^{t_1} T\, dt = \frac{1}{2}C^2 \int_{t_0}^{t_1} dt = \frac{1}{2}C \int_{s_0}^{s_1} ds;$$

the stationary character of Hamilton's integral implies that the length of path is stationary.

2. Let $t$ be a parameter along the curve $C$. On the geodesic perpendicular to $C$ at a point of $C$ with parameter $t$, we use arc length $s$ as parameter, counting $s$ from the point on $C$. Then $x = x\,(s, t)$, $y = y\,(s, t)$, $z = z\,(s, t)$ shall represent the curve obtained by laying off a fixed geodesic distance $s$ along each geodesic perpendicular to $C$ at the point with parameter $t$. Here, since $s$ is arc length, we have $x_s^2 + y_s^2 + z_s^2 = 1$; moreover, by formula (19), p. 765, $x_{ss}$, $y_{ss}$, $z_{ss}$ are proportional to $G_x$, $G_y$, $G_z$, and $G(x, y, z) = 0$ for all $s$, $t$ in question. On $C$(i.e., for $s = 0$) we have by assumption $x_s x_t + y_s y_t + z_s z_t = 0$. Then,

$$\frac{d}{ds}(x_s x_t + y_s y_t + z_s z_t) = \lambda(G_x x_t + G_y y_t + G_z z_t) + x_s x_{st} + y_s y_{st} + z_s z_{st}$$

$$= \lambda\frac{dG}{dt} + \frac{1}{2}\frac{d}{dt}(x_s^2 + y_s^2 + z_s^2) = 0.$$

Hence, $x_s x_t + y_s y_t + z_s z_t = \text{constant} = 0$ for all $s$, which proves that the curves $C'$ for which $s = \text{constant}$ are perpendicular to the geodesics.

## Exercises 7.4b (p. 767)

1. From the differential equations for geodesics (p. 765) we find that for a cylinder (i.e., if $G$ does not depend on $z$) $dz/dt$ is constant; hence, the geodesics on a cylinder make a constant angle with the $x$, $y$ -plane.

2. (a) $g(x) - \dfrac{y''}{\sqrt{(1 + y'^2)^3}} = 0.$

   (b) $g(x) - \dfrac{6y''(y''^2 + 4y'y''')}{(1 + y'^2)^4} + \dfrac{2y''''}{(1 + y'^2)^3} + \dfrac{48y'^2y''^3}{(1 + y'^2)^5} = 0.$

   (c) $y + y'' + y'''' = 0.$

   (d) $(2 - y'^2)\, y'' = 0.$

3. (a) $\phi d = (a_x + b_y)\phi_x + (b_x + c_y)\phi_y + a\phi_{xx} + 2b\phi_{xy} + c\phi_{yy}.$

   (b) $\Delta^2\phi = 0.$

   (c) $\Delta^2\phi = 0.$

4. $\dfrac{au'' + a'u' + u(b' - c)}{u} = \lambda = \text{constant}.$

5. (a) Euler's equation gives

$$f + 2\lambda u = 0;$$

   from this equation and $\int_0^1 \phi^2\, dx = K^2$, we have

$$\lambda = \pm \frac{\sqrt{\int_0^1 f^2\, dx}}{2K}, \qquad u = \frac{\pm Kf}{\sqrt{\int_0^1 f^2\, dx}}.$$

   (b) For any continuous admissible $\phi$ we have

$$I = \int f\phi\, dx \leq \sqrt{\int_0^1 f^2\, dx}\,\sqrt{\int_0^1 \phi^2\, dx} = K\sqrt{\int_0^1 f^2\, dx},$$

   the equality sign holding for $\phi = u$.

8. From the necessary condition (6b), p. 742, we find that

$$\int_{x_0}^{x_1} (F_{yy}\eta^2 + 2F_{yy'}\eta\eta' + F_{y'y'}\eta'^2)\, dx \geq 0$$

   for any $\eta(x)$ vanishing at $x = x_0, x_1$. Let $h$ and $\xi$ be such that $x_0 < \xi - h < \xi < \xi + h < x_1$. Define $\eta(x)$ to be $[(x - \xi)^2 - h^2]^2 h^{-7/2}$ for $|x - \xi| < h$, and to be 0 elsewhere. For $h \to 0$, the integral tends to $cF_{y'y'}(\xi, u(\xi), u'(\xi))$, where $c$ is a positive constant.

9. Problem really identical to standard isoperimetric problem. Solution is a circular arc, but since solutions are functions of $x$, there is an upper bound on permissible lengths in this problem, namely,

$$\frac{2[(x_1 - x_0)^2 + (y_1 - y_0)^2]}{x_1 - x_0} \arctan \frac{x_1 - x_0}{|y_1 - y_0|}.$$

**Exercises 8.1** (p. 777)

1. (a) Set $\alpha = a_1 + ia_2$, $\beta = b_1 + ib_2$.

   For the example of multiplication,

$$\overline{\alpha\beta} = (a_1b_1 - a_2b_2) - i(a_1b_2 + a_2b_1) = \bar{\alpha}\bar{\beta}.$$

   (b) Follows directly from part (a) on passage to the limit of the real and imaginary parts of the partial sums.

2. (a) From Exercise 1, $\overline{P(\alpha)} = P(\bar{\alpha})$; hence, $P(\alpha) = 0$ implies $P(\bar{\alpha}) = 0$, and conversely.

   (b) By long division express $P(z)$ in the form

$$P(z) = (z^2 - 2az + a^2 + b^2)\, Q(z) + cz + d,$$

   where $Q(z)$ is a polynomial with real coefficients and $c$ and $d$ are real. Setting $z = \alpha$ in this equation, obtain $c\alpha + d = 0$; whence,

$$ca + d = 0 \qquad \text{and} \qquad icb = 0.$$

   Since $b \neq 0$, $c = 0$, and hence, $d = 0$.

3. (a) Use the equation of a circle in the form

$$(z - z_0)(\bar{z} - \bar{z}_0) = r^2.$$

   Then $z_0 = \alpha - \lambda^2\beta$, $r^2 = z_0\bar{z}_0 - \alpha\bar{\alpha} + \lambda^2\beta\bar{\beta}$.

   If $\lambda = 1$, $z = x + iy$, the equation becomes that of a straight line, $ax + by = c$, where $a = 2Re\,\alpha$, $b = 2Im\,\beta$, $c = |\alpha|^2 - |\beta|^2$.

   (b) Invert the transformation to obtain

$$z = \frac{\beta - \delta z'}{\gamma z' - \alpha};$$

   then show that

$$|z - z_1| = \lambda|z - z_2|$$

   becomes

$$|z' - z_1'| = \lambda\left|\frac{\gamma z_1' - \alpha}{\gamma z_2' - \alpha}\right||z' - z_2'|.$$

4. For $x \geqq 0$.

5. Use the comparison test.

6. The coefficient of $z^n$ in the expansion of $\cos^2 z + \sin^2 z$ for $n > 0$ is

$$(-1)^{n/2}\sum_{\nu=0}^{n}\frac{(-1)^\nu}{\nu!(n-\nu)!} = \frac{(-1)^{n/2}}{n!}\sum_{\nu=0}^{n}(-1)^\nu\binom{n}{\nu} = 0$$

   [cf. Volume I, p. 110, Exercise 1 (b)].

7. The series is convergent if, and only if, $|z| < 1$, for if $|z| = \theta < 1$, then

$$\left|\frac{z^\nu}{1-z^\nu}\right| \leqq \frac{\theta^\nu}{1-\theta^\nu} \leqq \frac{1}{1-\theta}\theta^\nu$$

and we may compare with the geometric series. If $|z| > 1$, then $z^\nu/(1-z^\nu)$ tends to $-1$ as $\nu$ increases, whereas in a convergent series the terms must tend to 0. If $|z| = 1$, each term of the series either is undefined or has absolute value $\geq \frac{1}{2}$ and the series cannot converge.

## Exercises 8.2 (p. 786)

1. Set $f(z) = u + iv$, $g(z) = s + it$. Taking the product, for example, we find for

$$U(x, y) = \text{Re } \{f(z)\, g(z)\} = us - vt$$
$$V(x, y) = \text{Im } \{f(z)\, g(z)\} = ut + vs$$

that

$$U_x = u_x s + u s_x - (v_x t + v t_x)$$
$$= v_y s + u t_y + u_y t + v s_y$$
$$= u t_y + u_y t + v_y s + v s_y = V_y,$$

and so on.

2. For $f(z) = u + iv$, on differentiating $u^2 + v^2 = \text{constant}$, we obtain the pair of equations

$$u u_x + v v_x = 0, \quad u u_y + v v_y = 0.$$

Replacing the second equation through the Cauchy-Riemann equations by one in derivatives with respect to $x$ alone, we obtain a system with only the solution $u_x = v_x = 0$ (unless we are dealing with the trivial case $u^2 = v^2 = 0$). Consequently, $u_y = v_y = 0$ and the result follows.

3. (a) $-$(c) Everywhere continuous; not differentiable.

   (d) Continuous for $z \neq 0$: not differentiable.

4. If $z = re^{i\phi}$, $\zeta = \xi + i\eta$, then

$$\xi = \frac{1}{2}\left(r + \frac{1}{r}\right) \cos \phi$$

$$\eta = \frac{1}{2}\left(r - \frac{1}{r}\right) \sin \phi.$$

If $r = \text{constant} = c$, then

$$\frac{\xi^2}{\frac{1}{4}(c + 1/c)^2} + \frac{\eta^2}{\frac{1}{4}(c - 1/c)^2} = 1;$$

if $\phi = \text{constant} = c$, then

$$\frac{\xi^2}{\cos^2 c} + \frac{\eta^2}{\cos^2 c - 1} = 1$$

(cf. p. 256, Exercise 8).

5. From 8.1, Exercise 3b we know that the transformation maps circles into circles. Since the two points are fixed, circles through them map into

circles of the same family in both the transformation and its inverse. Since the mapping is conformal, the same is true of the orthogonal family of circles.

6. Set $z = x + iy$, $\zeta = 1/z = \xi + i\eta$. Thus,

$$\xi = \frac{x}{x^2 + y^2}, \quad \eta = \frac{-y}{x^2 + y^2}$$

and we recognize inversion as the composition $gf(z)$ of $1/z$ and reflection in the x-axis, $g(\zeta) = \bar{\zeta}$. Since reflection is conformal—with reversal of the sense of angles—and $1/z$ is analytic, inversion is conformal. Reflection maps circles into circles, and $1/z$, a general linear transformation (see Exercise 5), does the same; hence, inversion does the same. The Jacobian of inversion is the product of those for reflection and for $1/z$, hence, for inversion it is

$$-|f'(z)|^2 = -\frac{1}{|z|^2} = \frac{-1}{(x^2 + y^2)^2}.$$

7. $|\zeta|^2 = \zeta\bar{\zeta} = \dfrac{\alpha\bar{\alpha}z\bar{z} + \beta\bar{\beta} + (\alpha\bar{\beta}z + \bar{\alpha}\bar{\beta}\bar{z})}{\beta\bar{\beta}z\bar{z} + \alpha\bar{\alpha} + (\alpha\bar{\beta}z + \bar{\alpha}\bar{\beta}\bar{z})}$

Now for $\alpha\bar{\alpha} - \beta\bar{\beta} = 1$ the difference between the numerator and the denominator is

$$z\bar{z} - 1;$$

so the numerator is greater than the denominator for $|z| > 1$, and smaller for $|z| < 1$. If $\beta\bar{\beta} - \alpha\bar{\alpha} = 1$, the converse is the case.

8. First transform, by putting $\zeta = az + b$, into the unit circle; then apply the transformation

$$\zeta' = i\frac{1 + \zeta}{1 - \zeta}.$$

9. Use $\zeta_i - \zeta_j = \dfrac{(\alpha\delta - \beta\gamma)(z_i - z_j)}{(\gamma z_i + \delta)(\gamma z_j + \delta)}.$

## Exercises 8.3 (p. 796)

1. (a) Write the integrand in the form

$$\frac{1}{2}\left(\frac{1}{z - 1} + \frac{3}{z + 1}\right).$$

The first term in parentheses is analytic in the neighborhood of $z = -1$; hence, its integral around a small circle centered at $-1$ is 0. Similarly, the integral of the second term around a small circle centered at 1 is 0. To evaluate the integral in the circle about 1, set $z = re^{i\theta}$ to obtain $\pi i$. Similarly, for the small circle about $-1$, the integral is $3\pi i$.

   (b) Take a path circling 1 in one sense three times as many times as it circles $-1$ in the other; for example, (see Fig. 8.12).

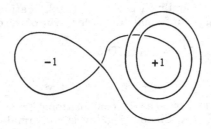

**Figure 8.12**

2.  $$\alpha^z \alpha^\zeta = \exp[z(\log \alpha + 2n\pi i)] \exp [\zeta(\log \alpha + 2m\pi i)],$$

whereas

$$\alpha^{z+\zeta} = \exp[(z + \zeta) (\log \alpha + 2k\pi i)].$$

Thus, addition of exponents is valid, provided the same branch of the logarithm is used throughout; that is, $n = m = k$. Note that this is the best one can do except in very special cases, for if the addition theorem is valid, then

$$k(z + \zeta) = nz + m\zeta + p,$$

where $p$ is some integer. If $z$ and $\zeta$ are linearly independent when considered as two-component vectors and $n \neq m$, the components of $z = a + ib$ and $\zeta = \alpha + i\beta$ are restricted by

$$\frac{(n - m) (a\beta - \alpha b)}{\beta + b} = p,$$

an integer, and if $n = m \neq k$, then $\beta + b = 0$. Neither condition is generally satisfied.

For the second law,

$$z^\alpha \zeta^\alpha = \exp [\alpha(\log z + 2n\pi i)] \exp [\alpha(\log \zeta + 2m\pi i)]$$

$$= \exp \{\alpha[\log z + \log \zeta + 2(n + m)\pi i]\},$$

whereas

$$(z\zeta)^\alpha = \exp \{\alpha[\log(z\zeta) + 2k\pi i]\}.$$

Here, equality need not even hold if $k = n + m$ because if $z = re^{i\theta}$ and $\zeta = \rho e^{i\phi}$, the conditions $-\pi < \theta \leq \pi$, $-\pi < \phi \leq \pi$ do not force $\theta + \phi$ to satisfy the same inequalities.

For the third law,

$$(\alpha^z)^\zeta = e^{\zeta \log \alpha^z} = \exp \{\zeta[z(\log \alpha + 2n\pi i) + 2m\pi i]\}$$

$$= \exp (z\zeta \log \alpha + 2z\zeta n\pi i + 2\zeta m\pi i).$$

Similarly,

$$(\alpha^\zeta)^z = \exp (z\zeta \log \alpha + 2z\zeta p\pi i + 2zq\pi i)$$

and

$$\alpha^z \zeta = \exp(z\zeta \log \alpha + 2z\zeta r\pi i),$$

where $m$, $n$, $p$, $q$, $r$ are arbitrary integers. Thus, we generally expect equality to hold only if $m = q = 0$ and $n = p = r$.

The best one can say is that it is possible to pick branches of the many-valued functions involved so that the laws of exponents hold, but we must be cautious about choosing them properly.

3. (a) The values of $i^i$ are $\exp\left[(2n - \dfrac{1}{2})\pi\right]$, for integral $n$.

   (b) Set $\zeta = \xi + i\eta$, $z = re^{i\theta}$, $-\pi < \theta \le \pi$ and $a = \log r = \log|z|$. Then,

   $$z^\zeta = \exp[a\xi - (\theta + 2k\pi)\eta]\, \exp\{i[a\eta + \xi(\theta + 2k\pi)]\}.$$

   The condition is that $a\eta + \xi(\theta + 2k\pi)$ be an integral multiple of $\pi$ for each choice of integral $k$. Setting $k = 0$, $1$, we obtain the condition $\xi = j/2$, where $j$ is any integer and, hence, for $a \ne 0$ $(r \ne 1)$,

   $$\eta = (l\pi - \frac{1}{2}j\theta)\big/a,$$

   where $l$ may be any integer. Thus, for any $z$ not on the unit circle, there exists an exponent $\zeta(j, l)$ for each pair of integers $j$, $l$ such that all values of $z^\zeta$ are real. If $a = 0$, the foregoing condition on $\eta$ above is replaced by the condition $\xi\theta = p\pi$, where $p$ may be any integer, and $\eta$ is now arbitrary. If $p \ne 0$, we see that $\theta = 2\pi p/j$ must be a rational multiple of $2\pi$. If $p = 0$, $\xi$ may be zero and then $\theta$ may be arbitrary.

   (c) Yes. Set $z = x + iy$, $\zeta = \xi + i\eta$, where $y = \eta = 0$. If $x > 0$, the solution of part (b) yields $\xi = j_2$, where $j$ is any integer. If $x < 0$, part (b) yields only integral values of $\xi = n$.

4. For $z = x + iy$, we may certainly differentiate under the integral sign with respect to $x$ and $y$, since these derivatives are continuous with respect to the parameters and convergence of the integrals of the derivatives at the lower limit $t = 0$ is uniform for $x > \varepsilon > 0$. Since the Cauchy-Riemann equations hold for the integrand, they must then hold for the integral. Integration by parts yields the functional equation.

5. Use the theorem in Volume I, p. 525, to show that the series is absolutely convergent.

6. (a) The value of the integral round the small circular detour tends to zero as the circle becomes smaller. If we put $z = e^{i\theta}$ on the unit circle and $z = x$, $z = iy$, respectively, on the axes, Cauchy's theorem gives

   $$0 = \int_0^1 \left(x + \frac{1}{x}\right)^m x^{n-1}\, dx + i \int_0^{\pi/2} (e^{i\theta} + e^{-i\theta})^m\, e^{in\theta}\, d\theta$$

   $$- i \int_0^1 \left(iy + \frac{1}{iy}\right)^m (iy)^{n-1}\, dy$$

$$= \int_0^1 \left(x + \frac{1}{x}\right)^m x^{n-1}\, dx + i \cdot 2^m \int_0^{\pi/2} \cos^m\theta\; e^{in\theta}\, d\theta$$

$$- e^{i\pi(n-m)/2} \int_0^1 \left(-y + \frac{1}{y}\right)^m y^{n-1}\, dy;$$

by equating the imaginary parts of this equation, we get

$$2^m \int_0^{\pi/2} \cos^m\theta \cos n\theta\, d\theta = \sin\frac{\pi(n-m)}{2} \int_0^1 \left(-y + \frac{1}{y}\right)^m y^{n-1}\, dy$$

$$= \frac{1}{2}\sin\frac{\pi(n-m)}{2} \int_0^1 (1 - \eta)^m\, \eta^{(n-m-2)/2}\, d\eta$$

$$= \frac{1}{2}\left(\sin\frac{\pi}{2}(n-m)\right) B\left(m + 1, \frac{n-m}{2}\right)$$

(cf. p. 508).

(b) Use the relation

$$\left(\sin\frac{(n-m)\pi}{2}\right)\Gamma\left(\frac{n-m}{2}\right) = \frac{\pi}{\Gamma[1-(n-m)/2]}$$

(cf. p. 508).

## Exercises 8.4 (p. 805)

1. The integrand has a continuous derivative with respect to $z$; consequently, differentiation under the integral sign is permissible. See Section 1.8b.

2. It is easily seen that

$$h(z) = \frac{1}{2\pi i} \int \frac{f(\zeta)}{\zeta - z} \frac{z^n}{\zeta^n}\, d\zeta$$

is an analytic function of $z$. By differentiating under the integral sign and using Leibnitz's rule (cf. Volume I, p. 203), we find that $h^{(\mu)}(z)$ is

$$\frac{1}{2\pi i} \sum_{\nu=0}^{\mu} \binom{\mu}{\nu} \nu!\, n \cdot (n-1) \cdots (n-\mu+\nu+1) \int \frac{f(\zeta)}{(\zeta - z)^{\nu+1}} \frac{z^{n-\mu+\nu}}{\zeta^n}\, d\zeta$$

$$= \frac{\mu!}{2\pi i} \sum_{\nu=0}^{\mu} \binom{n}{\mu - \nu} \int \frac{f(\zeta)}{(\zeta - z)^{\nu+1}} \frac{z^{n-\mu+\nu}}{\zeta^n}\, d\zeta.$$

Only the terms with $\mu - \nu \leqq n$ differ from zero, as otherwise $\binom{n}{\mu - \nu}$ vanishes. On the other hand, a term with $\mu - \nu < n$ vanishes for $z = 0$; if $\mu < n$, there are no other terms, so that $h^{(\mu)}(0) = 0$. If $\mu \geqq n$, there remains only the term with $\mu - \nu = n$, so that

$$h^{(\mu)}(0) = \frac{\mu!}{2\pi i} \int \frac{f(\zeta)}{(\zeta - z)^{n+1}}\, d\zeta = f^{(\mu)}(0).$$

3. By the Cauchy-Riemann equations the partial derivatives $v_x$ and $v_y$ of $v$ are given; a function $v$ with these derivatives does exist, since the

condition of integrability $u_{xx} + u_{yy} = 0$ is satisfied [see p. 104. formulae (75a, b)]; $v$ is uniquely determined apart from an additive constant $c$ and is given by the curvilinear integral

$$v(x,y) = \int_{(x_0,y_0)}^{(x,y)} (v_y \, dy + v_x \, dx) + c.$$

It also follows from the Cauchy-Riemann equations that $v$ is a potential function.

4. At $z = 1$, $\pi i$; at $z = -1$, $3\pi i$ (Section 8.3, Exercise 1).

5. Choose a circle of radius R centered at 0, with $R = |\zeta|$ so large that $R > 2|z|$. Then,

$$\left| \frac{1}{\zeta - z} - \frac{1}{\zeta} \right| = \frac{|z|}{|\zeta|^2 |1 - z/\zeta|} < \frac{2|z|}{R^2}.$$

Consequently, for the integral, obtain the bound

$$|f(z) - f(0)| \leqq 2M|z|/R.$$

Pass to the limit as $R$ tends to $\infty$.

6.
$$|a_\nu| = \left| \frac{1}{2\pi i} \int_C \frac{f(t)}{t^{\nu+1}} \, dt \right| \leqq \frac{1}{2\pi} \frac{M}{\rho^{\nu+1}} 2\pi\rho,$$

where $C$ is the circle of radius $\rho$ about the origin.

7. By assumption $|\alpha_n| > 0$. Consequently,

(i)
$$|P(z)| = |z|^n \left| \alpha_n + \frac{\alpha_{n-1}}{z} + \cdots + \frac{\alpha_0}{z^n} \right|$$

$$> \frac{1}{2} |z|^n |\alpha_n|,$$

provided we take

$$|z| > \max\left\{ 1, \, 2 \frac{|\alpha_{n-1}| + \cdots + |\alpha_0|}{|\alpha_n|} \right\};$$

for, then,

$$\left| \alpha_n + \frac{\alpha_{n-1}}{z} + \cdots + \frac{\alpha_0}{z^n} \right| \geqq |\alpha_n| - \left\{ \frac{|\alpha_{n-1}|}{|z|} + \cdots + \frac{|\alpha_0|}{|z^n|} \right\}$$

$$\geqq |\alpha_n| - \frac{|\alpha_{n-1}| + \cdots + |\alpha_0|}{|z|} > \frac{|\alpha_n|}{2}.$$

Now, since $P(z)$ has no roots, $f(z)$ is defined everywhere. But, since $|z| > 1$,

$$|f(z)| < \frac{2}{|\alpha_n| |z|^n} < \frac{2}{|\alpha_n|}.$$

Consequently, $f(z)$ is bounded and therefore constant. We conclude from the first of the foregoing inequalities that $f(z) = 0$, which contradicts $f(z) P(z) = 1$.

8. (a)–(b) The residue of $f'/f$ at $\alpha$ is $2\pi i I$. Set $f(z) = (z - \alpha)^p \phi(z)$, where

$\phi$ is analytic, $\phi(\alpha) \neq 0$, and $p$ represents either the order $n$ of the zero or $-m$ for the pole for parts (a) and (b), respectively. Then

$$\frac{f'(z)}{f(z)} = \frac{p\phi(z) + (z - \alpha)\, \phi'(z}{(z - \alpha)\, \phi(z)}.$$

Cauchy's integral formula then shows that $I$ is the value of $[p\phi(z) + (z - \alpha)]\, \phi'(z)/\phi(z)$ when $z = \alpha$; that is $p$.

(c) Apply the theorem of residues (p. 805).

9. (a) The number of roots of the equation $P(z) + \theta Q(z) = 0$, by Exercise 8, is

$$\frac{1}{2\pi i} \int_C \frac{P'(z) + \theta Q'(z)}{P(z) + \theta Q(z)}\, dz.$$

The denominator differs from zero for every $\theta$ for which $0 \leqq \theta \leqq 1$ at any point of $C$; the whole integral is therefore a continuous function of $\theta$. As its value is always an integer, it is constant and, hence, the same for $\theta = 0$ and $\theta = 1$.

(b) If

$$|a| < r^4 - \frac{1}{r},$$

then $r > 1$; so the equation $z^5 + 1 = 0$ has five roots inside the circle $|z| = r$; if we put $P(z) = z^5 + 1$, $Q(z) = az$, we have on the circle $|z| = r$,

$$|Q(z)| = |a|\, r < r^5 - 1 < |z^5 + 1| = |P(z)|.$$

10. From the lower bound (i) in Exercise 7 for $|P(z)|$, no root can lie outside or on a sufficiently large circle about 0. Applying the technique of estimation used in (i) in Exercise 7, we find

$$\frac{f'(z)}{f(z)} = \frac{n}{z} + R(z),$$

where the remainder $R(z)$ satisfies $|R(z)| < M/|z|^2$ outside a circle of sufficiently large radius $r$. Take $r$ so large that all the roots of $P$ lie in its interior. Applying the result of Exercise 8(c), we obtain for the number of roots, the integral about the circle of radius $r$

$$\frac{1}{2\pi i} \int \frac{f'(z)}{f(z)}\, dz = n + \frac{1}{2\pi i} \int R(z)\, dz.$$

Since

$$\left| \frac{1}{2\pi i} \int R(z)\, dz \right| < \frac{M}{r},$$

the remainder integral tends to zero as $r \to \infty$.

11. (a) Follow the method of solution for Exercise 8(a).

(b) If the roots are $\alpha_1, \alpha_2, \ldots, \alpha_j$, if the poles are located at $\beta_1, \beta_2, \ldots, \beta_k$, and if these have multiplicities $n_1, n_2, \ldots, n_j$ and $m_1, m_2, \ldots, m_k$, respectively, the integral has the value

$$n_1\alpha_1 + n_2\alpha_2 + \cdots + n_j\alpha_j - m_1\beta_1 - m_2\beta_2 - \cdots - m_k\beta_k.$$

12. Since $f(z) = e^z$ is everywhere analytic, since $f'(z)/f(z) = 1$, and since the integral $I$ of Exercise 8(a) must therefore vanish on any circle, no matter how large, $f(z)$ can have no roots.

## Exercises 8.5 (p. 814)

1. (a) Expressing the functions in the neighborhood of $\alpha$ by
$$f(z) = a_0 + a_1(z - \alpha) + \cdots + a_{n-1}(z - \alpha)^{n-1} + \cdots$$
and
$$g(z) = (z - \alpha)^{-n}[c_{-n} + c_{-n+1}(z - \alpha) + \cdots + c_{-1}(z - \alpha)^{n-1} + \cdots],$$
we obtain the residue
$$2\pi i \sum_{\nu=0}^{n-1} a_\nu c_{-\nu-1}.$$

(b) In the foregoing solution, use $c_k = 0$ for $k > -n$ and $a_{n-1} = f^{(n-1)}(\alpha)/(n-1)!$.

2. Set
$$f(z) = (z - \alpha)^2 \phi(z) = (y - \alpha)^2 \left[ \frac{f''(\alpha)}{2} + \frac{f'''(\alpha)}{6} (z - \alpha) + \cdots \right]$$
and determine the first-order coefficient in the expansion of $1/\phi(z)$.

3. (a) $\pi/\sqrt{2}$.

(b) Use the result of Exercise 2 for the residues at $e^{i\pi/4}$ and $e^{3i\pi/4}$ to obtain $3\pi/4\sqrt{2}$. Here, for $f(z) = (1 + x^4)^2, f''(z) = 24x^2(1 + x^4) + 32x^6$ and $f'''(z) = 48x(1 + x^4) + 9\cdot 32x^5$.

(c) The integrand has simple poles at the points $z_k = \omega^{2k-1}$ ($k = 1$, $2, \ldots, 2n$), where $\omega = e^{i\pi/2n}$ is the principal $(4n)$-th root of unity. For $k \leq n$, the poles are in the upper half-plane. Thus, from formula (8.21b) the integral is equal to
$$I = 2\pi i \sum_{k=1}^{n} \frac{z_k^{2m}}{2nz_k^{2n-1}} = -\frac{\pi i}{n} \sum_{k=1}^{n} z_k^{2m+1},$$
where we have used $z_k^{2n} = -1$. Entering the expression for $z_k$ in this last sum, we obtain $I$ in the form of a geometric series and then sum to obtain the result:
$$I = -\frac{\pi i}{n\omega^{2m+1}} \sum_{k=1}^{n} [\omega^{4m+2}]^k = -\frac{\pi i \omega^{2m+1}}{n} \frac{1-(\omega^{4m+2})^n}{1 - \omega^{4m+2}}$$
$$= \frac{\pi}{n} \frac{2i}{\omega^{2m+1} - \omega^{-(2m+1)}} = \frac{\pi}{n \sin[(2m + 1/2n)\pi]}.$$

4. The left-hand side of the formula is the sum of the residues of the function $z^k/f(z)$ divided by $2\pi i$ and is therefore equal to
$$\frac{1}{2\pi i} \int \frac{z^k}{f(z)} \, dz$$

round a circle enclosing all the roots $\alpha_v$. But this integral tends to zero as the radius of the circle tends to infinity (the center remaining fixed).

5. Because $x \cos x$ is odd and $x \sin x$ is even, the integral is equal to

$$\frac{1}{2i} \int_{-\infty}^{\infty} \frac{xe^{ix}}{x^2 + c^2} \, dx.$$

The residue in the upper half-plane of $ze^{iz}/2i(z^2 + c^2)$ is $\frac{1}{2}\pi e^{-|c|}$. Take $z = r(\cos \theta + i \sin \theta)$ and integrate over the closed path $C$ from $-r$ to $r$ along the $x$-axis and over the semicircle $|z| = r$ in the upper half-plane. We need only prove the part of the integral over the semicircle tends to zero in the passage to the limit as $r \to \infty$. We find for the integral over the half circle $0 \leq \theta \leq \pi$,

$$J = \int_0^\pi \frac{r^2 e^{i\theta} e^{-r \sin \theta} e^{ir \cos \theta}}{r^2 e^{2i\theta} + c^2} \, d\theta.$$

Choose $r$ so large that $|r^2 e^{2i\theta} + c^2| > \frac{1}{2} r^2$; for example, choose $r^2 > 2c^2$.

It follows that

$$|J| < 4 \int_0^{\pi/2} e^{-r \sin \theta} \, d\theta < 4 \int_0^{\pi/2} e^{-2r\theta/\pi} \, d\theta < \frac{2\pi}{r}.$$

## Miscellaneous Exercises 8 (p. 818)

1. $(z_1 - z_3)/(z_2 - z_3)$ must be real.

2. Let arg $z$ be the argument of $z = re^{i\theta}$; that is, arg $z = \theta + 2n\pi$. The directed angle from the segment $\overrightarrow{\alpha\beta}$ to the segment $\overrightarrow{\alpha\gamma}$ is

$$\arg \frac{\gamma - \alpha}{\beta - \alpha} + 2p\pi,$$

where $p$ is an integer. The given equation tells us that

$$\arg \frac{\gamma - \alpha}{\beta - \alpha} = -\arg \frac{\gamma - \beta}{\alpha - \beta} + 2n\pi.$$

Thus, taking the segment joining $\alpha$ and $\beta$ as the base of the triangle, we see that the angles from the base to the sides are equal and opposite in sign. Conversely, equality of the base angles yields the given equation.

3.
$$\Delta = \frac{(z_1 - z_3)/(z_2 - z_3)}{(z_1 - z_4)/(z_2 - z_4)}$$

must be real, for if $C$ is the circle through $z_1$, $z_2$, $z_3$, we may transform $C$ by a linear transformation $\zeta = (\alpha z + \beta)/(\gamma z + \delta)$ into the real axis (cf. Section 8.2, Exercise 8). By Section 8.2, Exercise 9, $\Delta$ is unchanged. Then a necessary condition that the image of $z_4$ shall lie on the same circle as the images of $z_1$, $z_2$, $z_3$ is that it be real, which is equivalent to $\Delta$ being real.

4. The equality to be proved is

$$\sqrt{|z_1 - z_2||z_3 - z_4|} + \sqrt{|z_2 - z_3||z_1 - z_4|} = \sqrt{|z_1 - z_3||z_2 - z_4|}$$

or

$$1 + \sqrt{\left|\frac{(z_1 - z_2)(z_3 - z_4)}{(z_2 - z_3)(z_1 - z_4)}\right|} = \sqrt{\left|\frac{(z_1 - z_3)(z_2 - z_4)}{(z_2 - z_3)(z_1 - z_4)}\right|}$$

Now the expressions under the square roots are invariant in a linear transformation (cf. Section 8.2, Exercise 8, 9). If by a suitable linear transformation we transform the circle into the real axis, we have only to prove the relation $AB \cdot CD + BC \cdot AD = AC \cdot BD$ for four points on a straight line, where it is trivial.

5. $\zeta = e^{iz}$ takes every value except $\zeta = 0$, as is easily seen from the relation $e^{iz} = e^{-y}(\cos x + i \sin x)$. Now we have to choose $\zeta$ so that

$$c = \cos z = \frac{1}{2}\left(\zeta + \frac{1}{\zeta}\right);$$

this quadratic equation always has a solution

$$\zeta = c \pm \sqrt{c^2 - 1}.$$

and this solution is not zero, so that a corresponding $z$ exists.

6. Cf. Exercise 5. If $\zeta = e^{iz}$, then

$$\tan z = \frac{1}{i}\frac{\zeta - (1/\zeta)}{\zeta + (1/\zeta)} = c$$

or

$$\zeta = \sqrt{\frac{1 + ic}{1 - ic}};$$

there is a finite $\zeta \neq 0$ only when $c \neq \pm i$; hence, $\tan z = c$ only has a solution if $c$ is neither $+i$ nor $-i$.

7. If $z = x + iy$, $\cos z$ is real if $x = \pi n$ or $y = 0$, and $\sin z = 0$ if $x = \pi n + \pi/2$ or $y = 0$ (where $n$ is an integer).

8. (a) $r = 1$ (for $|z| > 1$ the individual terms tend to $\infty$; for $|z| < 1$ compare with the geometric series).

   (b) $r = 0$.

   (c) $r = 1$.

9 (a) Integrate $e^{iz}/(1 + z^4)$ over upper semicircle:

$$\frac{\pi\sqrt{2}}{4}e^{-\sqrt{2}/2}\left(\sin\frac{\sqrt{2}}{2} + \cos\frac{\sqrt{2}}{2}\right).$$

   (b) Integrate $z^2 e^{iz}/(1 + z^4)$ over upper semicircle:

$$\frac{\pi\sqrt{2}}{4}e^{-\sqrt{2}/2}\left(\cos\frac{\sqrt{2}}{2} - \sin\frac{\sqrt{2}}{2}\right).$$

   (c) Integrate $e^{iz}/(q^2 + z^2)$ over upper semicircle:

$$\frac{\pi}{2q}e^{-q}.$$

(d) Integrate $x^{\alpha-1}/[(x+1)(x+2)]$ over a region bounded by a large circle about the origin and slit along the positive real axis:

$$\frac{\pi(2^{\alpha-1}-1)}{\sin \pi\alpha}.$$

10. (a) $+2\pi i$ at $z = 2n\pi$, $-2\pi i$ at $z = (2n+1)\pi$.
   (b) $+2\pi i$ at $z = 2n\pi + 3\pi/2$, $-2\pi i$ at $z = 2n\pi + \pi/2$.
   (c) Use the functional equation $\Gamma(z) = \Gamma(z+\nu+1)/z(z+1)\cdots(z+\nu)$;

$$\frac{(-1)^n}{n!}\,2\pi i \text{ at } z = -n.$$

   (d) $2\pi i$ at $z = n\pi i$.

11. $$|\sinh(x+iy)|^2 = \left(\frac{e^{x+iy}-e^{-x-iy}}{2}\right)\left(\frac{e^{x-iy}-e^{-x+iy}}{2}\right)$$

$$= \frac{1}{2}(\cosh 2x - \cos 2y)$$

$$\geq \frac{1}{2}(\cosh 2x - 1).$$

Integrate along the boundary of a square with sides $x = \pm \pi(n+\frac{1}{2})$ and $y = \pm(n+\frac{1}{2})$, where $n$ is an integer. As $n \to \infty$, the integral tends to zero; hence, the sum of the residues tends to zero.

12. Write

$$\frac{\cot \pi t}{t-z} = \frac{\cot \pi t}{t} + \frac{z \cot \pi t}{t(t-z)};$$

$\cot \pi t$ is bounded on the square $C_n$, and the integrals of $(\cot \pi t)/t$ over opposite sides of the square almost cancel one another; hence,

$$\lim_{n\to\infty}\int_{C_n}\frac{\cot \pi t}{t-z}\,dt = \lim_{n\to\infty}\int_{C_n}\frac{z \cot \pi t}{t(t-z)}\,dt = 0.$$

If we put together residues of opposite poles, the sum of the residues converges and we obtain

$$\cot \pi x = \frac{2x}{\pi}\left(\frac{1}{2x^2}+\frac{1}{x^2-1^2}+\frac{1}{x^2-2^2}+\cdots\right)$$

(cf. Volume I, p. 602).

13. $$\frac{1}{1+t} = 1 - t + t^2 - + \cdots \pm t^{n-1} + (-1)^n\frac{t^n}{1+t}.$$

Hence,

$$\log(1+z) = z - \frac{z^2}{2} + \frac{z^3}{3} - \cdots \pm \frac{z^n}{n} + R_n,$$

where

$$R_n = (-1)^n \int_0^z \frac{t^n}{1+t}\, dt.$$

If we take $z = e^{i\theta}$ and the straight line from 0 to $e^{i\theta}$ as path of integration, we have, for $e^{i\theta} \neq -1$,

$$|R_n| = \left| \int_0^1 \frac{t^n}{1 + e^{i\theta}t}\, dt \right| \leq \frac{1}{m} \int_0^1 t^n\, dt = \frac{1}{m(n+1)},$$

where $m$ denotes the minimum of $|1 + e^{i\theta}t|$ for $0 \leq t \leq 1$. Hence, if $z = e^{i\theta} \neq -1$, $R_n$ tends to 0.

14. If $x \neq 0$ and if $C'$ is a contour in the region in which $f$ is regular and contains $y$ but not 0, then, by p. 801,

$$\frac{d^n}{dy^n} \frac{yf(y)}{(y-a)^{n+1}} = \frac{n!}{2\pi i} \int_{C'} \frac{tf(t)}{(t+a)^{n+1}(t-y)^{n+1}}\, dt.$$

If we put $a = y = \sqrt{x}$, the latter integral becomes

$$\frac{n!}{2\pi i} \int_{C'} \frac{tf(t)}{(t^2 - x)^{n+1}}\, dt.$$

If we then substitute $t^2 = \tau$, the integral becomes

$$\frac{n!}{2\pi i} \int_C \frac{f(\sqrt{\tau})}{(\tau - x)^{n+1}}\, d\tau,$$

where $C$ is a contour containing $x$ but not 0; the integral is equal to

$$\frac{1}{2} \frac{d^n}{dx^n} f(\sqrt{x}).$$

15. (a)
$$f(z) = \sum_{\nu=1}^{\infty} \left( \frac{1}{(2\nu - 1)^z} - \frac{1}{(2\nu)^z} \right);$$

now

$$\frac{1}{(2\nu - 1)^z} - \frac{1}{(2\nu)^z} = z \int_{2\nu-1}^{2\nu} \frac{1}{y^{z+1}}\, dy \leq \frac{|z|}{|(2\nu-1)^{z+1}|} = \frac{|z|}{(2\nu - 1)^{1+z}},$$

and the series $\sum\limits_{\nu} 1/(2\nu - 1)^{1+z}$ is absolutely convergent for $x > 0$.

(b) $(1 - 2^{1-z})\zeta(z) = 1 + \dfrac{1}{2^z} + \dfrac{1}{3^z} + \dfrac{1}{4^z} + \cdots - \dfrac{2}{2^z} - \dfrac{2}{4^z} - \dfrac{2}{6^z} - \cdots$

$$= 1 - \frac{1}{2^z} + \frac{1}{3^z} - \frac{1}{4^z} + \cdots = f(z).$$

(c) $\lim\limits_{z \to 1} (z - 1)\, \zeta(z) = f(1) \cdot \lim\limits_{z \to 1} \dfrac{z-1}{1 - 2^{1-z}} = \dfrac{f(1)}{g'(1)} = 1,$

where

$$g(z) = 1 - 2^{1-z}.$$

# List of Biographical Dates

Abel, Niels Henrik (1802-1829)
Amsler, Jakob (1823-1912)
Archimedes (287?-212 B.C.)
Bernoulli, Jakob (1654-1705)
Bernoulli, John (1667-1748)
Bessel, Friedrich Wilhelm (1784-1846)
Birkhoff, George David (1884-1944)
Bohr, Harald (1887-1951)
Bolzano, Bernhard (1781-1848)
Borel, Felix Édouard Émile (1871-1956)
Brouwer, Luitzen Egbertus Jan (1881-1966)
Cauchy, Augustin (1789-1857)
Cavalieri, Francesco Bonaventura (1598-1647)
Chebyshev, Pafnuti Lvovich (1821-1894)
Clairaut, Alexis Claude (1713-1765)
Cramer, Gabriel (1704-1752)
Coulomb, Charles Augustin de (1736-1806)
De Moivre, Abraham (1667-1754)
Descartes, (Cartesius) René (1596-1650)
Dirac, Paul Adrien Maurice (1902-    )
Dirichlet, Gustav Lejeune (1805-1859)
Du Bois-Reymond, Paul (1831-1889)
Euler, Leonhard (1707-1783)
Fermat, Pierre de (1601-1665)
Fourier, Joseph (1768-1830)
Frenet, Fréderic-Jean (1816-1900)
Fréchet, Maurice René (1878-    )
Fresnel, Augustin Jean (1788-1827)
Gauss, Carl Friedrich (1777-1855)
Gram, Jörgen Pederson (1850-1916)
Green, George (1793-1841)
Guldin, Paul (1577-1643)
Hamilton, Sir William Roan (1805-1865)
Heine, Heinrich Eduard (1821-1881)
Helmholtz, Hermann Ludwig Ferdinand von (1821-1894)
Hermite, Charles (1822-1901)
Heron (of Alexandria) (third century A.D.)
Hölder, Otto (1860-1937)
Holditch, Hamnet (1800-1867)

Huygens, Christian  (1629-1695)
Jacobi, Carl Gustav Jacob  (1804-1851)
Kepler, Johannes  (1571-1630)
Lagrange, Joseph Louis  (1736-1813)
Laplace, Pierre Simon  (1749-1827)
Lebesgue, Henri  (1875-1941)
Legendre, Adrien-Marie  (1752-1833)
Leibnitz, Gottfried Wilhelm von  (1646-1716)
Lipschitz, Rudolf Otto  (1832-1903)
Lissajous, Jules Antoine  (1822-1880)
Maxwell, James Clerk  (1831-1879)
Möbius, August Ferdinand  (1790-1868)
Mollerup, Peter Johannes  (1872-1937)
Morera, Giacinto  (1856-1909)
Morse, Marston Harold  (1892-    )
Newton, Isaac  (1642-1727)
Parseval-Deschènes, Marc Antoine  (B?  -1836)
Plateau, Joseph Antoine Ferdinand  (1801-1883)
Poincaré, Henri  (1854-1912)
Poisson, Siméon Denis  (1781-1840)
Riccati, Jacopo Francesco  (1676-1754)
Riemann, Bernhard  (1826-1866)
Schuler, Maximilian Joseph Johannes Eduard  (1882-    )
Schwarz, Hermann Amandus  (1843-1921)
Steiner, Jacob  (1796-1863)
Stokes, George Gabriel  (1819-1903)
Taylor, Brook  (1685-1731)
Wallis, John  (1616-1703)
Weierstrass, Karl  (1815-1897)
Wronski, (Hoene), Jozef Maria  (1778-1853)

# Index